CWAP™

Certified Wireless Analysis Professional™ Official Study Guide

(Exam PW0-205)

FIRST EDITION

Planet3 Wireless

McGraw-Hill/Osborne
New York Chicago San Francisco Lisbon London Madrid
Mexico City Milan New Delhi San Juan Seoul Singapore Sydney Toronto

McGraw-Hill/Osborne
2100 Powell St. 10th Floor
Emeryville, CA 94608
U.S.A.

To arrange bulk purchase discounts for sales promotions, premiums, or fund-raisers, please contact McGraw-Hill/Osborne at the above address. For information on translations or book distributors outside the U.S.A., please see the International Contact Information page immediately following the index of this book.

CWAP Certified Wireless Analysis Professional
Official Study Guide (Exam PW0-205) First Edition

1234567890 JPI JPI 019876543

ISBN 0-07-225585-4

Publisher
Brandon A. Nordin

Vice President & Associate Publisher
Scott Rogers

Acquisitions Editor
Timothy Green

Authors
Devin Akin
Jim Geier

Editorial Director
Gareth Hancock

Technical Editors
Criss Hyde

Copy Editor
Kevin Sandlin

Proofreaders
Kevin Sandlin
Criss Hyde

Indexer
Jack Lewis

Computer Designers
Scott Turner

Illustrator
Scott Turner

Series Design
Scott Turner

CWNP® Certification Program

The CWNP Program is the industry standard for wireless LAN training and certification, and is intended for individuals who administer, install, design, troubleshoot, and support IEEE 802.11 compliant wireless networks. Achieving CWNP Certification will help you to design, install, and maintain wireless LANs that are more secure, cost-effective, and reliable.

The CWNP Program has three levels of knowledge and certification covering all aspects of wireless LANs.

Foundation

CWNA covers a broad range of wireless networking topics. CWNA brings those of you who are new to wireless networking up to speed quickly. For those of you already familiar with wireless LANs, earning the CWNA certification fills in any gaps in your knowledge, and officially proves your expertise to help your competitive edge.

Advanced

CWSP ensures that you understand how to secure a wireless LAN from hackers and protect the valuable information on your network. CWSP offers the most thorough information available on how attacks occur and how to secure your wireless network from them.

CWAP focuses entirely on the analysis and troubleshooting of wireless LAN systems. The CWAP certified individual will be able to confidently analyze and troubleshoot any wireless LAN system using any of the market leading software and hardware analysis tools.

Expert

CWNE credential is the final step in the CWNP Program. By successfully completing the CWNE practical examination, network engineers and administrators will have demonstrated that they have the most advanced skills available in today's wireless LAN market.

We at The CWNP® Program would like to dedicate this book to our Lord Jesus Christ, our friend, our Savior. We have experienced His presence daily since the inception of this company, and owe everything to Him. We would like to thank Jesus for the incredible blessings we have received, including our friend and brother Mark Elliott whose dedication to serving our Lord on the Mercy Ships has inspired us all. We thank Him for our new brothers Scott Williams and Scott Daniel who both recently came to know the Lord. We thank Him for providing for every need since He created this adventure back in 1999. We thank Him for bringing together a group of brothers, each with complimentary talents, who strengthen and support each other in our personal and professional lives. We thank Him for making the seemingly impossible not only possible, but unfolding reality, right before our eyes. Each time He shuts a door, He opens another - encouraging us to boldly walk through it. We don't know what tomorrow holds, but we do know the Lord, who holds tomorrow in His hands. We're excited to see what He's going to do in each of our lives and those lives who we touch by His grace. We claim no honor, but give Him all the glory.

2 Corinthians 12:9 – But he said to me, "My grace is sufficient for you, for my power is made perfect in weakness." Therefore I will boast all the more gladly about my weaknesses, so that Christ's power may rest on me.

Acknowledgements

Planet3 Wireless, Inc. would like to acknowledge and thank the following people for their contributions to the CWAP Study Guide:

Devin Akin, Author: Devin is the Chief Technology Officer of Planet3 Wireless, Inc. Devin has over 10 years of IT experience and holds Cisco's CCNP, CCDP, and CCSP, Microsoft's MCSE, Novell's MCNE, and the esteemed NSA/CNSS INFOSEC certifications among many others. He is the primary author of the CWNA and CWSP Study Guides and CWNA, CWSP, and CWAP Courseware. He was the primary subject matter expert for all CWNP exams and practice exams, holds all CWNP certifications, and teaches wireless courses around the world. Devin has previously worked as a Senior Network Access Design Engineer with EarthLink and BellSouth, and a Senior Systems Engineer for Foundry Networks and Sentinel Technologies.

Jim Geier, Co-Author: Jim Geier, Co-Author: Jim Geier is the principal consultant of Wireless-Nets, Ltd. (www.wireless-nets.com), where he provides independent analysis, design, and planning of wireless LANs to product developers, system integrators and large end users throughout the world. Jim is a voting member within the Wi-Fi Alliance, certifying interoperability of 802.11 (Wi-Fi) wireless LANs. He served as Chairman of the IEEE Computer Society, Dayton Section, and Chairman of the IEEE International Conference on Wireless LAN Implementation. He has been an active member of the IEEE 802.11 Working Group, developing international standards for wireless LANs. Jim is author of several books on wireless networks, and his education includes a bachelors and masters degree in electrical engineering and a Master's degree in business administration. You can contact Jim Geier at jgeier@wireless-nets.com.

Criss Hyde, Technical Editor: Criss has over 25 years of IT experience. He has contributed substantially to all the Planet3 products published or republished during 2004. His attention to technical detail simply astounds us. He holds the CWAP and CWSP certifications, as well as Cisco Certified Network and Design Professional, Cisco Wireless LAN Design and Support Specialist, and Sun Certified System and Network Administrator. Criss has earned Engineering and a Law degree from Penn State and George Mason Universities, respectively, worked for 13 years for Raytheon, and is a member of the Virginia Bar. Criss is married, and is the father of eight home schooled children. Criss worked briefly for the Executive Office of the White House.

Eric Geier, for testing many of the RF and protocol concepts explained in this book. Eric is a member of the technical staff of Wireless-Nets, Ltd., where he researches and analyzes wireless network technologies, performs wireless LAN analysis, and develops training media.

Scott Daniel, for relentlessly configuring endless complex lab scenarios to prove concepts explained in this book. Scott is a lab engineer and courseware designer at the CWNP Program with a diverse background in systems, network, and security engineering across a wide variety of platforms.

Contents At a Glance

Contents

Forward

Data communication networks are, at their most fundamental level, the exchange of bits, grouped into units and sub-units, and exchanged in accordance with a predefined set of rules we call protocols. Understanding the operation of any data communication system may be decomposed into understanding the bits, the units, and the protocols. Troubleshooting, optimizing, and securing a network is accomplished by comparing and contrasting the observed behavior of these bits and data units with the established protocol rules, restrictions, and capacity requirements. The process of observing the internal operations in a communication network is the process of "protocol analysis", and, in the realm of the IEEE 802.11 wireless LAN (WLAN) standards, that's what this book is all about.

My first exposure to WLAN technology was through the draft 802.11 standards document from the Institute of Electrical and Electronics Engineers (IEEE). The 802.11b draft standards soon followed, and were ratified in the second half of 1999. Since there were few practical implementations of the 802.11 standard it was too early to know what was going to be important, what was going to be found to be flawed, and what would never be widely implemented. Those of us in the RF engineering space watched this new "wireless LAN" technology explode into the marketplace with unprecedented growth. Today you have the benefit of the past five years of industry experience with 802.11 technology, and The CWAP Study Guide has distilled key concepts and facts into a very readable volume. You're not going to be struggling to understand a sometimes seemingly cryptic IEEE standard. You're going to learn how this new, pivotal technology works and how you can confirm its efficiency, security, and correct operation. You're going to be able to demonstrate your knowledge and expertise through professional certification in the Certified Wireless Network Professional (CWNP®) program, the accepted industry standard for 802.11 engineering expertise.

When we consider the bits in a WLAN, we're thinking in terms of the binary 1's and 0's that are represented by the wiggling and jiggling of electromagnetic energy transmitted through space. There are many different ways that a radio circuit and an antenna can cause an electromagnetic wave to wiggle, and different bit representations offer different advantages along with different disadvantages. The consideration of the representation of bits based on some particular way of jiggling the electromagnetic wave is the physical (PHY) layer of communication. You'll be introduced to the various 802.11 PHY standards and you'll see how they've evolved and how they differ.

There are a number of behavioral rules that an 802.11 WLAN communicator must follow, and these rules, along with various PHY standards, are stipulated by working groups within the IEEE. The 802.11 group is subdivided with letters of the alphabet to form the 802.11a, b, c, d, e, f, etc., all the way up to 802.11k and beyond! To be an expert in the analysis of WLAN communication, it's necessary to understand the expected rules of protocol behavior, and you will be introduced to those rules, and the associated 802.11 standards, as you read through this book.

Of course, a WLAN doesn't operate apart from some type of wired network infrastructure. 802.11 radios ("access points") are typically connected together through Ethernet cables, switches, and routers. There can be wireless "backhaul" connections between access points, but you're going to encounter a cable somewhere along the line. As you read, you'll be introduced to these infrastructure components, and you'll see how the whole system fits together.

The very thing that makes a wireless network attractive also creates a security exposure. Since users don't need to physically "plug in" (an attractive feature), it's possible for an unauthorized person to gain access to the wireless network (a security exposure). Prior to 2004 the only security options that were part of the 802.11 standard ("Wired Equivalent Privacy, WEP") were determined to be flawed. Today, new recommendations and standards ("Wi-Fi Protected Access, WPA" and the 802.11i standards) have provided the capability to effectively secure a wireless network. Security is a crucial part of wireless network design and administration and awareness of security issues is a key part of any WLAN engineer's education.

Today there are two groups of engineers in the communications arena who are converging into the WLAN space. Neither group brings with them the knowledge necessary to successfully design, implement, secure, manage, and troubleshoot an 802.11 wireless LAN. Both groups have things to learn, and both groups are actively moving towards WLAN expertise. The first group consists of the legacy Ethernet LAN engineers. These are the "wireline" folks who know about switches, routers, and firewalls. They know how to move data through wires, but moving it through the air is new territory. The second group consists of the telephony folks who have been creating the cellular phone network for many years. They know how to move signals through the air, but moving data is new territory.

Today there are a number of tools available to help engineers observe the behavior of communicating devices in the air. These "wireless LAN analyzers" offer a broad range of features and capabilities, and a WLAN engineer must be knowledgeable in the use of an analysis tool, and in the interpretation of the packet level decodes that the analyzer presents. In this book you'll see examples of frame level traces that will show the behavior of wireless communicators in action.

Why should you learn about the 802.11 standards? Why should you challenge yourself to demonstrate your knowledge through professional certification? The answer is that 802.11 WLAN technology is, today, the focal point in an almost unimaginable global technology convergence, and, without an 802.11 foundation, communication engineers (LAN and telephony) are going to be at a shocking disadvantage in the years to come. Now, this statement should not be taken to mean that the 802.11 standards are going to become universal and pervasive the way TCP/IP and the Internet did during the 1990's. Perhaps they will, and perhaps they won't. There are many other competing standards in the field today: 802.16 WiMAX, 802.20, 3G cellular with EV-DO, UltraWideband, and more. However, as of the publication of this book, the 802.11 standards have risen to a position of prominence and they're going to serve, at the least, as a springboard into an evolving set of future communication systems.

We're going to see a convergence of cell phone and WLAN networking over the next several years. When you're inside a building your cell phone will roam onto the in-building 802.11 network and realize high-speed data transfer capabilities. In your car you'll roam back onto the cellular network. This is the realm of Voice-over-WLAN (also called wireless Voice-over-IP). It's here today as a proprietary offering from many vendors and convergence with the global cellular network is right around the corner. You've got to be on top of the 802.11 engineering issues to be on top of wireless voice.

In the retail and commercial sector, Radio Frequency Identification (RFID) is emerging as an alternative to bar code scanning. Management of the inventory supply chain from manufacture, through shipping, through merchandising, to ultimate product end-of-life can be tracked with tiny RFID "tags". A little bit of web searching on "RFID" will reveal just how pervasive this new technology is going to be. Where is 802.11 in all this? It's the 802.11 WLAN that's going to connect the hand-held RFID scanners and "smart shelves" back to the store database. We're talking about every retail store that uses bar code scanning today – it's going to be RFID right around the corner. You've got to be on top of the 802.11 engineering issues to be on top of RFID.

There are very few times in the course of human history when it's been evident that a new technology would dramatically change the way society functions. What if you knew where the automotive industry was heading when Carl Benz championed the internal combustion engine in his 3-wheeled car in 1885. What if you could have foreseen the aviation industry when the Wright's flew at Kitty Hawk? How about a crystal ball on the computer industry in the 1960's or 1970's? What if you could have foreseen the Internet? Well, you're right there again, and this time the technology falls under the umbrella of wireless convergence: data, voice, video - and today they all revolve around 802.11 WLAN technology. You've got to be on top of the 802.11 engineering issues to be on top in technology in the coming years.

Between the covers of this book you'll find a wealth of information that will be a basis for your understanding of wireless data networking, 802.11 protocol analysis, and WLAN management. You may find, as I have often discovered, that "the more you know, the more you know that you don't know", and, at the end of the day, have fun learning new things and never stop challenging yourself.

-Joseph Bardwell

Chief Scientist and President of Connect802 Corporation

Preface

First, I would like to thank Criss Hyde for constantly staying on my case about details. This book is all about details, and let me say right now in writing that there are *lots* of them in the 802.11 series of standards. They are written for rocket scientists who have law degrees. Luckily, Criss is one of those people.

I have learned more technical details while writing this book than I ever wanted to know about anything, ever. When I began writing this book, I only thought I knew this material. Now, I know enough to know that I'll never be able to keep all of it straight in my head. Referencing back to the standards from time to time is necessary unless you've opted for Google appliance integration surgery.

This whole experience has been mostly one of learning and re-learning. There's nothing worse than committing something to memory – only to find out it is wrong and you have to start again from scratch. I have done exactly that over and over throughout the writing (and rewriting) of this book. I equate the current series of 802.11 standards to a mystery novel. As you're reading, there are many twists and turns, but it's not until you've read the last page of the last standard that the light comes on in your head. Of course, to be totally honest, you have to read the novel (meaning the standards) several times to get the light to come on, and even then, the light is somewhat dim at times.

My wish for all of the readers of this book is to suffer less, if only slightly, than I have in learning all of this material. I can't imagine anything better than having all of the 802.11 facts in one text in a digestible format. That's what this book is meant to be. The 802.11 series of standards is often called the "Mother of All Standards." I think that nickname is very appropriate.

I have trained a great number of the CWNTs out there, and a trait common to most of them is constantly asking, "How does it work?" and "Why does it work that way?" In the CWNA and CWSP classes, there's a modest amount of answers for those types of questions, but extreme detail is missing due to time constraints. To some degree, the same is true of the CWAP class – time is against you all week. With this book, all of the details are here if you can keep them straight in your head.

Have you ever performed a site survey, security audit, or other common WLAN function to realize that something is happening on the network, and you simply do not know why? In fact, you don't even know where to start looking! After reading and memorizing this book, you can kiss those days goodbye. The knowledge you'll receive from reading this book will allow you to design, troubleshoot, and test the most complex and comprehensive wireless LANs on the market today. That's real job security for real people in the real world.

Just the other day, a systems engineer from a WLAN product manufacturer asked me, "What determines whether or not you're a WLAN analyst?" That was a surprisingly good question. My answer was a question. I asked, "Have you attempted the CWAP exam yet?"

-Devin Akin

CTO and co-founder of the CWNP Program

Introduction

This official CWAP Study Guide is written for two purposes. First, the book is your indispensable resource for understanding the MAC & PHY frame formatting and frame exchange processes behind every 802.11 wireless LAN. In other words, you can memorize the 802.11 standard, as well as every addition and amendment to that standard, or you can make the CWAP Study Guide a permanent part of your network analysis toolkit. Second, the book is the official self-study resource for helping you prepare for the CWAP exam (PW0-205).

If you are pursuing the CWAP certification, you are required to pass the CWNA exam as a pre-requisite. The knowledge that you will acquire through achievement of the CWNA certification will serve as the foundation for learning what has been called, "the most technically advanced" subject matter currently taught in the IT industry.

If you are reading this book, then you undoubtedly have either a solid knowledge of network analysis and are now endeavoring to take on the wireless end of that trade, or you have earned your CWNA (and possibly also CWSP) certification and are ready to take on the next challenge in your wireless LAN-focused career. Whichever description is more accurate, the outcome will be the same. You will learn more about the technologies that drive 802.11 wireless LANs than you have ever seen before. In addition, you have begun the process of setting yourself further apart from the rest of the IT industry by pushing yourself to learn some of the most complex and technically challenging information in all of IT.

Your study of wireless LAN analysis will enable you to troubleshoot any wireless LAN scenario, improve the performance of any wireless LAN, and add a whole new dimension to your ability to secure a wireless LAN. Furthermore, because the CWAP curriculum and exam are supported by the industry's leading wireless LAN analysis software makers, you will also have a thorough understanding of how to use all of the leading wireless LAN analysis software, which software is most appropriate for your organization, and which software is best for any given situation. As a wireless LAN analyst, these skills will make you more valuable to your organization and to your clients.

How is this book organized?

The CWAP Official Study Guide is organized into two major units:

- MAC and PHY Layers
- Applied Analysis

The IEEE 802.11 series of standards focuses only on the MAC sublayer (the lower half of layer 2) and the physical layer options. Therefore, that is what is at the heart of this book and the CWAP certification. We divide this subject – wireless LAN analysis – up into sections taking the reader step by step through what is specified at the MAC layer (frame formats, frame types, frame exchange processes, etc). Then, we move on to the PHY layers, differentiating between the most commonly used PHY layers and how they work. From there, we spend the rest of the book talking about the types of analyzers, common features, how to use them, how to interpret traces, giving examples of troubleshooting, etc.

Why should I pursue the CWAP certification?

Protocol analysis itself is an entire industry, both for wired and wireless. Network analysis these days is primarily Ethernet analysis because almost everything on the LAN is Ethernet. Ethernet Analysis is relatively uncomplicated because Ethernet framing is relatively simple. The 802.11 standards introduce an overwhelmingly popular (as popular as Ethernet) layer 2 protocol with *very* complicated framing and frame exchange processes.

Since 802.11 (Wi-Fi®) has become ubiquitous, any hopes of it going away are futile. For this reason, network analysts *must* learn wireless LAN analysis to the same degree of understanding as they have with Ethernet. With 802.11, there are so many security protocols and designs that it's difficult to keep up with them. A detailed understanding of the construction of these security protocols and frame exchanges is critical to designing and troubleshooting enterprise wireless networks. So, how can an IT professional prove their knowledge of this complex set of frame exchange rules, security protocols, and the 802.11 frame format? CWAP.

How do I earn the CWAP certification?

To earn the CWAP certification, you have to pass 2 exams. First, you must pass exam PW0-100, which will earn you the CWNA certification. Then you must pass exam PW0-205, and then you will have earned the CWAP certification.

The best ways to prepare for both exams, ranked by importance, are:

1. Attend an authorized course delivered by a CWNP Education Center
2. Get some hands on experience with wireless LAN gear and, specifically for the CWAP exam, with wireless LAN analyzers
3. Use the CWAP Official Study Guide for self-studying and review for the CWAP exam, as this book is *the* textbook for reviewing for the CWAP exam
4. Use the official practice test for each exam

All of these resources can be found at cwnp.com.

Free Trials of WLAN Analysis Software

Each of the wireless LAN analysis product manufacturers who have supported the creation and distribution of the CWAP courseware are offering the readers of this book free trial ware of their products. To download the software, visit www.cwnp.com/cwap/vendor_trials.html

Who is this book for?

This book builds upon the knowledge that is gained through preparing for and earning the CWNA certification. This book is focused entirely on the analysis and troubleshooting of wireless LAN systems, and is intended for three main audiences: CWNAs, IT professionals who analyze and troubleshoot wireless LANs, and people who want to become experts in all aspects of wireless LANs.

For CWNAs, this book represents a starting point towards achieving an advanced CWNP certification. For those who have set themselves apart by proving their knowledge of the foundations of 802.11 wireless networks, the next logical step is to progress towards CWNE by learning how to analyze and troubleshoot the wireless networks they have learned to install, manage, and secure.

For IT professionals, this book may represent an opportunity to achieve another certification. The CWAP Study Guide contains a significant amount of new information presented in an easy to understand format. You will gain an understanding of the use of common tools found in wireless LAN protocol analyzers and detailed knowledge of appropriate application of a wireless protocol analyzer.

For people who want to become wireless LAN experts, the CWAP Study Guide will give you intimate knowledge of the inter-workings of IEEE 802.11 family of standards and a thorough understanding of wireless LAN troubleshooting from performance and security perspectives. The CWAP certification is without equal or competition and is recognized and endorsed by nearly all of the leading wireless LAN analysis vendors in the market today.

Exam Objectives

The CWAP certification, covering the current CWAP exam objectives, will certify that the successful candidate understands the frame structures and exchange processes for each of the 802.11 series of standards and how to use the tools that are available for analyzing and troubleshooting today's wireless LANs. The CWAP candidate must have obtained the CWNA certification prior to taking the CWAP certification exam.

The skills and knowledge measured by this examination are derived from a survey of wireless networking professionals and analyzer product manufacturers from around the world. The results of this survey were used in weighing the subject areas and ensuring that the weighting is representative of the relative importance of the content.

This section outlines the exam objectives of the CWAP exam.

802.11 MAC Frames and Exchange Processes – 42%

1.1. Distinguish the intended purpose of each 802.11 MAC layer frame type

- 1.1.1. Control frames
- 1.1.2. Management frames
- 1.1.3. Data Frames

1.2. Explain the structure of each 802.11 MAC layer frame type

- 1.2.1. MAC Layer terminology used in the 802.11 series of standards
- 1.2.2. Header fields and subfields
- 1.2.3. Control, Management, and Data frame payload contents and sizes
  - Fixed fields
  - Information elements
- 1.2.4. Frames sizes
- 1.2.5. MAC layer addressing
- 1.2.6. Modifications for 802.11e

1.3. Explain and describe the frame exchange processes involved in:
   - 1.3.1. Authentication and Association
   - 1.3.2. Disassociation, Reassociation, and Deauthentication
   - 1.3.3. Active and Passive Scanning
   - 1.3.4. Roaming within an ESS
   - 1.3.5. Power Management mode operation
   - 1.3.6. Fragmentation
   - 1.3.7. 802.11b/g mixed mode environments and protection mechanisms
   - 1.3.8. Security mechanisms
     - WEP
     - WPA / WPA2 / 802.11i
     - 802.1x/EAP

1.4. Contrast the differences between the frame exchange processes in:
   - 1.4.1. Infrastructure vs. Independent Service Sets
   - 1.4.2. PCF vs. DCF access methods

802.11 Physical Layer Technologies – 20%

2.1. Explain PHY Layer terminology used in the 802.11 series of standards

2.2. Explain Interframe Spacing
   - 2.2.1. SIFS
   - 2.2.2. PIFS
   - 2.2.3. DIFS
   - 2.2.4. EIFS

2.3. Explain Slot Times
   - 2.3.1. 802.11a
   - 2.3.2. 802.11b
   - 2.3.3. 802.11g
   - 2.3.4. Special operation in 802.11b/g mixed mode environments

2.4. Explain 802.11 Contention (DCF Mode)
   2.4.1. CSMA/CA (half duplex)
   2.4.2. Backoff timer operation
   2.4.3. Virtual Carrier Sense (NAV)
   2.4.4. Physical Carrier Sense (CCA)
   2.4.5. Contention Window operation
2.5. Describe the PLCP Sublayer (802.11a/b/g)
   2.5.1. Purpose
   2.5.2. Preambles and Headers for each PHY
   2.5.3. Payloads
2.6. Describe the PMD Layer

802.11 Wireless LAN Protocol Analyzer Use and Trace Interpretation – 38%

3.1. Demonstrate appropriate application of an 802.11a/b/g protocol analyzer for:
   3.1.1. Troubleshooting
   3.1.2. Performance testing
   3.1.3. Security analysis
   3.1.4. Intrusion analysis
   3.1.5. Distributed analysis
3.2. Interpret 802.11a/b/g protocol traces when performing:
   3.2.1. Troubleshooting
   3.2.2. Performance testing
   3.2.3. Security analysis
   3.2.4. Intrusion analysis
3.3. Apply generic features common to most 802.11a/b/g protocol analyzers (not limited to those listed)
   3.3.1. Protocol decodes
   3.3.2. Peer map functions
   3.3.3. Conversation analysis
   3.3.4. Expert functions

Contact Information

We are always eager to receive feedback on our courses, training materials, study guides, practice tests, and exams. If you have specific questions about something you have read in this book, please use the information below to contact the CWNP Program

The CWNP® Program
Planet3 Wireless, Inc.
PO Box 20063
Atlanta, GA 30325
866.GET.CWNE
866.422.8354 fax
www.cwnp.com

Direct feedback via email:
feedback@cwnp.com

CHAPTER

1

Introduction to Wireless LAN Analysis

CWAP Exam Objectives Covered:

- Demonstrate appropriate application of an 802.11a/b/g protocol analyzer

In This Chapter

What is Wireless LAN Analysis?

Why Analyze Wireless LANs?

When Should You Apply Analysis?

Types of Wireless LAN Analysis

Wireless LAN Analysis Tools

What is Wireless LAN Analysis?

Wireless LAN analysis entails the inspection of a wireless LAN and the assessment of performance, security, RF coverage, and root causes of problems. The steps of analysis include measuring, interpreting, and reporting. A wireless LAN analyst uses protocol analyzers, RF analyzers, and simulation in order to produce data that will enable a better understanding of the network. With this information, IT staff and system integrators are able to effectively interpret and report underlying issues and solutions.

In the early 1990s, wireless LANs were proprietary, standalone systems that someone from operations, such as the warehouse manager or manufacturing VP, would purchase directly from the vendor. Most of these earlier wireless solutions would include bar code-based applications. For example, a warehouse would commonly install an inventory management system or a retail store would deploy a price marking application. The only data traffic on these networks would be the occasional transmission of tiny bar codes that consume little bandwidth. As a result, there was no critical need for analysis of the wireless LANs because very seldom did anything go wrong.

Wireless LANs, though, have been evolving rapidly over the past decade, especially within the past few years. Wireless LANs now operate at much higher data rates and simultaneously support a multitude of applications, such as web browsing, large file transfers, custom applications, voice communications, and video streaming. Some companies are even deploying wireless LANs to support both public and private users, which significantly complicates the implementation. With these demands on the wireless LAN, problems will inevitably occur. Consequently, it has become very important for IT professionals to become proficient at analyzing wireless LANs.

Why Analyze Wireless LANs?

Any good business will continually analyze its operations. Tiger Woods is constantly analyzing his golf swing. Shaquille O’Neal analyzes his powerful post-up moves. Analysis is meant to continually improve performance, and this methodology holds true for today's wireless LANs. Analyzing your wireless LAN and improving its performance increases return on investment because of higher up-time and ability to effectively support applications.

Maximize Performance

With a growing use of higher end applications over wireless LANs, it is becoming crucial to maximize performance. Users want to stream video from websites and send large email attachments. End users typically do not consider that the wireless LAN does not have the speed and capacity of the wired LAN they are accustomed to using. This misunderstanding puts high expectations and a big strain on the wireless network.

RF interference may also crop up from time-to-time, which increases overhead on the network and reduces throughput. This somewhat silent issue causes applications to run slowly or even to fail altogether. Unfortunately, most companies take a reactive approach, waiting until users complain before investigating what might be causing the problem. In order to realize the ROI of a wireless LAN, it is imperative to optimize it using a more proactive approach. Wireless protocol analyzers allow you to take this proactive approach.

Because of the large number of wireless-enabled laptops and PDAs, companies are finding that more users than expected are utilizing the wireless LAN. Once it is known that a wireless LAN exists, people think of new applications as well. In addition, hotspot operators, such as airports and hotels, are striving to share their wireless LANs for both public and private applications. As a result, there are many more users active on the wireless LAN in any give area.

The increase in utilization of wireless LANs and varying presence of RF interference is demanding thorough performance analysis to ensure that all users can continually make use of their applications without significant delays and frustrations. This performance analysis involves carefully designing the wireless LAN and testing actual capacity after installing the network. Also, regular performance analysis is necessary to ensure the network keeps up with demands as the number of users increase and applications change. For example, analysis would indicate whether or not

802.11b is getting the job done or if 802.11g or 802.11a might be needed to handle the current capacity adequately.

Improve Security

Passive monitoring of data frames, unauthorized access to resources, and denial of service attacks are major security issues of wireless LANs. The fact that a wireless network uses radio waves for connecting users makes wireless LANs vulnerable to attack by hackers. If a company ignores these threats, that company is risking substantial losses.

The only way that a company can guard against potential security attacks is to implement effective security mechanisms. Solid encryption and authentication go a long way for securing a wireless LAN, but it is important to perform regular security analysis. Regular security analysis ensures that changes made to the network, such as updating firmware, do not leave the wireless network open to attacks. Today's leading-edge wireless LAN intrusion detection systems (IDS) are basically wireless protocol analyzers with detailed alarms and notification procedures. Thoroughly understanding protocol analysis will allow you to be proficient at using both stand-alone analyzers and distributed analyzers aimed at intrusion detection.

Improve RF Coverage

When deploying a wireless LAN, installers install access points in strategic locations in order to provide required RF coverage. A problem with this type of configuration is that the coverage pattern generally changes as the layout of the facility changes and users find new areas to make use of the wireless LAN. The result is spotty RF coverage, which disrupts connections to network applications. In some cases, the loss of connectivity is acceptable, such as when browsing the Internet. The user can readjust the position of their laptop or PDA until they receive an acceptable signal and association with an access point. However, many corporate applications, especially those developed initially to operate over wired networks, may incorporate errors in databases if wireless connections are lost in the middle of a transaction. Other applications are session-based, such as Telnet, and may time out (breaking the session) with the loss of only a few frames.

Because of these problems, it is paramount to maintain continuous RF coverage through the areas of the facility where users need access to wireless applications. Regular RF coverage analysis, as part of the installation process of access points, and periodically measuring and analyzing coverage throughout the life of the network will make continuous RF coverage a reality. If RF coverage analysis is not part of the regular routine, the wireless LAN will likely cause end-user headaches on a regular basis.

Learn How Wireless LANs Operate

The analysis of wireless LANs leads to a much better understanding of how the protocols and RF propagation work. After taking measurements using protocol and RF analyzers and interpreting the results, you gain a much richer understanding of wireless LANs. The data and corresponding reports enable you to clearly see the frame exchange sequences and statistics between nodes on a wireless LAN.

A more in-depth appreciation and understanding of how wireless LANs function helps you better support them. When symptoms occur, you are able identify the underlying problems much faster, and thus reduce downtime and increase the availability of the system to users.

When Should You Apply Analysis?

Analysis applies to many aspects of the life of a wireless LAN. Be sure to incorporate analysis into every phase, from design to operational support.

System Design

You should perform analysis when designing the system in order to determine whether the chosen wireless LAN technologies, architecture and components will adequately support application and user requirements. Design analysis is especially necessary when the business requirements will likely stretch the limits of the wireless LAN. The goal here is to determine whether the wireless system will operate adequately before your client spends lots of money on hardware and installation of the system. This pre-installation analysis significantly reduces risk for complex implementations.

For example, the deployment of a wireless LAN to support 3,000 users in a stock exchange facility would entail simulation or careful analysis of a working prototype before investing money to purchase the necessary hardware. Practical utilization models for this level of overall utilization do not exist for wireless LANs. Some analysis and simulation, however, will help verify that there is a quantity and placement of access points that will actually satisfy requirements. In such cases, it is common to use a protocol analyzer and a device capable of emulating many clients and traffic types.

RF Site Surveys

One of the first steps of deploying a wireless LAN where you will probably utilize analysis is when performing a RF site survey. The site survey is done prior to installing the wireless LAN because the site survey discovers potential sources of interference and identifies the optimum installation location for access points. The measurements of RF signals and proper interpretation of signal quality and throughput tests, that is part of "analysis" as a whole, is critical for properly defining and qualifying range boundaries.

Some installers of access points simply use the data rate at which the client is associated with an access point as the basis for the range boundary. For example, an installer may interpret that the range boundary has been reached when the data rate on a laptop with association with an 802.11g access point drops to 24 Mbps. The problem with this approach, however, is this data rate is merely the data rate that the 802.11 data frames are sent. Access delays, collisions with nearby stations, and errors due to interference may prompt significantly high transmission retry rates. These events in turn lower actual throughput dramatically. For example, a user next to a microwave oven cooking popcorn and trying to browse the web will experience significant page load delays, even though the associated data rate is 54 Mbps.

Thus, analysis of range boundaries is absolutely necessary when performing site surveys. Analysis of range boundaries includes using parameters such as retry rates, throughput values, and interaction between protocols when defining range boundaries. You will need to determine what is acceptable for these parameters, but that is what wireless LAN analysis skills will enable you to do.

Acceptance Testing

After installing a wireless LAN, more analysis is needed to ensure that the installed system offers required levels of performance, coverage, and security. If a discrepancy is found, the corrective actions can be made before users begin sending heavy traffic across the network. For example, acceptance testing may find that collision rates in a particular area of the facility are very high. Through analysis, the source of the problem, such as low signal levels in that area, can be found.

It is very common for a site surveying firm not to return to the site for acceptance testing. Rather, the site survey experts perform the site survey, assume that it is correct given all variables involved, and then move on to the next site. Such a lack of follow through places the performance analysis squarely on the shoulders of the IT staff, unless there is enough budget to hire yet another firm to perform acceptance testing. Regardless of the situation, analysis is necessary to validate the site survey and to optimize the network given its current set of variables.

Baseline Analysis

What is a Baseline?

A baseline is a process for studying the network at regular intervals to ensure that the network is working as designed. It is more than a single report detailing the health of the network at a certain point in time. Data networks have been around for many years. Until recently, keeping the networks running has been a fairly forgiving process, with some margin for error. With the increasing acceptance of latency-sensitive applications such as Voice over IP (VoIP), the job of running the network is becoming harder and requires more precision. In order to be more precise and to give a network administrator a solid foundation upon which to manage the network, it is important to have some idea of how the network is running. To do this, you must go through a process called a baseline. By following the baseline process, you can obtain the following information:

- Gain valuable information on the health of the hardware and software
- Determine the current utilization of network resources
- Make accurate decisions about network alarm thresholds
- Identify current network problems
- Predict future problems

By performing a baseline on a regular basis, you can find out the current state and extrapolate when failures will occur and prepare for them in advance. This also helps you to make more informed decisions about when, where, and how to spend budget money on network upgrades.

Why a Baseline?

A baseline process helps you to identify and properly plan for critical resource limitation issues in the network. These issues can be divided into two groups, each described as either "control plane" resources or "data plane" resources. Control plane resources are unique to the specific platform and device and can be impacted by a number of issues including:

- Data link utilization
- Features enabled
- Network design

Control plane resources include parameters such as:

- CPU utilization
- Memory utilization
- Buffer utilization

Control plane resources would be relevant to access points, wireless LAN switches, wireless gateways, client devices, etc. Data plane resources are impacted only by the type and quantity of traffic and include, at a minimum, link utilization. By baselining resource utilization for critical areas, you can avoid serious performance issues, or worse, a network failure. With the introduction of latency-sensitive applications such as voice and video, baselining is now more important than ever. Traditional

TCP/IP applications are forgiving and allow for a certain amount of delay. Voice and video are typically UDP based and do not allow for retransmissions or network congestion.

Due to the new mix of applications, baselining helps you to understand both control plane and data plane resource utilization issues and to proactively plan for changes and upgrades to ensure that the network continues to function without interruption.

Baseline Objective

The objective of a baseline is to:

- Determine the current status of network devices
- Compare that status to standard performance guidelines
- Set thresholds to alert you when the status exceeds those guidelines

Due to the large amount of data and the amount of time it takes to analyze the data, you must first limit the scope of a baseline to make it easier to learn the process.

Continual Monitoring

After the wireless LAN is operational, it is important for support staff to be proactive with analysis by continually monitoring the system for indications that identify potential problems. This regular analysis includes measuring traffic flows and throughput over time to spot trends that could warrant redesigning the network. Regular performance monitoring is especially important with wireless LANs because usage generally increases over time as the company learns about and deploys new wireless applications. Security assessments should also be conducted periodically using tools to detect incorrectly configured access points and access control vulnerabilities. Intrusion detection systems should be kept up-to-date to stay abreast of new attacks against wireless LANs.

Troubleshooting Problems

Through analysis techniques, it is possible to systematically troubleshoot problems that may be occurring on the wireless LAN. Symptoms offer clues of what the underlying problem may be. Capturing frames and measuring RF signals provide data that you can analyze in order to determine the root causes of the problems.

Wireless LANs are generally part of a much larger enterprise information system. In addition to wireless LAN access points and radio cards in client devices, the system may consist of a wired LAN, servers, routers, and WAN communications links. An administrator usually must manage the larger system, not just the wireless LAN. As a result, it is often necessary to determine whether problems are resulting from the wireless or wired side of the network.

With effective analysis, it is possible to determine whether the wireless LAN is the problem. An administrator armed with wireless LAN analysis skills can take some measurements and determine whether issues with low performance, security threats, or limited RF coverage are causing the problems. If not, then the problems likely fall within the wired network. Thus, analysis can isolate network segments when diagnosing problems on the network.

Types of Wireless LAN Analysis

Analysis of wireless LANs falls into several categories. Each one, however, follows the same steps of measuring, interpreting, and reporting results.

RF Propagation Analysis

Propagation analysis verifies the coverage of the wireless LAN, which is critical in order for users to associate with an access point from any applicable location. Propagation will change over time because of reconfiguration of offices and presence of new machinery. This constant change makes it imperative to measure and assess propagation to avoid coverage holes in a proactive manner. Proactive analysis reduces frustration when users make use of applications requiring a high quality wireless link.

Performance Analysis

Performance analysis involves the measurement of throughput and delays in order to determine how well the wireless LAN is supporting data traffic. The goal of performance analysis is to minimize the impacts of the wireless LAN on the speed at which the user can utilize network applications. For example, monitoring a part of the network may reveal that utilization is approaching the capacity of a particular access point. Users in this area will likely be experiencing delays when using any network-intensive applications, and access to the network may not be possible if utilization increases any further. Provisions can be made to either add more access points in that area or offer a form of bandwidth control for each user or application.

Performance analysis may involve protocol interaction due to multiple 802.11 modulations in use in the same area. For example, 802.11g access points and stations may be configured to be backwards compatible with 802.11b stations and access points, but the reverse is not true. There will be environments in which 802.11g radios are configured to work with 802.11b radios, and there will be environments where 802.11g radios are configured to ignore 802.11b radios. Both of these situations have advantages and disadvantages that the analyst should be aware of.

Security Analysis

Companies must define security policies for wireless LANs. The problem is that installers and administrators sometimes make mistakes when configuring access points, wireless LAN switches, wireless gateways, and even client devices. For example, someone may forget to enable encryption for one particular access point, which makes the access point vulnerable to hackers. The hacker can associate with this access point from outside the facility and attack corporate resources. As a result, companies must periodically perform security assessments to ensure that all security policies are being met. Performing a manual security walk-through on a regular basis is incredibly time-consuming, so administrators should consider automated, distributed systems such as a wireless LAN

IDS. Automated systems with fixed policies are less likely to make mistakes than administrators performing manual scans.

Troubleshooting

Even well-designed wireless LANs will eventually have trouble meeting the needs of users. For example, spotty coverage can crop up after an incomplete, inaccurate, or poorly done site survey or when changes are made to the facility. If proactive analysis is not done regularly, then users will ultimately complain about coverage issues. Or, defective access point firmware may disallow some users from associating. This problem would also prompt users to complain to the help desk. Support staff must then reactively troubleshoot the problems and define a solution. As the wireless LAN administrator, you can analyze before the problem occurs and prevent it, or you can analyze after the end-users have become frustrated and start complaining.

Wireless LAN Analysis Tools

There are several types of tools available that aid in the analysis of wireless LANs, including protocol analyzers, RF analyzers, and simulation tools.

Protocol Analyzers

A protocol analyzer, sometimes referred to as a packet analyzer, is the primary tool for analyzing the protocols of wireless LANs. In general, protocol analyzers focus on measuring and displaying 802.11 MAC Layer frames, fields, and parameters (see Figure 1.1). However, many of the protocol analyzers also implement some basic form of RF signal measurement, such as signal amplitude, noise amplitude and channel number (see Figure 1.2). The protocol analyzer is an invaluable tool for the wireless LAN analyst.

FIGURE 1.1 A sample 802.11 frame trace using AirMagnet's Laptop Trio.

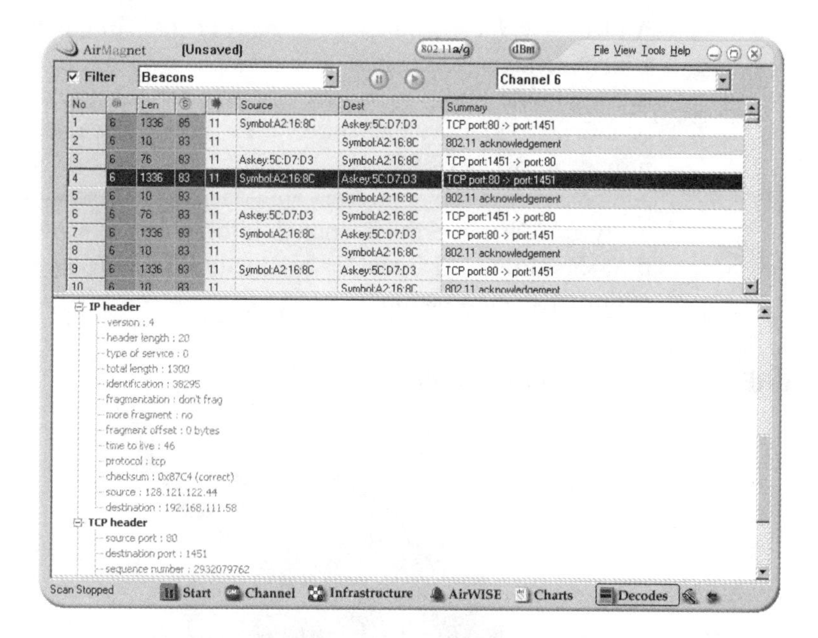

FIGURE 1.2 A sample RF signal display using AirMagnet's Laptop Trio.

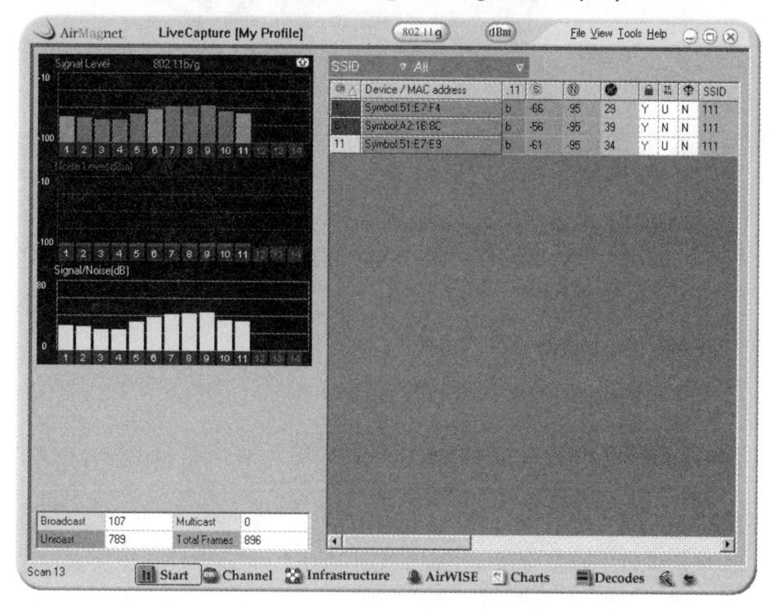

RF Analyzers

An RF analyzer measures and displays details of the 802.11 PHY Layer, which includes RF signals. For example, a spectrum analyzer displays the amplitude of the signal over the frequency spectrum, which is beneficial to identify the source of RF interference. Other RF analyzers provide a variety of RF signal information, such as delay spread, jitter, and phase shift. RF analyzers are generally necessary when the protocol analyzer does not provide enough information to solve an RF signal problem. Figures 1.3 and 1.4 are pictures of and some screenshots from a popular RF analysis tool.

FIGURE 1.3 An RF Analysis Tool

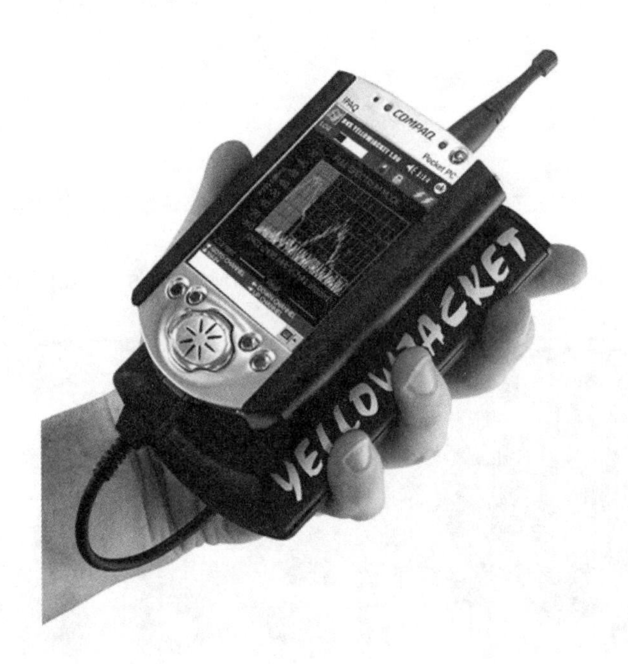

FIGURE 1.4 Screenshots from an RF Analysis Tool

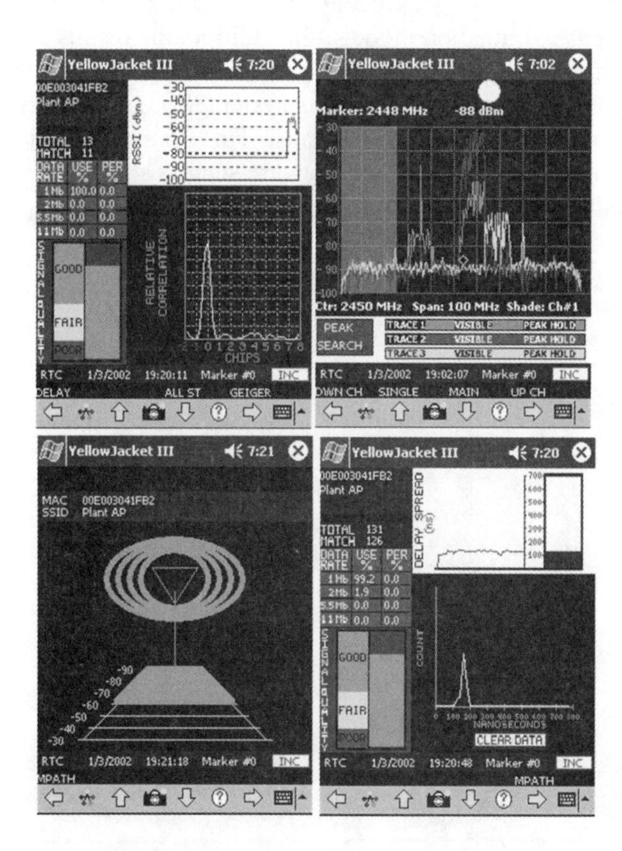

Simulation Tools

Simulation software enables the analyst to create a model of a wireless LAN that runs on a PC. By plugging in parameters, such as utilization, number of users, number of access points, and channel settings, a simulation run produces reports that estimate throughput and delays.

The advantage of simulation is that it can be accomplished before installing the network, making it an effective tool for use during the design stage of a wireless LAN deployment project. Continual use of simulation throughout the operational phase of the wireless LAN can also lend value when expanding the network.

A disadvantage of using simulators is that they are relatively expensive. They also require significant training before someone is able to develop effective simulation models. In addition, the results are only as accurate as the predictions you make for traffic loads and numbers of users. Figure 1.5 is a picture of an emulation system that can emulate up to 64 simultaneous 802.11a/b/g clients. This system in particular can use external traffic generation software such as shown in Figure 1.6 to produce real-world traffic flows.

FIGURE 1.5 WLAN Simulation System

FIGURE 1.6 Traffic Generation Software

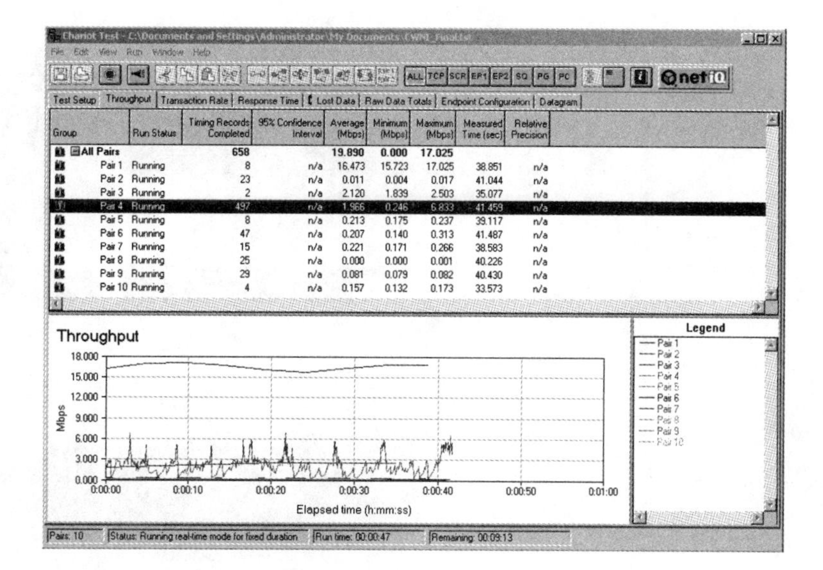

Intrusion Detection Systems

Many protocol analyzers have been significantly expanded to analyze not only the packets, but the events happening on the network as well. These events could be broadly categorized as security events and performance events. Wireless IDS may perform baseline testing, inventory the devices on the wireless network, and track all performance and security related events. Most early models of protocol analyzers were limited to software running on a single device. Recently, many manufacturers have released distributed systems that use remote hardware and/or software probes reporting to a centralized console.

In today's security- and performance-conscience organizations, it is hard to justify not having such protocol and event analysis information available to engineers and administrators. Figure 1.7 below illustrates a typical Wireless IDS console monitoring multiple sensors.

FIGURE 1.7 Wireless Intrusion Detection System Console

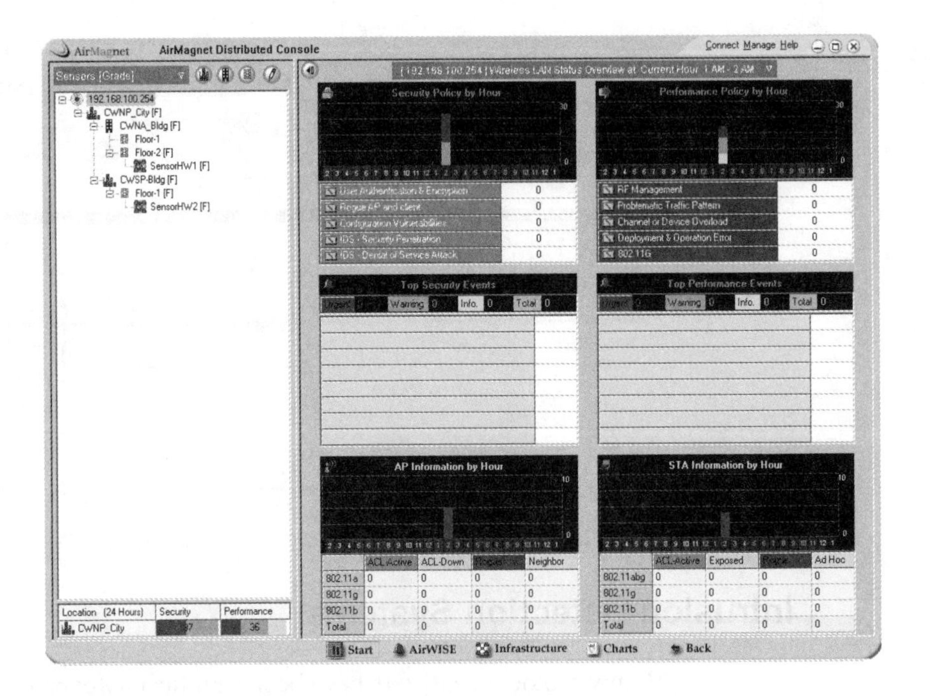

Summary

Wireless LAN analysis is certainly beneficial for improving the coverage, performance, and security of wireless LANs. Analysis comes in handy when trying to troubleshoot problems that occur during installation and operations of wireless LANs. In order to perform analysis, you should make use of protocol and RF analyzers to measure signals and monitor protocol operation.

Key Terms

Before taking the exam, you should be familiar with the following terms:

protocol analyzer

performance analysis

security analysis

Wireless Intrusion Detection System

Review Questions

1. What are three important reasons to implement wireless LAN protocol analysis?

2. What is a baseline analysis?

3. What is the difference between RF analysis and protocol analysis?

4. What are two types of analysis that may be performed with a wireless IDS?

5. What are three specific tasks that can be performed with wireless protocol analyzers?

CHAPTER

2

802.11 Protocol Architecture

CWAP Exam Objectives Covered:

- Explain and describe the frame exchange processes involved in:
  - Active and Passive Scanning
  - 802.11b/g mixed mode environments and protection mechanisms
- Explain Interframe Spacing
  - SIFS, PIFS, DIFS, EIFS
- Contrast the differences between the frame exchange processes in:
  - PCF vs. DCF access methods
- Explain 802.11 Contention
  - CSMA/CA (half duplex)
  - Backoff timer operation
  - Virtual Carrier Sense (NAV)
  - Physical Carrier Sense (CCA)
  - Contention Window operation
- Explain Slot Times
  - 802.11a/b/g
  - Special operation in 802.11b/g mixed mode environments
- Contrast the differences between the frame exchange processes in:
  - Infrastructure vs. Independent Service Sets

In This Chapter

802.11 Services

The 802.11 standard defines an architecture that satisfies various wireless requirements. A basic service set (BSS) involves one or more wireless stations (hereafter referred to as, "mobile stations," "client stations," or just, "stations") that share a single coordination function. An independent BSS (IBSS), which is often referred to as an Ad Hoc or peer-to-peer network, enables connections between stations with relatively little planning or provisioning. For example, two business people in a meeting room can switch their wireless clients to Ad Hoc mode and transfer documents to each other's laptops. 802.11 data frames in this form of network flow directly from one wireless client to another.

The advantage of the IBSS is that you can form the network quickly because such a network does not require an access point. An access point is a station that has Distribution System (DS) services including the distribution service responsible for delivering information between access points. Access points relay information within 802.11 data frames between stations on the same BSS and provide connections to other BSSs through a DS (generally Ethernet). Access points may also, and usually do, have an integration service to provide a portal function so that the access point can bridge data frames in and out of the BSS.

The 802.11 extended service set (ESS) is made up of one or more BSSs interconnected via access points. These types of networks are referred to as infrastructure wireless LANs and are the most common implementation of 802.11 networks in enterprises, homes, and public hotspots.

Each access point in an ESS contains an 802.11 station, an 802.3 (or similar) portal, and DS services. The access points enable stations to send data frames to both wireless stations located within the ESS and wired stations beyond the ESS. For example, a message being sent from one station to another will traverse an access point. The access point receives a data frame from User A and forwards its information to User B in a different 802.11 data frame.

Because of the connection to a wired network, an ESS architecture enables users to access a much wider range of services, such as web browsing, email, and corporate servers. In addition, roaming between access points is usually implemented by connecting each access point to Ethernet.

The deployment of wireless LANs using access points is much more common than Ad Hoc networks. As a result, this book focuses on describing 802.11 functions as they relate to ESSs. When necessary, however, this book describes variations corresponding to an IBSS.

Here we reiterate that the 802.11 standard refers to stations as either access point (AP) stations or Non-AP stations and uses the word "station" to refer to both. In this text we will refer to Non-AP stations as client stations, mobile stations, or simply stations, such as laptops using wireless PC cards for connectivity both to each other and to access points.

System Services

The IEEE 802.11 architecture allows for the possibility that the DS medium may not be identical to an existing wired LAN. A DS medium may be created from many different technologies including current IEEE 802 wired LANs. The 802.11 series of standards does not constrain the DS medium to be either data link or network layer based, nor do the 802.11 standards constrain a DS to be either centralized or distributed in nature. IEEE 802.11 explicitly does not specify the details of DS implementations. Instead, 802.11 specifies *services*. The services are associated with different components of the architecture. There are two categories of an IEEE 802.11 service - the station service (SS) and the Distribution System Service (DSS). Both categories of service are used by the 802.11 MAC sublayer.

The service provided by stations is known as the *station service*. The SS is present in every 802.11 station (including access points, because access points include station functionality). The SS is specified for use by MAC sublayer entities. All conformant stations provide SS. The station services are as follows:

- Authentication
- Deauthentication
- Frame Protection

- MSDU delivery

The services provided by the distribution system are known as *distribution system services*. The physical embodiment of various services may or may not be within a physical access point. These services are accessed via a station, and a station that is providing DS services is an access point. The DS services are as follows:

- Association
- Disassociation
- Distribution
- Integration
- Reassociation

Distribution System (DS)

The DS interconnects a set of Infrastructure Basic Service Sets (BSS) and portals to create an Extended Service Set (ESS). The DS consists of a set of DS Services (DSS) that allows any two stations in the ESS to communicate indirectly. In practice, the DS services are implemented in access points and in a medium that connects multiple access points to each other.

An Infrastructure BSS is a set of stations that share an instance of the Distributed Coordination Function (DCF) and is characterized by the data found in a stream of beacons produced by its access point. A portal is a logical connecting point between a DS and a network outside the ESS. While conceptually portals are distinct from access points and only communicate with the ESS via the DS, in practice portals are implemented in access points in the form of bridged IEEE 802.3 Ethernet ports. An access point is a station that includes DS services as well as Station Services (SS) and supports indirect communication between 802.11 stations.

An ESS that consists of one BSS has a relatively simple DS. The DS is entirely in the access point. The DS enables the access point to deliver messages:

- Between stations in the BSS (for communication within the BSS)
- Between stations in the BSS and the portal (for communication with devices outside the ESS)

Figures 2.1 illustrates Intra-BSS Communication.

FIGURE 2.1 Intra-BSS Communication

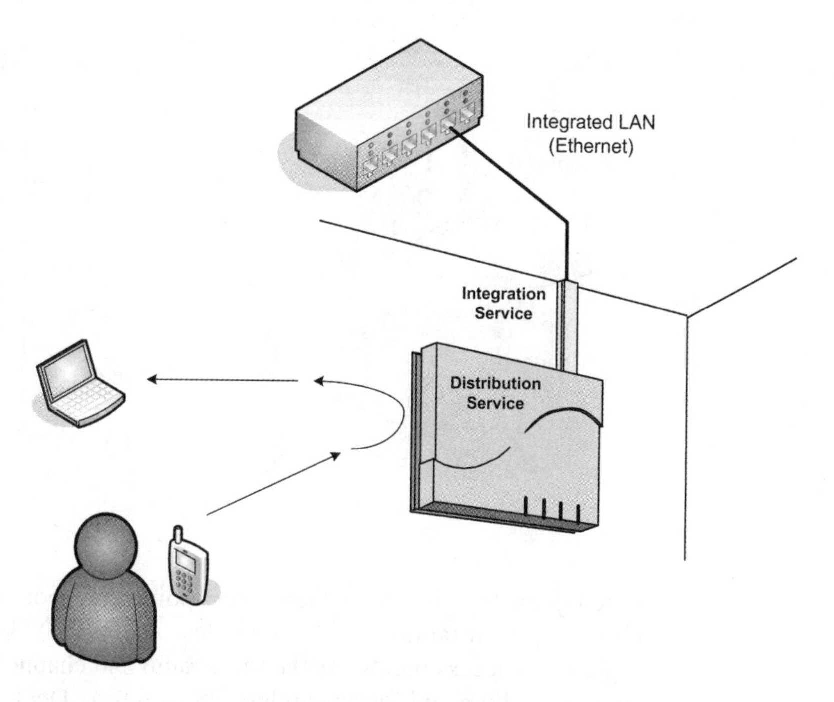

An ESS that consists of multiple BSSs has a more complex DS. The DS is in each access point and in the wired or wireless DS medium (DSM) across which the access points distribute messages to each other. The DS medium is sometimes incorrectly equated to the DS itself.

If the DS medium is Ethernet, then access points transport messages to each other using their portals and Ethernet ports (called the Integrated LAN). While the messages distributed between access points actually leave the ESS while traversing the integrated LAN, abstractly the messages are carried by a medium devoted to the DS and the messages do not leave the ESS. Figure 2.2 illustrates this concept.

FIGURE 2.2 Ethernet Distribution System Medium

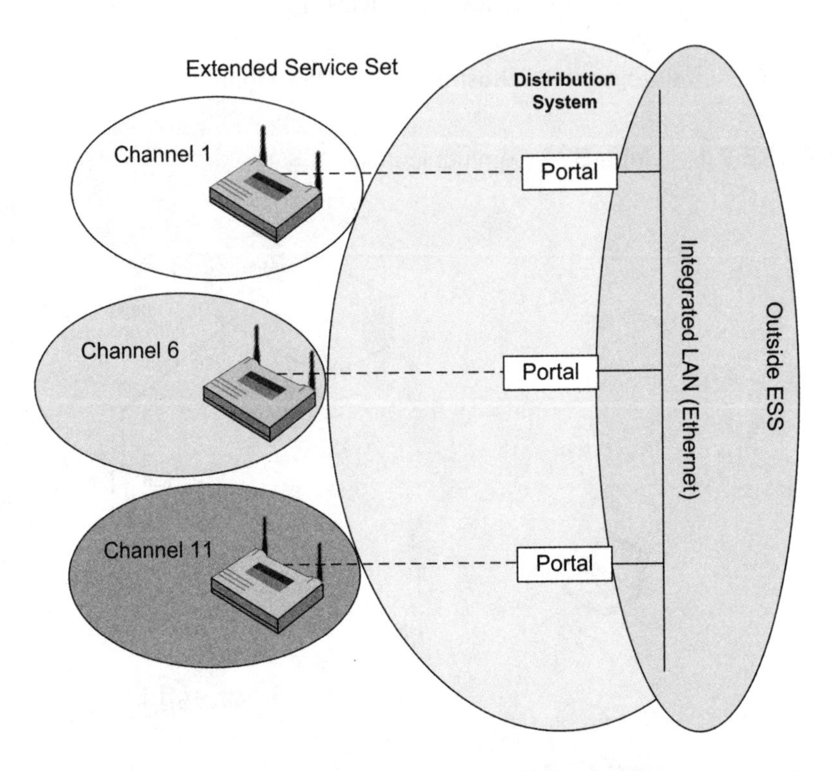

If the DS medium is 802.11 then access points transport messages to each other using their radios. This is a wireless DS (WDS). In a wireless DS single-radio access points use the same radio and channel for its BSS wireless medium and for its wireless DS meditum. Dual-radio access points may use one radio for the WDS, and another radio on a different channel for the BSS wireless medium. The wireless DS is a special case for transporting messages between access points of an ESS and is supported by the otherwise unused Address 4 field in the 802.11 MAC header. Figure 2.3 illustrates the physical construction of a WDS.

FIGURE 2.3 Wireless Distribution System

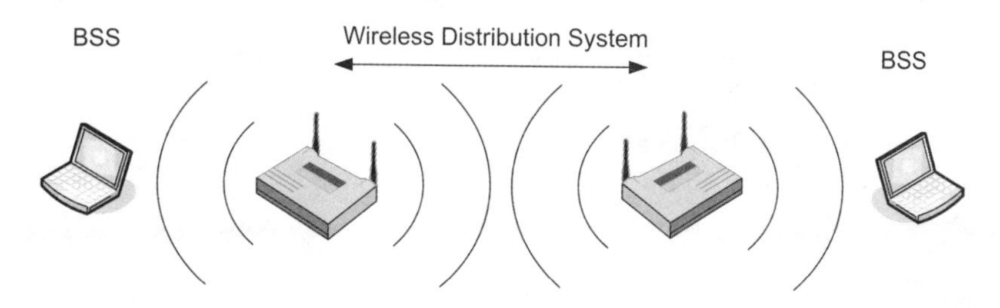

The DS medium may be a combination of wired and wireless segments, all bridged together as one data-link broadcast domain, and configured as one IP subnet. Marketing departments have given access points that support special media combinations and feature sets peculiar names to fit their applications such as, "wireless bridge", "wireless repeater", and "wireless workgroup bridge."

In summary, a DS brings indirect message delivery and portals to a set of basic service sets and creates an extended service set. Without a DS, all you have are Independent Basic Service Sets (IBSS) whose stations have to communicate with each other directly, if at all.

Integration Service

If the distribution service, one part of the DS, determines that the intended recipient of a message is a member of an integrated LAN, the "output" point of the DS would be a portal instead of an access point. Messages that are distributed to a portal cause the DS to invoke the Integration function (conceptually after the distribution service). The integration service is a DS service. Messages received from an integrated LAN (via a portal) for an 802.11 station will invoke the Integration function before the message is distributed by the distribution service. The details of an integration function are dependent on a specific DS implementation and are outside the scope of the 802.11 standard. As illustrated in figure 2.4 below, the distribution service comprises a thin layer above the MAC sublayer, and the integration service is between the distribution service and the integrated LAN.

FIGURE 2.4 802.11 Distribution System Services

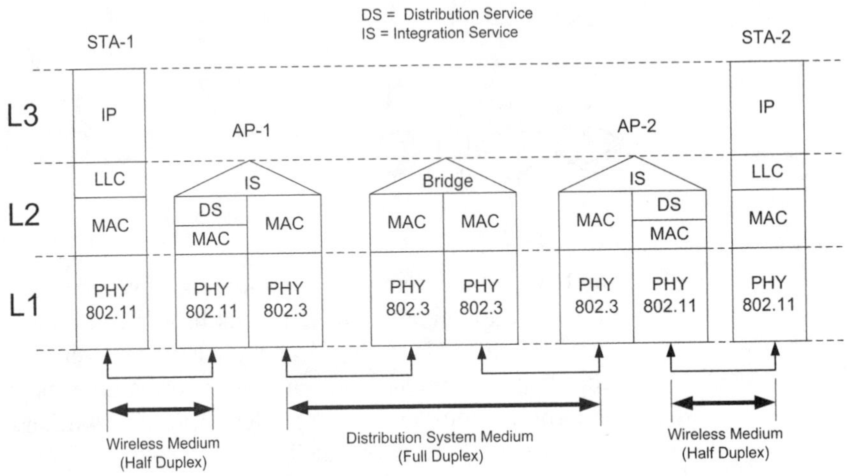

802.11 MAC Layer Overview

As with any IEEE 802 network, the 802.11 standard defines a common MAC Layer that controls the operation of stations (STAs) and access points (APs). Much of the MAC Layer functionality resides in firmware, and developers and end users have access to many of the attributes that change the configuration and operation of the wireless LAN.

The 802.11 MAC Layer provides the following functions and elements:

Scanning

Before authenticating to a BSS, a station may scan for all available BSSs (as advertised by an access point or station). The scanning station will find valid BSSs as those having the same service set identifier (SSID) as the station that is scanning. After association, the station may continue to scan when conditions occur that indicate the need to find another BSS. This functionality is implemented differently among chipset vendors, so it is common to see that one vendor's card may scan after association while another vendor's may not.

Synchronization

Some of the 802.11 functions require all stations to have synchronized clocks. Each access point (or master station in an Ad Hoc network) periodically broadcasts beacon frames at specific points in time based on a configurable interval. Stations update their clocks with a time value found in each beacon.

Frame transmission

A station must follow specific medium access procedures prior to frame transmission with the goal of only transmitting a frame when the medium is idle. Special timing elements enhance this process. A station receiving a unicast data or management frame must respond immediately with a positive acknowledgement frame if no errors are found.

Authentication

When a BSS is selected, the station goes through an authentication process to prove that it is a member of a set of stations authorized to join this BSS.

Association

If authentication is successful, the station can complete the association process. The association process involves exchange of capability information between the station and the BSS and establishes an access point/station mapping. In the case of an ESS, the access point allocates memory and assigns an association ID (AID) to the station requesting association. After association, the station can send and receive 802.11 data frames through the access point (a.k.a. invoking the DSS) unless an additional security mechanism is in place.

Reassociation

As users roam about a facility, they often travel out of range of the access point to which they are associated. In this case, the station reassociates with another access point. Generally, the decision to reassociate depends on received signal strengths of the corresponding beacon frames. These values are known as Received Signal Strength Indicators (RSSIs).

Data Protection

Optional encryption alters the payload bits of each data frame to prevent hackers from understanding the contents of frames in-transit between stations and access points.

Power management

Stations include transceivers that consume significant power that can drain batteries relatively quickly. The optional 802.11 power-save function allows a station to periodically enter and exit a doze state, which uses much less power and increases battery life.

Fragmentation

Sometimes, such as in the presence of RF interference, it is beneficial to break larger frames into smaller frames before transmission across a wireless link. The 802.11 standard defines fragmentation as an optional function, and it can be initiated on either the client station or the access point or both. Fragmentation is unique to each wireless link.

Request-to-Send / Clear-to-Send (RTS/CTS)

Hidden nodes may exist on the network that can significantly increase collisions, which reduces performance. The 802.11 RTS/CTS mechanism can optionally be used by stations and access points to announce the intended transmission of unicast data and management frames over a certain threshold size in order to limit retransmission time when hidden nodes are present. RTS/CTS may also used as an optional protection mechanism by 802.11g stations to announce the intended transmission of ERP-OFDM frames in an 802.11b/g mixed mode environment.

Management

The 802.11 standard defines a Management Information Base (MIB) that contains many parameters that impact the operation of the wireless LAN. It is possible to interface with this MIB through Simple Network Management Protocol (SNMP) over the IP network.

The remainder of this chapter will describe these functions and elements in greater detail.

Scanning

Each 802.11 station periodically scans each RF channel in order to find a BSS to join. The process of scanning is critical when a station is first activated. After powering up, the station will initiate scanning to find an initial BSS to join. As RF conditions change, the station will periodically scan and possibly reassociate with another BSS.

Before a station switches to a different channel than the channel on which it is associated, the station may (with some vendors) send a Null Function data frame with the power management bit in the frame control field of the MAC header set to "1" to inform the access point to buffer frames intended for the station. Once the station has retuned to its associated channel, the station will send another Null Function data frame with the power management bit set to "0" to inform the access point to begin forwarding frames again. This behavior prevents missed frames during scanning.

There are two forms of scanning: passive scanning and active scanning.

Passive Scanning

Passive scanning is the process through which a station listens to each channel (or set of channels) for a specific period of time. The station waits for the transmission of beacon management frames (a.k.a. beacons) having the SSID of the network that the station is configured to join. Beacons contain fixed fields and information elements that hold information about the BSS which are used by stations to determine whether or not the station may associate. Some vendors allow configuration of access points to remove the SSID value from the beacon so that the access point is not "announcing" its SSID to nearby stations. The passive scanning process is illustrated in Figure 2.5a.

FIGURE 2.5a Passive Scanning

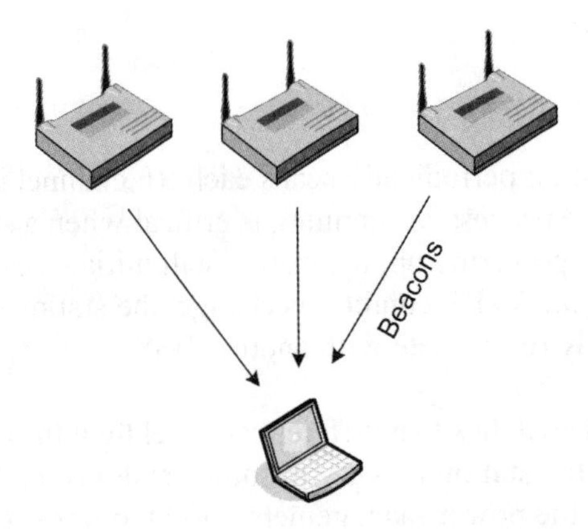

Once the station detects beacons from one or more access points, the station will decide which access point with which to associate based on a vendor-proprietary algorithm. The station will negotiate a connection on the applicable channel by proceeding with authentication and association processes. An advantage of passive scanning is that it does not require the transmission of any additional frames, which reduces overhead traffic on the wireless medium and improves overall network throughput.

Active Scanning

Active scanning requires that a station broadcast probe request frames indicating the SSID of the network that the station is configured to join. The station that sends the probe request frames will receive probe response frames from access points within range and having the specified SSID. This process, like that of passive scanning, provides information that the station can use to determine the access point with which to associate. Alternately, a station can send probes containing a broadcast SSID (a null value) that causes all access points within reach to respond. The active scanning process is illustrated in Figure 2.5b.

FIGURE 2.5b Active Scanning

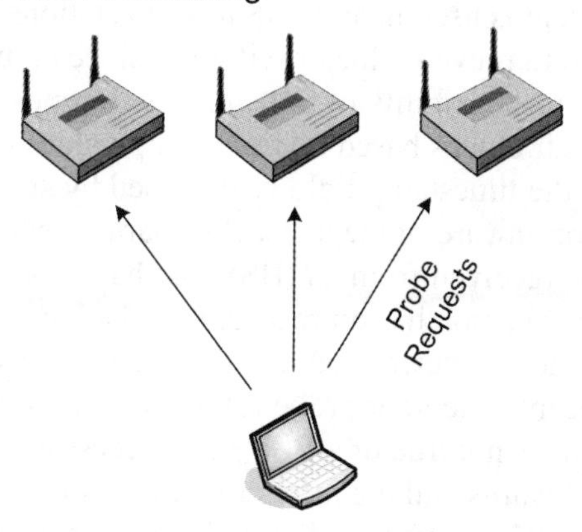

An access point must reply to all probes that contain the broadcast SSID or an SSID that matches its own. This standard is ignored when the vendor provides a proprietary mechanism allowing the network administrator to disable probe responses to probes with broadcast SSIDs. This feature is very common in today's access points and wireless LAN switches. With Ad Hoc networks, the station that generated the last beacon frame will respond to probes. The advantage of active scanning is that it identifies potential access points faster, which may be necessary if the client station is experiencing a rapid decrease in received signal strength from frames.

Synchronization

Stations within a BSS must remain in time synchronization with the access point (or each other in the case of an Ad Hoc network) to ensure that they are able to effectively implement all functions. Specifically, time synchronization in a BSS is needed for PCF and Power-Save modes. To support this, the access point (or individual station in an IBSS) periodically transmits beacon frames. In addition to announcing the presence of the BSS, a beacon contains the access point's clock value in the timestamp field of the beacon's frame body. Each station receiving the beacon from the access point to which they're associated uses the timestamp information to update its own clock accordingly. The timestamp is used by stations to update their clocks regardless of any other factors.

The timestamp field is used in an IBSS to keep stations' clocks synchronized and to prevent clock oscillation since all participating stations have the responsibility of transmitting beacons. IBSS stations will only reset their clocks based on timestamp values greater than their own clocks, and the timestamp field is also used by stations in an IBSS to make other stations aware of the latest BSS parameters. If a beacon or probe response is received from an IBSS that has parameters that differ from those that the station has on record, it will adopt the new values only if the timestamp field is newer than the timestamp that was associated with the most recent beacon or probe response data of which the station has a record. This is not true of a BSS. The access point could literally roll its clock backwards and the infrastructure would still work.[1] A sample protocol analyzer trace in Figure 2.6 shows the timestamp field below.

[1] IEEE 802.11 - 1999 (R2003) – Section 11.1.1.1, 11.1.4

FIGURE 2.6 Timestamp field in the beacon

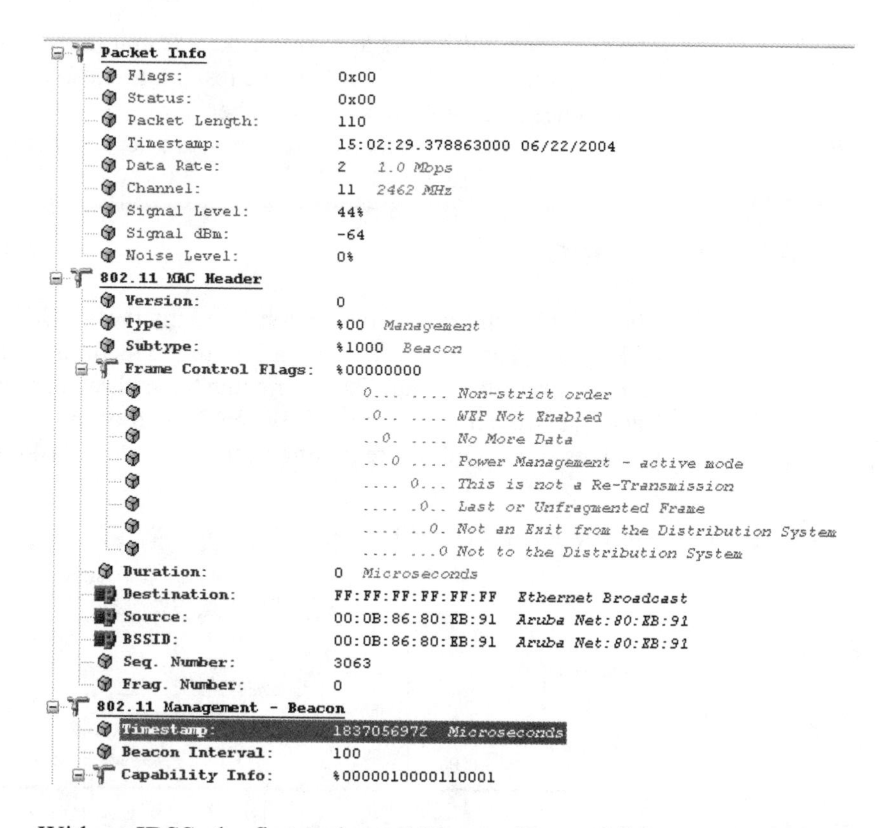

With an IBSS, the first active Ad Hoc station establishes the RF channel and starts sending beacons. Other Ad Hoc stations can join the network after receiving a beacon and accepting the IBSS parameters (e.g., beacon interval) found in the beacon frame. All stations that join the IBSS participate in the sending of beacons. Each station schedules itself to send a beacon a very short random delay interval after each Target Beacon Transmission Time (TBTT). TBTTs are the exact points in time that each access point has scheduled a beacon to be sent onto the wireless medium. TBTTs are based on a zero reference time and spaced according to the beacon interval. Each station that receives a beacon during its delay cancels sending its beacon this time and schedules one for a new random delay interval after the next TBTT.

Frame Transmission

The purpose of 802.11 is to move packets from higher layers across the wireless interface. This involves encapsulating these packets into 802.11 data frames. In addition, a host of other frames, such as beacons, association requests, and probe requests, are necessary to manage and control the wireless link.

Access Timing

The 802.11 standard defines four spacing intervals – SIFS, PIFS, DIFS, and EIFS – that defer a station's access to the medium and provide various levels of priority. Figure 2.7 below illustrates three of these intervals. Each interval defines the time between the end of the last symbol of the previous frame and the beginning of the first symbol of the next frame.[1]

FIGURE 2.7 Basic Medium Access Procedure

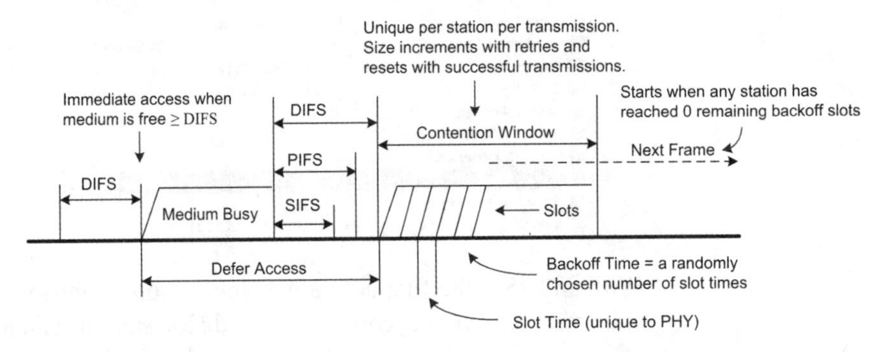

Short IFS (SIFS)

The SIFS is the shortest of the interframe spaces, providing the highest priority level by allowing some frames to access the medium before others. The following frames use the SIFS interval because they require expedient access to the network:

[1] IEEE 802.11 – 1999 (R2003) – Section 9.2.3

- Acknowledgement (ACK) frame
- Clear-to-Send (CTS) frame
- The second (or subsequent) fragment of a fragment burst

FIGURE 2.8a Use of Short Interframe Spaces (SIFS)

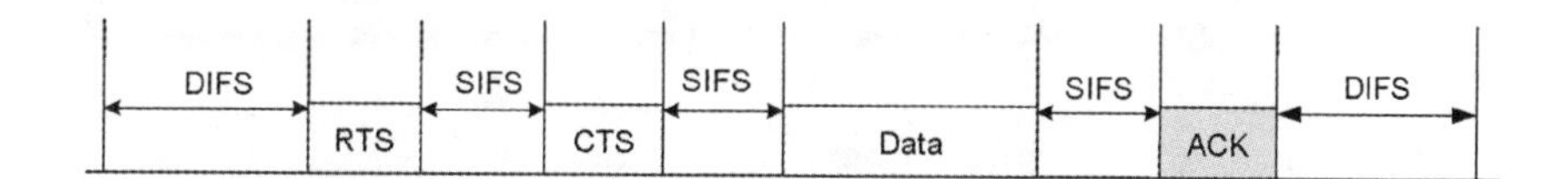

PCF IFS (PIFS)

The PIFS is the interval that an access point operating under PCF mode uses to gain access to the medium before stations using DCF mode. PIFS provides priority over DIFS because of its shorter time period. Access points only use PIFS during a CFP to send the initial beacon and to retransmit frames for which they did not receive an ACK frame.

FIGURE 2.8b Use of PCF Interframe Spaces (PIFS)

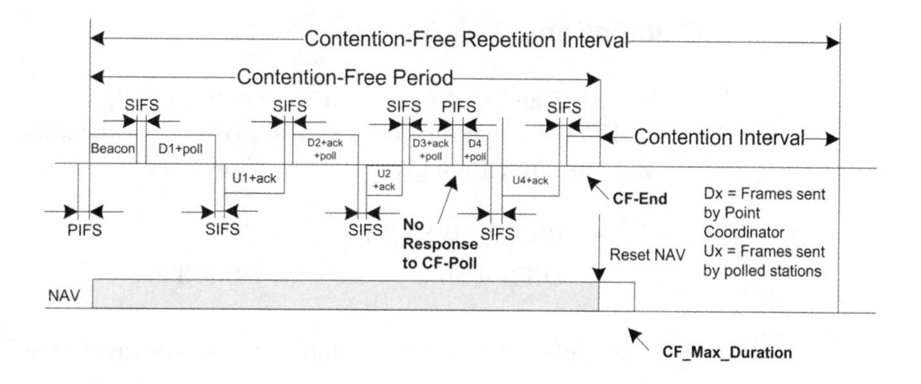

DCF IFS (DIFS)

All stations operating according to the distributed coordination function use the DIFS interval for transmitting data frames and management frames. This spacing makes the transmission of these frames lower priority than PCF-based transmissions.

FIGURE 2.8c Use of DCF Interframe Spaces (DIFS)

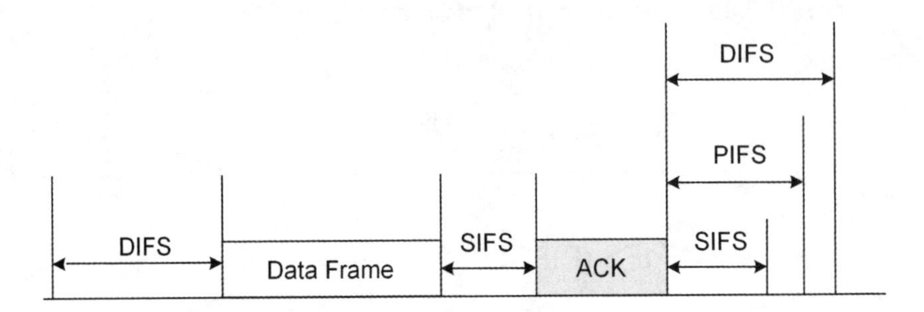

Extended IFS (EIFS)

The EIFS is used by DCF stations whenever the PHY has indicated to the MAC that a frame transmission began but did not result in the correct reception of a complete MAC frame with a correct Frame Check Sequence (FCS) value.[1]

Calculations

The SIFS and Slot Time values are defined in the relevant physical layer (PHY) specification. The PIFS and DIFS are calculated based on SIFS and Slot Time for a given PHY as shown below:

- PIFS = SIFS + Slot Time
- DIFS = SIFS + Slot Time + Slot Time

The EIFS is derived from a number of factors as shown in the formulas below:

```
EIFS (µs) = aSIFSTime + (8 x ACKSize) + aPreambleLength +
aPLCPHeaderLngth + aSIFSTime + (2 x aSlotTime)
```

[1] IEEE 802.11 – 1999 (R2003) – Section 9.2.3.4

or, you could simplify the formula a little to say the following:

```
EIFS (µs) = (2 x aSlotTime) + (2 x aSIFSTime) +
aPreambleLength + aPLCPHeaderLngth + (8 x 14)
```

ACKSize is the length, in bytes, of an ACK frame. The 802.11-1999 standard does not intend for the 8xACKSize to be converted to bits, but rather that the number yielded by 8xACKSize (8x14=112) be taken as units of microseconds (since 1 Mbps was the lowest mandatory rate of 802.11-1999). The same is true of PLCP Preamble Length and PLCP Header Length.

FIGURE 2.9 EIFS Reference Chart

802.11 FH	**Section 14.9 FH PHY characteristics:**	
	aSlotTime:	50 μs
	aSIFSTime:	28 μs
	aPreambleLength:	96 μs
	aPLCPheaderLength:	32 μs
	EIFS:	28 + 112 + 96 + 32 + 28 + 100 = 396 μs
802.11 DS	**Section 15.3.3 DS PHY characteristics (long preamble):**	
	aSlotTime:	20 μs
	aSIFSTime:	10 μs
	aPreambleLength:	144 μs
	aPLCPheaderLength:	48 μs
	EIFS:	10 + 112 + 144 + 48 + 10 + 40 = 364 μs
802.11a	**Section 17.4.4 OFDM PHY characteristics:**	
	aSlotTime:	9 μs
	aSIFSTime:	16 μs
	aPreambleLength:	20 μs
	aPLCPheaderLength:	4 μs
	EIFS:	16 + 112 + 20 + 4 + 16 + 18 = 186 μs
802.11b	**Section 18.3.4 DS High Rate PHY characteristics (long preamble):**	
	aSlotTime:	20 μs
	aSIFSTime:	10 μs
	aPreambleLength:	144 μs
	aPLCPheaderLength:	48 μs
	EIFS:	10 + 112 + 144 + 48 + 10 + 40 = 364 μs
or	**Section 18.3.4 DS High Rate PHY characteristics (short preamble):**	
	aSlotTime:	20 μs
	aSIFSTime:	10 μs
	aPreambleLength:	72 μs
	aPLCPheaderLength:	24 μs
	EIFS:	10 + 112 + 72 + 24 + 10 + 40 = 268 μs
802.11g	**Section 19.8.4 OFDM Extended Rate PHY characteristics:**	
	aSlotTime:	20 μs (long slot time)
	aSIFSTime:	10 μs
	aPreambleLength:	20 μs
	aPLCPheaderLength:	4 μs
	EIFS:	10 + 112 + 20 + 4 + 10 + 40 = 196 μs
or	**Section 18.3.4 DS High Rate PHY characteristics (short preamble):**	
	aSlotTime:	9 μs (short slot time)
	aSIFSTime:	10 μs
	aPreambleLength:	20 μs
	aPLCPheaderLength:	4 μs
	EIFS:	10 + 112 + 20 + 4 + 10 + 18 = 174 μs

In 802.11g basic service sets where legacy (NonERP) stations are present, 20μs slot times are used for backwards compatibility.

Medium Access

As part of transmitting a frame, the station must first gain access to the wireless medium, which all stations within the BSS share.

Distributed Coordination Function (DCF)

The DCF is a mandatory 802.11 access method and offers a distributed mechanism for sharing access to the common wireless medium. DCF makes use of a carrier-sense multiple access/collision avoidance (CSMA/CA) protocol, which is similar to what 802.3 (Ethernet) LANs utilize.

Carrier Sense

The DCF includes a combination of both virtual and physical carrier sense mechanisms and attempts to allow only one station to transmit on the wireless medium at any given time.

FIGURE 2.10 The 802.11 DCF access method

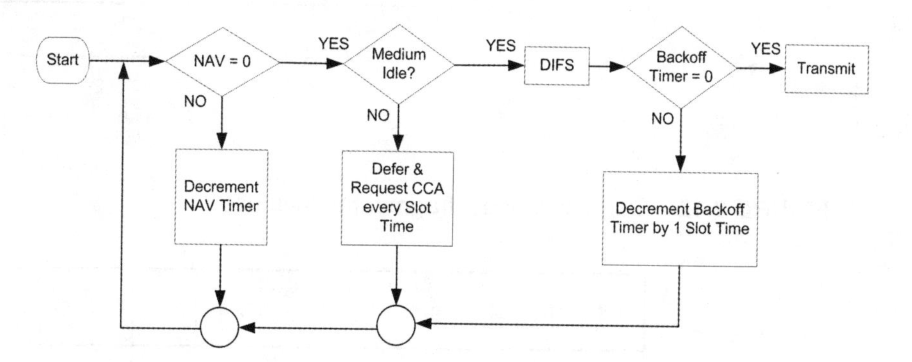

The DCF carries out a virtual carrier sense protocol based on reservation information found in the duration field of all frames' MAC headers. This process helps improve performance by reducing collisions and allowing stations to efficiently complete frame exchange sequences. The content of the duration field, which is a time value measured in microseconds (μs), announces impending use of the medium for the remaining frames in a frame exchange sequence.

Duration Values

For example, a station transmitting a data frame would include a duration value that includes the time needed for one SIFS and to receive the acknowledgement frame from the receiving station. Stations receive and monitor the duration field of all frames sent within their own BSS and by nearby stations, which might not be part of their BSS, on the same or overlapping channel. Stations place the duration field contents in their network allocation vector (NAV). The NAV operates like a timer, starting with a value equal to the duration field value of the last frame transmission sensed on the medium and counting down to zero. Once the NAV reaches zero, the station proceeds with the physical medium access process of sensing the wireless medium. Examples of the duration values carried by frames are shown in figures 2.11 and 2.12 below.

FIGURE 2.11 Duration values, no fragmentation in use

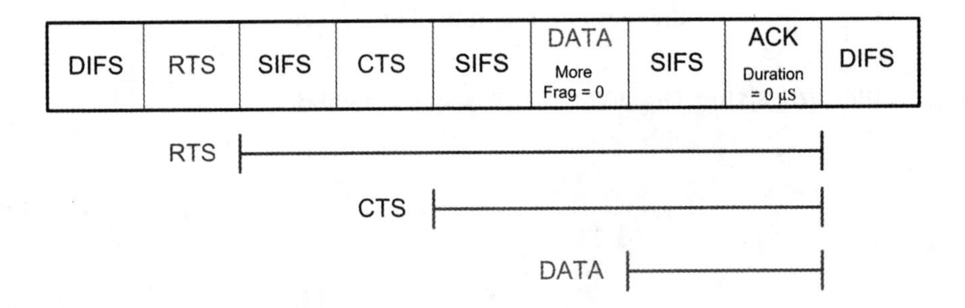

FIGURE 2.12 Duration values, fragmentation in use

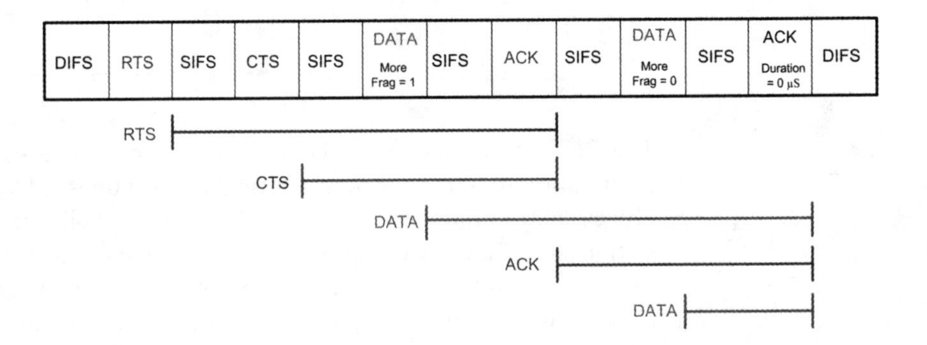

When the station needs to transmit a frame, the DCF determines whether the wireless medium is idle by monitoring the station's receiver. If the medium is not idle, which is the case when another station is transmitting or RF interference is present, then the station wanting to transmit the frame will defer (wait). Once the medium is idle, however, the station may transmit the frame provided its NAV is equal to zero and its backoff timer has expired.

If the transmitting station learns that its frame transmission was not successful (due the absence of an acknowledgement frame within a specific amount of time), then the station will execute the medium access process again in order to retransmit the frame. A station reduces the probability of collisions among stations sharing the medium by using a random backoff time. The period of time immediately following a busy medium is when the highest probability of collisions occurs, especially under high utilization. The reason for this occurrence is that many stations may be waiting for the medium to become idle and may transmit at the same time. Once the medium is idle, a random backoff time defers a station from transmitting a frame, minimizing the chance that stations will collide.

The MAC entity calculates the random backoff time using the following formula:

```
Backoff Time = Random() X aSlotTime
```

Random( ) is a pseudo-random integer drawn from a uniform distribution over the interval [0,CW], in which CW (contention window) is an integer within the range of values of the MIB (management information base) attributes aCWmin and aCWmax. The random number drawn from this interval should be statistically independent among stations. aSlotTime equals a constant value found in the station's MIB.

FIGURE 2.13 Contention Window and Slot Times

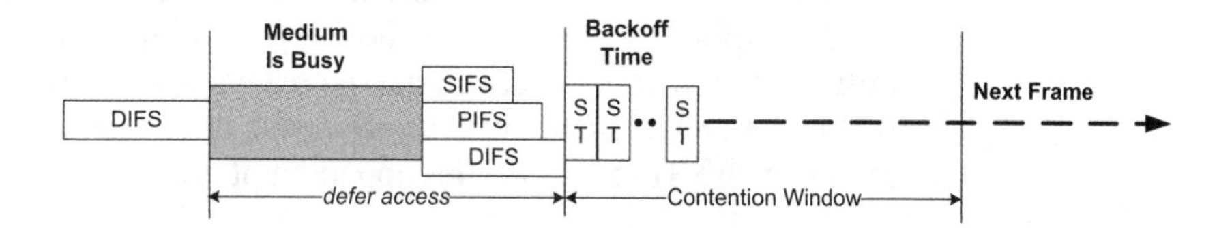

CW increases exponentially for retransmissions (unique per station) to minimize collisions and maximize throughput for both low and high network utilization. Under low utilization, stations are not forced to wait very long before transmitting their frame. On the first or second attempt, a station will make a successful transmission within a short period of time. If the utilization of the network is high, the protocol holds stations back for longer period of times to avoid the probability of multiple stations transmitting at the same time.

FIGURE 2.14 Contention Window values

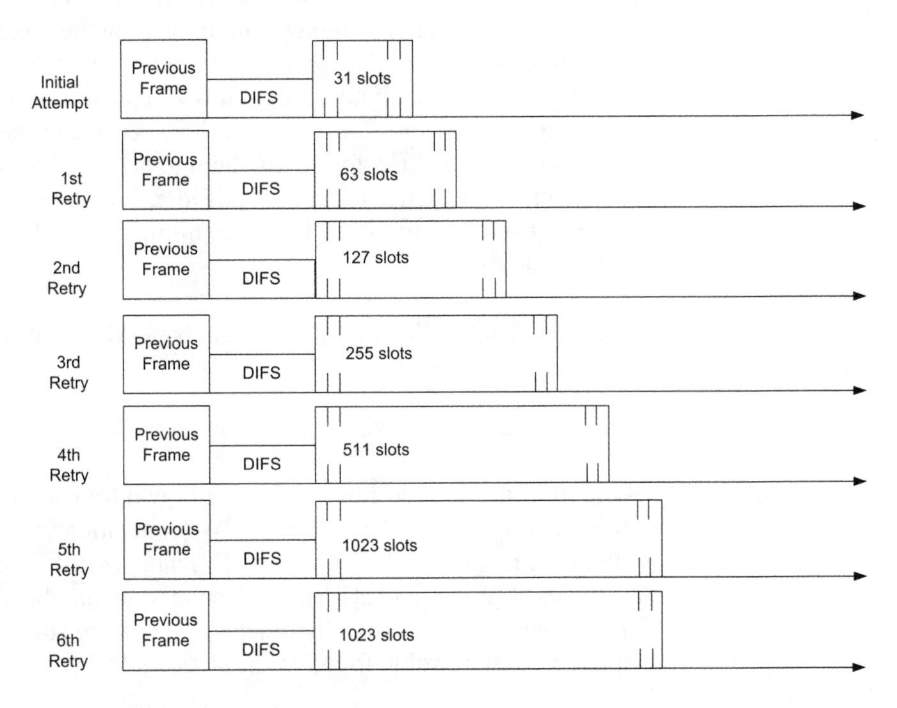

Under high utilization, the value of CW increases to relatively high values after successive retransmissions, providing substantial transmission spacing between stations needing to transmit. This mechanism does a good job of avoiding collisions; however, stations on networks with high utilization will experience substantial delays while waiting to transmit frames. In 802.11 and 802.11b networks, the values for aCWmin and aCWmax are 31 (first transmission attempt) and 1023 (5th and all subsequent retries until the retry limit is reached).[1] In 802.11g networks, the values for aCWmin and aCWmax are 15 (first transmission attempt) and 1023 (6th and all subsequent retries until the retry limit is reached).[2] This gives 802.11g a statistical transmission advantage in 802.11b/g mixed mode networks.

Extended Rate Physical (ERP)

ERP is a term used to specify PHYs able to both transmit and receive certain OFDM physical modulations defined by Clause 19, also known as 802.11g. In the case of ERP networks (i.e., where 802.11g access points are present), ERP-OFDM stations (also called ERP stations) may need to implement protection mechanisms (when mandated by the access point), which enable NonERP stations (e.g., 802.11b – commonly referred to as Clause 15 and 18 stations) to understand when ERP-stations are transmitting. If all of the stations in the area are ERP then protection mechanisms are unnecessary. If one or more of the stations in this scenario are NonERP types, a condition which is made known to the BSS by the access point using the beacon frame, then all ERP stations must use a protection mechanism that announces the pending use of the medium using a modulation that the NonERP client stations can understand.

---

[1] IEEE 802.11 - 1999 (R2003) – Sections 9.2.4 & 15.3.3
[2] IEEE 802.11g - 2003 – Section 19.8.4

FIGURE 2.15 Protection Mechanisms

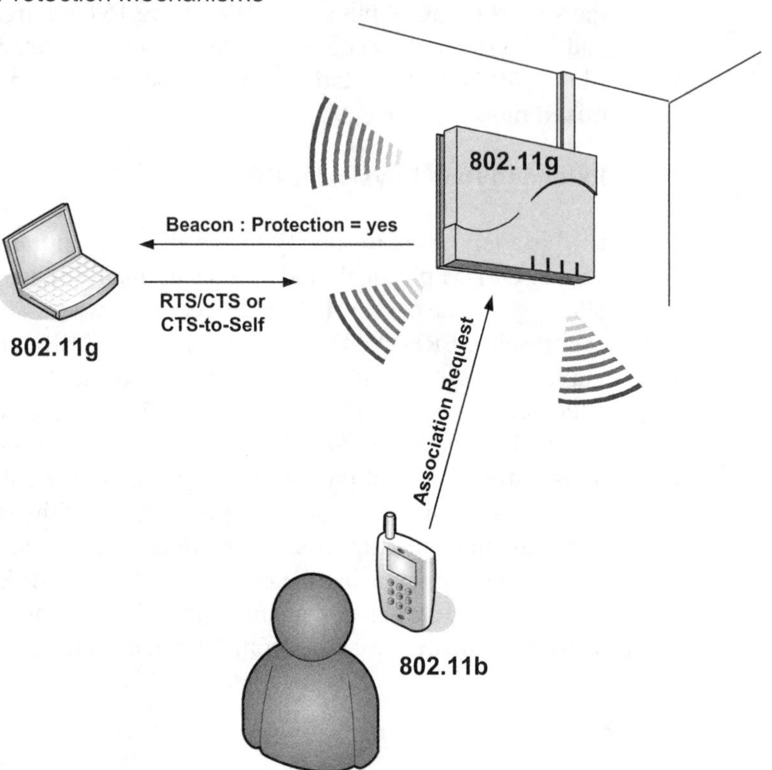

For example, when the access point allows a NonERP station to associate, it immediately tells all ERP stations in the BSS to use "protection." ERP stations must then use DSSS-modulated RTS/CTS and/or CTS-to-Self protection mechanisms. ERP stations must transmit the protection mechanism frames using a modulation understandable by all stations in the BSS, typically either BPSK or QPSK. The duration field value in the MAC frame header must be set to an appropriate value to reserve the medium for a period of time long enough for the entire transmission (when fragmentation is not in use). [1]

[1] IEEE 802.11g - 2003 – Section 7.3.2.13

FIGURE 2.16 Protection Announced in Beacons

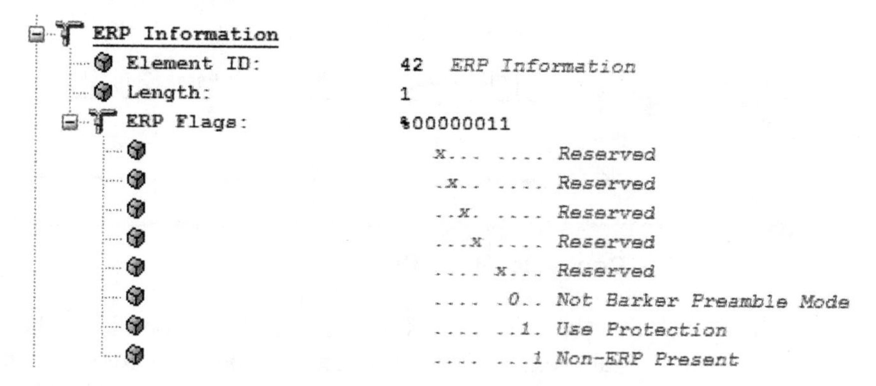

The duration value will be understood by NonERP stations so they will update their NAV, causing them to refrain from transmitting during the reservation period specified by the ERP station.[1]

Point Coordination Function (PCF)

The optional priority-based point coordination function provides contention-free frame transfer for processing time-critical information transfers via a station polling mechanism. There are very few vendors that support this access method, so this book only covers the basic operation. Using PCF, a point coordinator (a software function) resides in the access point to control the transmission of frames to and from stations.

Only Infrastructure Basic Service Sets may use PCF. Additionally, when PCF is in use, it is not actually PCF only, but rather a combination of PCF and DCF. PCF operates on top of DCF. Polling happens during the Contention-Free Period (CFP), and regular CSMA/CA access happens during the Contention Period (CP).

---

[1] For more information on Protection Mechanism use in ERP wireless LANs, see *Protection Ripple in 802.11 WLANs* (Devin Akin, June 2004) at www.cwnp.com

FIGURE 2.17 CFP/CP Alternation (a.k.a. Contention-Free Repetition Interval)

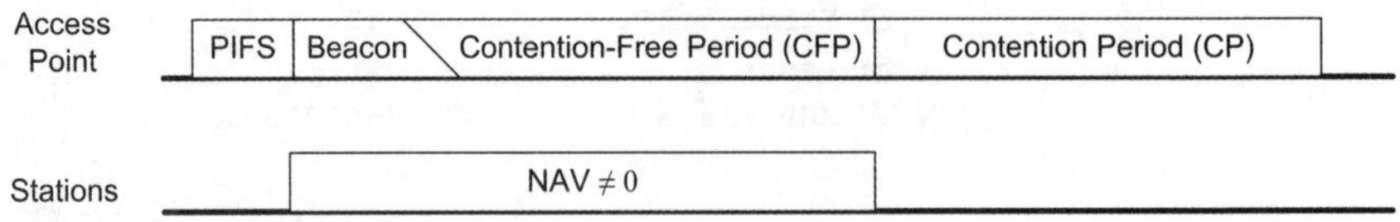

Access Point	PIFS	Beacon	Contention-Free Period (CFP)	Contention Period (CP)
Stations		NAV ≠ 0		

Contention-Free Period

At the beginning of the CFP, the point coordinator has an opportunity to gain control of the medium. The point coordinator senses the medium at the beginning of each CFP. If the medium is idle after a PIFS interval, the point coordinator sends a beacon frame that includes the CF Parameter Set element.

Before receiving a CFP beacon, stations set their NAVs to the CFPMaxDuration value found in the previous beacon's CF Parameter Set. Stations set their NAV at the expected Target Beacon Transmission Time (TBTT), which is the same time the beacon is expected to be transmitted. Since stations anticipate their backoff in this manner, there is a less likely chance of colliding with the beacon when it is sent by the point coordinator. Stations know at which TBTT the CFP beacon will be transmitted based on the CFPPeriod value found in the CF Parameter Set.

When stations receive the beacon, they see that the beacon contains a duration field value of 32,768 which is reserved for use in all frames sent during a CFP. This special value causes receiving stations to skip the normal NAV update procedure based on Duration field found in every received frame and continue counting down from the NAV value set by the CFPDurRemaining value in the CF Parameter Set found only in beacons.

FIGURE 2.18 Duration Field Value & CF Parameter Set

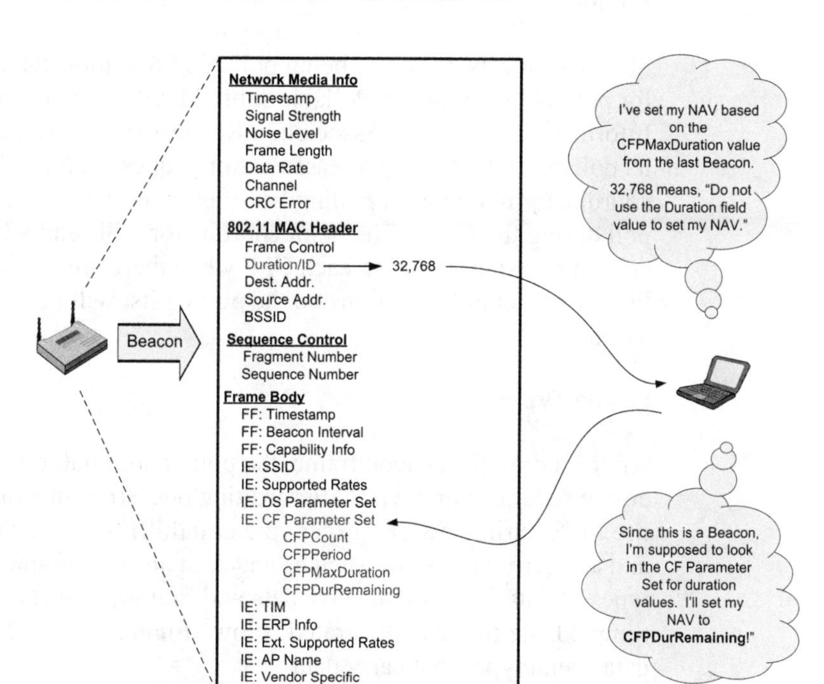

CFPDurRemaining indicates the estimated time, in Time Units (TUs), remaining in the present CFP, and is set to zero in CFP Parameter elements of beacons transmitted during the contention period. TUs are measured in Kilomicroseconds (Kμs). The CFPDurRemaining value is used by all stations to update their NAVs during CFPs. The CF Parameter Set structure is shown in figure 2.19 below.

FIGURE 2.19 CF Parameter Set Structure

	Element ID	Length	CFP Count	CFP Period	CFP MaxDuration (TU)	CFP DurRemaining (TU)
Octets:	1	1	1	1	2	2

Polling

Stations have an option of being pollable. A station can indicate its desire for polling using the CF Pollable subfield within the Capability Information field of an Association Request frame. A station can change its pollability by issuing a reassociation request frame. The point coordinator maintains a polling list of eligible stations that may receive a poll during the CFP. The point coordinator will send a CF-Poll frame to at least one station during each CFP when there are entries in the polling list. A subset of the stations on the polling list will be polled, in order, by ascending AID.

Frame Types

After sending the beacon frame, the point coordinator may then transmit any of several frame types after waiting one SIFS interval. All functions use SIFS during the CFP except the initial PIFS before the first beacon and any error conditions such as a lack of an ACK frame. There are four types of data frames that carry data and four types that do not. The Simple Data frame is illustrated below. Figures 2.20 – 2.23 show those data frame types that carry data:

- Simple Data
- Data+CF-Ack
- Data+CF-Poll
- Data+CF-Poll+CF-Ack

Figures 2.24 – 2.27 show those data frames that do not carry data:

- CF-Ack
- CF-Poll
- CF-Ack+CF-Poll
- Null Function

Simple Data frame: This point coordinator frame is directed from the access point's point coordinator to a particular station. If the point coordinator does not receive an acknowledgement frame from the recipient, the point coordinator can retransmit the unacknowledged frame during the CFP after a PIFS interval. A point coordinator can send unicast, broadcast, and multicast frames to all stations, including stations in Power-Save mode.

FIGURE 2.20 Simple Data frame (DATA)

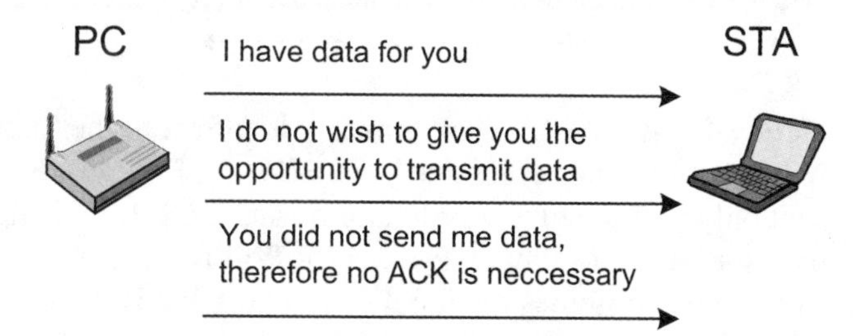

Data + CF-Ack frame: Identical to the Simple Data frame, with the following exceptions. The Data+CF-Ack frame may be sent only during a CFP. The acknowledgement carried in this frame is acknowledging the previously received data frame, which may have been sourced from some other client.

FIGURE 2.21 Data + CF-Ack frame (DATA)

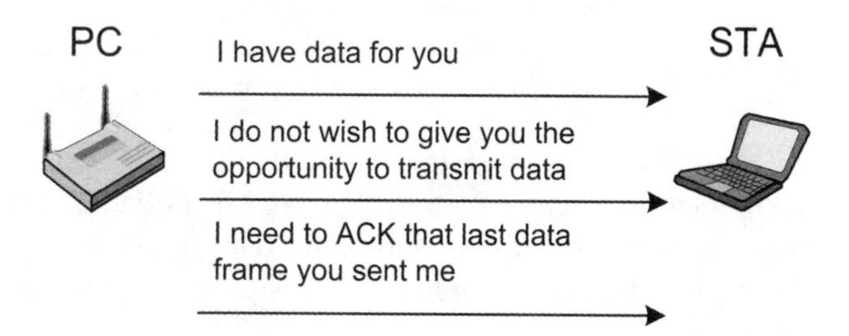

Data + CF-Poll frame: In this case, the point coordinator sends a data frame to a station and polls that same station for sending a contention-free frame. This is a form of piggybacking that reduces overhead on the network.

FIGURE 2.22 Data + CF-Poll frame (DATA)

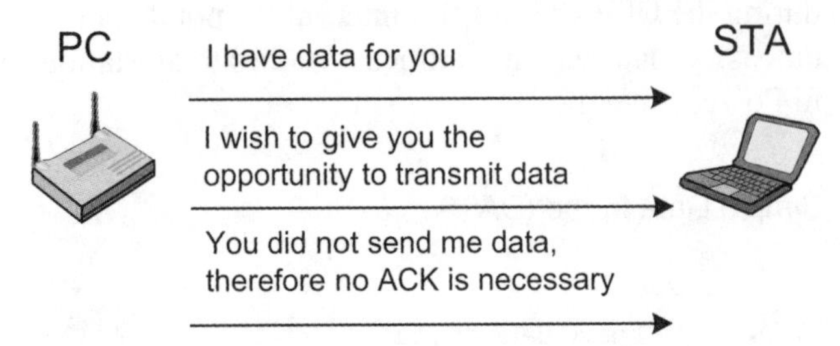

Data + CF-Poll + CF-Ack frame: Identical to the Simple Data frame, with the following exceptions. The Data+CF-Ack+CF-Poll frame may be sent only by the point coordinator during a CFP. This frame is never sent by a station. The Data+CF-Ack+CF-Poll frame is never used in an IBSS. This frame combines the functions of both the Data+CF-Ack and Data+CF-Poll frames into a single frame.

FIGURE 2.23 Data + CF-Poll + CF-Ack frame (DATA)

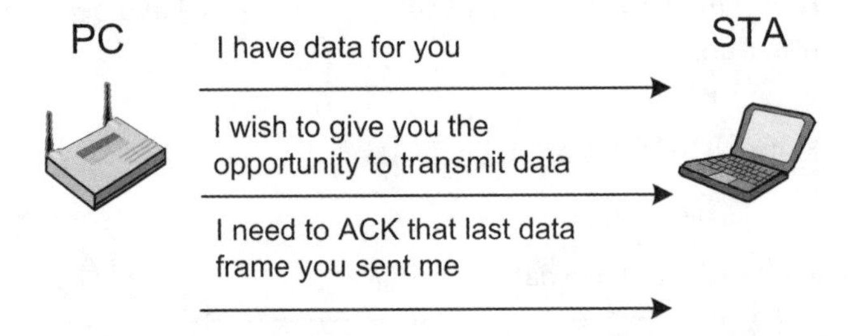

CF-Ack frame: This frame is used by the point coordinator to acknowledge receipt of a frame from a CF-Pollable station. This frame also means that the point coordinator does not have any data to send to the station and does not wish to give the station the right to transmit a frame (by polling).

FIGURE 2.24 CF-Ack frame (NO DATA)

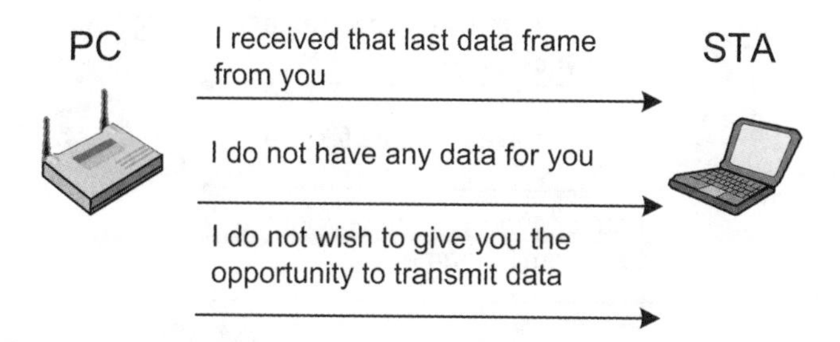

CF-Poll frame: The point coordinator sends this frame to a station granting the station permission to transmit a single frame to any destination. If the polled station has no frame to send, it must send a Null Function Data frame. If the sending station does not receive any frame acknowledgement, it cannot retransmit the frame unless the point coordinator polls it again. If the receiving station of the contention-free transmission is not CF Pollable, it acknowledges the reception of the frame using DCF rules. DCF rules specify that the receiving station wait one SIFS, and then transmits the ACK frame without regard to its own NAV.

FIGURE 2.25 CF-Poll frame (NO DATA)

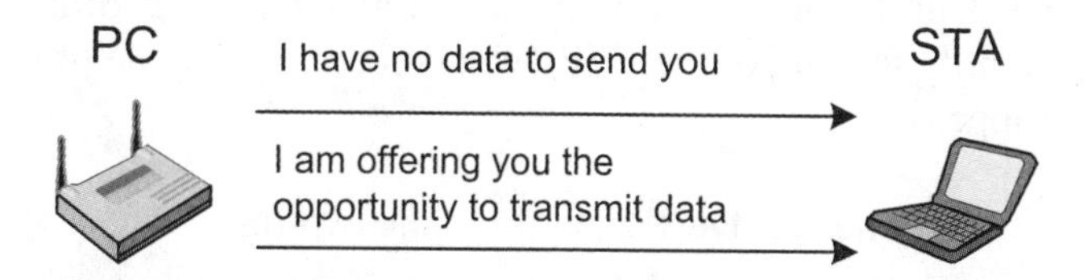

CF-Ack + CF-Poll frame: Used by the point coordinator to acknowledge a correctly received frame and to solicit a pending frame from a station.

FIGURE 2.26 CF-Ack + CF-Poll frame (NO DATA)

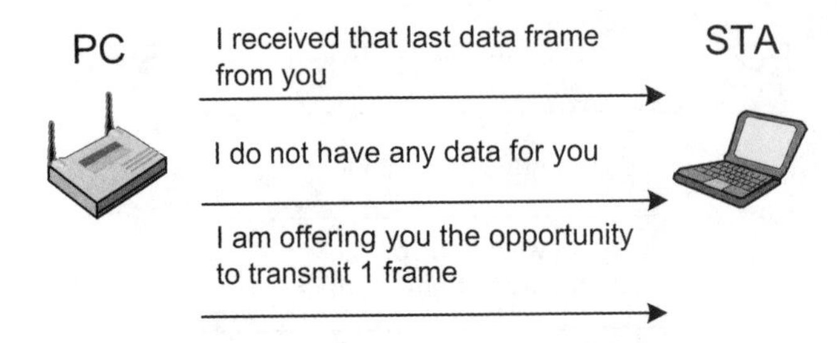

Null Function frame: This frame is used by a station to notify the point coordinator that it has not data to transmit. Also, this frame may be used by stations both in PCF and DCF modes to notify the access point of its intention to change its power state (awake or asleep).

FIGURE 2.27 Null Function frame (NO DATA)

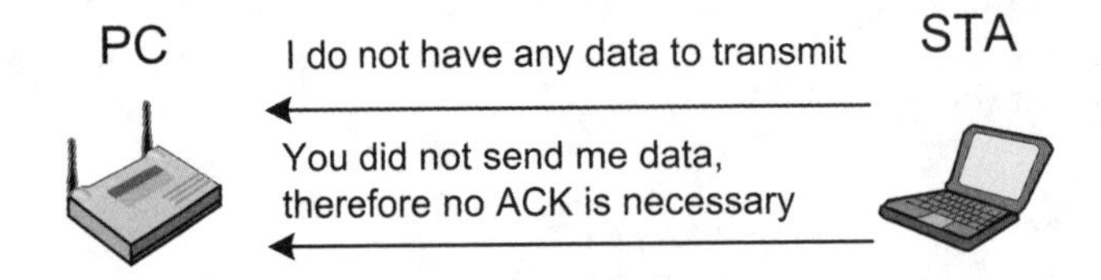

CF-End frame: This frame is sent by the point coordinator to the BSS to identify the end of the CFP, which occurs when one of the following happens:

- The CFPDurRemaining time expires
- The point coordinator has no frames to transmit and no stations to poll

FIGURE 2.28 CF-End frame

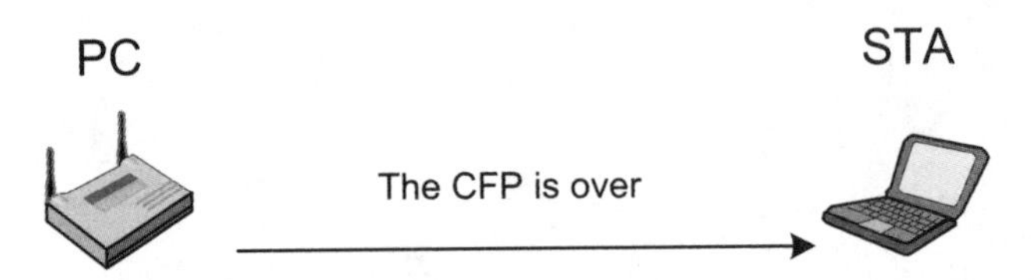

CF-End + CF-Ack frame: This frame is sent by the point coordinator to the BSS to identify the end of the CFP, and to acknowledge a data frame previously sent by a station. The CF-End + Ack frame is directed at the broadcast group address. It has a zero microsecond duration which resets all stations' NAVs to zero.

FIGURE 2.29 CF-End + CF-Ack frame

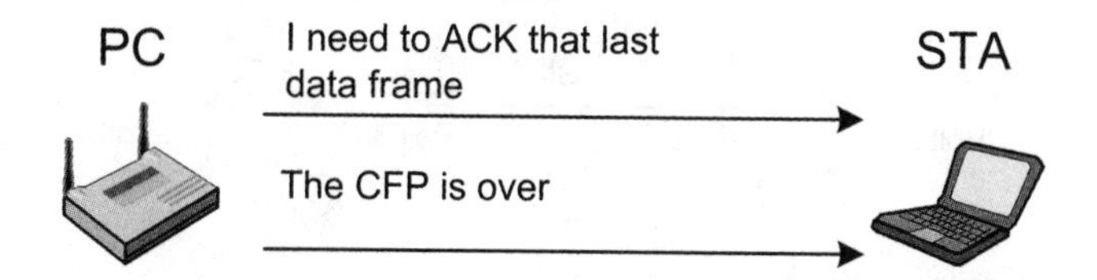

The PCF does not routinely operate using the backoff function associated with DCF mode; therefore, a risk of collisions exists when overlapping point coordinators are present on the same PHY channel. This may be the case when multiple access points form an infrastructure network (an ESS). To minimize collisions in a situation like this, the PC utilizes a random backoff time only if it experiences a busy medium when attempting to transmit the initial beacon.

PCF Summary

PCF has a complex method of providing priority access to a set of stations in a BSS. Without visualizing all of these intricacies, it is difficult to keep them straight. Few if any vendors currently implement PCF 802.11 wireless LANs. Review figures 2.30 – 2.31 below to help reiterate functionality both during the CFP and the transition between CFP and CP.

FIGURE 2.30 Contention-Free Period Medium Access

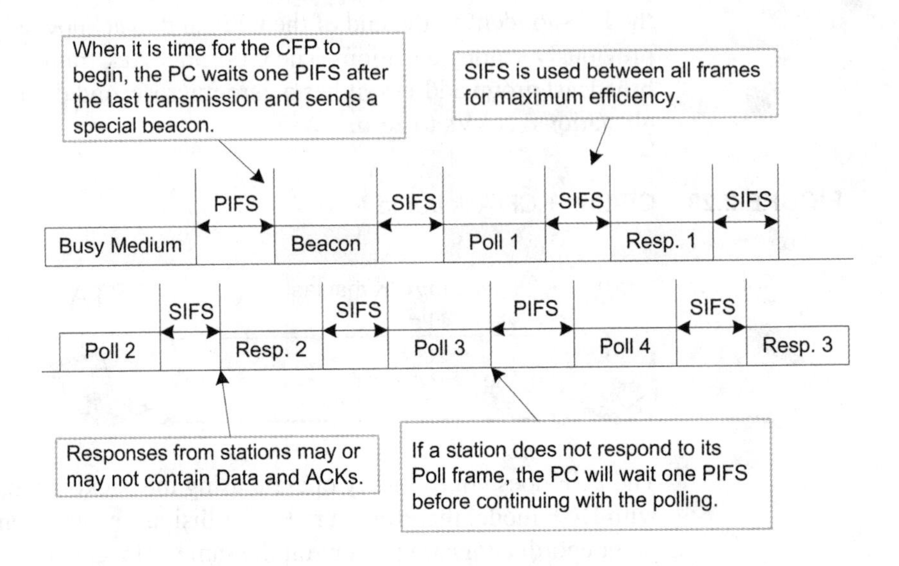

FIGURE 2.31 Ending the Contention-Free Period and Starting the Contention Period

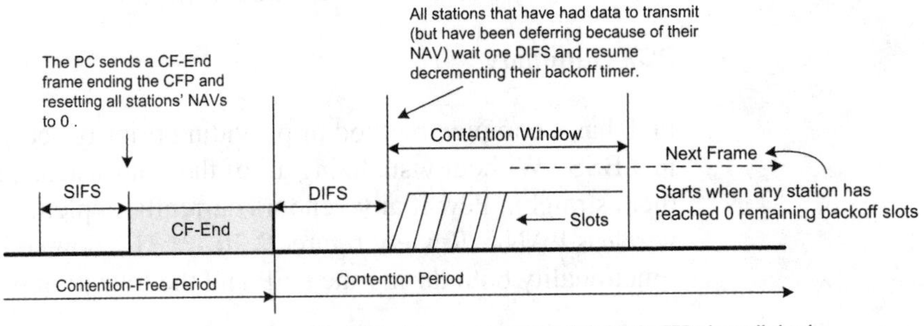

Error Control

Because of potential transmission impairments, such as RF interference and collisions, bit errors can disrupt the transmission of frames. For example, station A may send a Request-to-Send (RTS) frame and never receive the corresponding Clear-to-Send (CTS). Or, station A may send a data frame and never receive an acknowledgement. Because of these problems, the MAC Layer has error recovery mechanisms.

Stations initiating frame exchange sequences have the responsibility of error recovery. The station or access point receiving a data frame will respond with an acknowledgement frame if no errors in the frame are found. The error control function involves the retransmission of the data frame after a period of time if no acknowledgement frame is received from the destination station. This also takes into account that bit errors could have made the acknowledgement frame unrecognizable.

The MAC layer has limits on how many times it will retry an apparently failed frame exchange sequence. The relevant MIB variables are *dot11RTSThreshold*, *dot11ShortRetryLimit* and *dot11LongRetryLimit*, whose default values are 2347, 7 and 4 respectively. If a frame is less than or equal to the dot11RTSThreshold value in length (too short to be preceded by RTS/CTS control frames), then the MAC layer will transmit it dot11ShortRetryLimit times before discarding it. If a frame is long enough to be preceded by RTS/CTS control frames, then the MAC layer will transmit it dot11LongRetryLimit times before discarding it. Default values for short and long try limits are seven and four respectively.[1]

Unless the RTS threshold is lower than the default, all frames are "short." And although called "retry" limits they are really "try" limits.

Data Rate Shifting

The 802.11 standard recognizes that vendors may incorporate an automatic data rate shifting process that enables the station to adjust the data rate of the frame transmission to make efficient use of the medium. If the station senses too many retransmissions or the received signal strength decreases below a specific threshold, then the station will shift down to a lower data rate. This enables the station to maintain association with the access point at greater ranges and in the presence of RF interference.

[1] IEEE 802.11 - 1999 (R2003) – Section 9.2.5.3

Summary

The 802.11 MAC Layer specifies two primary medium access methods: PCF and DCF. PCF is polling based and DCF is purely contention based. Each medium access method has its advantages and disadvantages, but DCF is clearly the most popular implementation in the industry. Each access method has a set of frame types that are used to coordinate data transfer between stations and access points.

Each station uses passive or active scanning (or both) to identify a potential BSS with which to associate. After associating with an access point, a station can send and receive data frames and roam between access points. A solid understanding of duration values, backoff procedures, contention windows, and mixed mode environments is important to troubleshooting performance problems.

Key Terms

Before taking the exam, you should be familiar with the following terms:

active scanning

Distributed Coordination Function (DCF)

Distribution System (DS)

duration value

Extended Interframe Space (EIFS)

passive scanning

PCF Interframe Space (PIFS)

Point Coordination Function (PCF)

protection mechanisms

Short Interframe Space (SIFS)

Review Questions

1. When a station that is associated to an access point is performing active scanning, on what channel does it send probe request frames?

2. For a given PHY, what two parameters are used to calculate the value of DIFS?

3. Name two Distribution System Services.

4. In PCF mode, describe the situation that warrants the AP to send a CF-End+CF-Ack frame.

5. In PCF mode, what value found in the beacon is used to set (and reset) a station's NAV?

6. In an 8011g network, the contention window on a first transmission attempt has a maximum number of how many slots?

7. The first data fragment in a fragment burst carries a duration value sufficient to reserve the medium during the transmission of which frames?

8. In an IBSS (Ad Hoc) network, what qualifier determines whether or not a station will accept the updated parameters in a beacon?

9. In an 802.11g network, what causes the access point to enable *Use_Protection* in its beacons?

10. If an error is encountered during a Contention Free Period that causes an acknowledgement to be corrupted in transit, what interframe space is used by the point coordinator to maintain control of the medium?

11. A CTS frame always contains a duration value equal to the time it will take to transmit which frames?

12. Which service within an access point allows the access point to function as a portal device?

13. When sent during a CFP, what is the function of a Null Function data frame?

CHAPTER

3

Connectivity and Data Protection

CWAP Exam Objectives Covered:

- Explain and describe the frame exchange processes involved in:
  - Authentication and Association
  - Disassociation, Reauthentication, and Deauthentication
  - Roaming within an ESS
  - Security Mechanisms

In This Chapter

Authentication

Because of the open broadcast nature of wireless LANs, designers need to implement appropriate levels of security. The 802.11 standard describes the following two types of authentication services: Open System authentication and Shared Key authentication.

Open System Authentication

Open system authentication is the default authentication service that simply announces a station's desire to authenticate with the access point. This is a very simple two-step process. First, the station wanting to authenticate with an access point (or another station in the case of an IBSS) sends an authentication management frame containing the sending station's identity. The receiving access point (or station) then sends back an authentication management frame indicating success or failure of the authentication, as seen in Figure 3.1.

FIGURE 3.1 Authentication frame

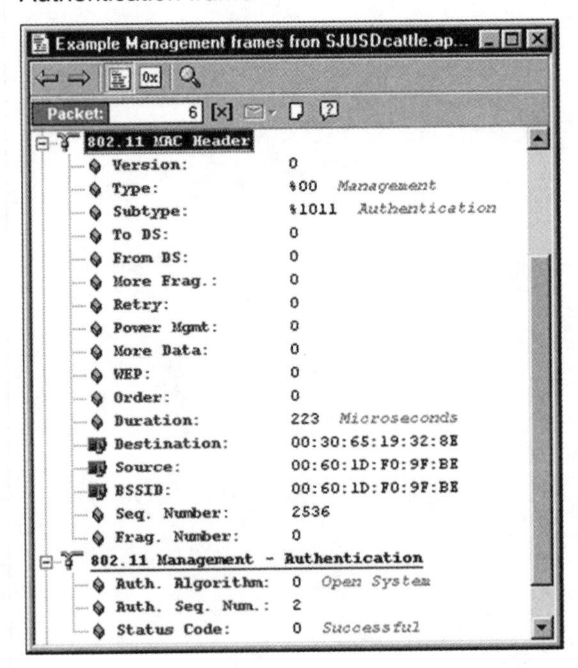

This means that Open System authentication is a null authentication. The identification of the station is not really checked. See Figure 3.2 to view the Open System authentication process.

FIGURE 3.2 Two-Step Open System Authentication Process

CH	Len	S	#	Source	Dest	Summary
1	30	48	1	Netgear:66:E5:F1	00:0D:ED:A5:4F:70	802.11 authentication
1	10	73	1		Netgear:66:E5:F1	802.11 acknowledgement
1	30	73	11	00:0D:ED:A5:4F:70	Netgear:66:E5:F1	802.11 authentication
1	10	81	11		00:0D:ED:A5:4F:70	802.11 acknowledgement
1	69	51	1	Netgear:66:E5:F1	00:0D:ED:A5:4F:70	802.11 association request
1	10	71	1		Netgear:66:E5:F1	802.11 acknowledgement
1	80	76	11	00:0D:ED:A5:4F:70	Netgear:66:E5:F1	802.11 association response
1	10	81	11		00:0D:ED:A5:4F:70	802.11 acknowledgement

Shared Key Authentication

Shared Key authentication is an optional mechanism that involves a more rigorous exchange of frames and attempts to determine whether the requesting station is authentic. Stations authenticate through shared knowledge of a secret 802.11 wired equivalent privacy (WEP) key.

FIGURE 3.3 Four-step Shared Key Authentication Process

CH	Len	S	#	Source	Dest	Summary
1	30	83	1	00:0C:30:52:93:6A	00:0D:ED:A5:4F:70	802.11 authentication
1	10	71	1		00:0C:30:52:93:6A	802.11 acknowledgement
1	160	61	11	00:0D:ED:A5:4F:70	00:0C:30:52:93:6A	802.11 authentication
1	10	83	11		00:0D:ED:A5:4F:70	802.11 acknowledgement
1	168	81	1	00:0C:30:52:93:6A	00:0D:ED:A5:4F:70	802.11 authentication
1	10	60	1		00:0C:30:52:93:6A	802.11 acknowledgement
1	30	61	11	00:0D:ED:A5:4F:70	00:0C:30:52:93:6A	802.11 authentication
1	10	81	11		00:0D:ED:A5:4F:70	802.11 acknowledgement
1	83	83	1	00:0C:30:52:93:6A	00:0D:ED:A5:4F:70	802.11 association request
1	10	60	1		00:0C:30:52:93:6A	802.11 acknowledgement
1	80	60	11	00:0D:ED:A5:4F:70	00:0C:30:52:93:6A	802.11 association response
1	10	83	11		00:0D:ED:A5:4F:70	802.11 acknowledgement

The Shared Key authentication service uses a four-step transmission of frames as follows.

1. A requesting station sends an authentication frame to the access point (or station in the case of an IBSS).
2. The access point replies with an authentication frame containing 128 octets of clear text (unencrypted) challenge text.
3. The requesting station copies the challenge text into a new authentication frame which it encrypts and sends to the access point.
4. The access point decrypts the authentication frame and compares the challenge text found there to the original. If a match occurs, then the access point replies with an authentication frame indicating a successful authentication. If not, the responding access point (or station) sends an authentication frame containing an unsuccessful indication.

Shared Key authentication is very easy to break because the process exposes both the encrypted and unencrypted form of the challenge text. Since the WEP key is used both for authentication and for data encryption, this is a severe security risk. As a result, you should avoid using Shared Key authentication. Open System authentication with WEP allows automatic connectivity to the access point, but the user must then have the correct WEP key configured in order to pass traffic through the access point. Instead of using either of these basic security methods, you should consider using WPA-PSK, 802.1X/EAP-WPA, IEEE 802.11i, or VPN technologies for authentication and encryption. These methods impose user-based authentication based on passwords, tokens, and/or digital certificates.

Association

The process of association means that a station joins a BSS and accepts operating parameters, such as available data rates. In order to associate with an access point, the station must be configured with the SSID of the access point or be using the broadcast SSID, which is often null (blank). A matching SSID is generally sufficient to continue with association.

After authenticating with the access point, a station must complete the association process by sending an Association Request frame to the access point. The access point then responds with an Association Response frame that includes information regarding the BSS. At this point, the client station has joined the BSS and is able to send and receive data frames. This process can be seen in Figures 3.2 and 3.3 above.

Use of 802.1X/EAP based authentication mechanisms are very common in today's wireless LANs. It is important to point out that 802.1X/EAP authentication is actually comprised of two separate authentication and association processes. First, Open System authentication followed by association is used to connect the station to the access point so that the access point can talk to the station for the purpose of the second authentication process: EAP authentication. Figure 3.4 illustrates the steps you might see in a wireless protocol analyzer while analyzing the authentication and association of a wireless station using 802.1X/EAP as the security protocol. Following this exchange, all data frames will be encrypted. 802.1X/EAP authentication is covered in more detail in the Data Protection portion of this chapter.

FIGURE 3.4 802.1X/EAP Authentication and Association Process

No	M	Time	Delta		Len			Source	Dest	Summary
10		5/5 20:53:51.449511	3.540251	1	30	48	1	Netgear:66:E5:F1	00:0D:ED:A5:4F:70	802.11 authentication
11		5/5 20:53:51.449822	3.540562	1	10	73	1		Netgear:66:E5:F1	802.11 acknowledgement
12		5/5 20:53:51.450089	3.540829	1	30	73	11	00:0D:ED:A5:4F:70	Netgear:66:E5:F1	802.11 authentication
13		5/5 20:53:51.450297	3.541037	1	10	81	11		00:0D:ED:A5:4F:70	802.11 acknowledgement
14		5/5 20:53:51.451488	3.542228	1	69	51	1	Netgear:66:E5:F1	00:0D:ED:A5:4F:70	802.11 association request
15		5/5 20:53:51.451798	3.542538	1	10	71	1		Netgear:66:E5:F1	802.11 acknowledgement
16		5/5 20:53:51.452242	3.542982	1	80	76	11	00:0D:ED:A5:4F:70	Netgear:66:E5:F1	802.11 association response
17		5/5 20:53:51.452352	3.543092	1	10	81	11		00:0D:ED:A5:4F:70	802.11 acknowledgement
18		5/5 20:53:51.452895	3.543635	1	36	50	11	Netgear:66:E5:F1	00:0D:ED:A5:4F:70	802.1x: EAPOL-Start
19		5/5 20:53:51.453005	3.543745	1	10	85	11		Netgear:66:E5:F1	802.11 acknowledgement
20		5/5 20:53:51.453321	3.544061	1	78	75	11	00:0D:ED:A5:4F:70	Netgear:66:E5:F1	802.1x:EAP ID/request
21		5/5 20:53:51.453434	3.544174	1	10	80	11		00:0D:ED:A5:4F:70	802.11 acknowledgement
22		5/5 20:53:51.453706	3.544446	1	78	75	11	00:0D:ED:A5:4F:70	Netgear:66:E5:F1	802.1x:EAP ID/request
23		5/5 20:53:51.453823	3.544563	1	10	78	11		00:0D:ED:A5:4F:70	802.11 acknowledgement
24		5/5 20:53:51.454286	3.545026	1	46	50	11	Netgear:66:E5:F1	00:0D:ED:A5:4F:70	802.1x:EAP ID/response
25		5/5 20:53:51.454400	3.545140	1	10	75	11		Netgear:66:E5:F1	802.11 acknowledgement
26		5/5 20:53:51.455086	3.545826	1	46	48	11	Netgear:66:E5:F1	00:0D:ED:A5:4F:70	802.1x:EAP ID/response
27		5/5 20:53:51.455200	3.545940	1	10	76	11		Netgear:66:E5:F1	802.11 acknowledgement
28		5/5 20:53:51.458244	3.548984	1	78	73	11	00:0D:ED:A5:4F:70	Netgear:66:E5:F1	802.1x:EAP LEAP/request
29		5/5 20:53:51.458357	3.549097	1	10	78	11		00:0D:ED:A5:4F:70	802.11 acknowledgement
30		5/5 20:53:51.458907	3.549647	1	73	50	11	Netgear:66:E5:F1	00:0D:ED:A5:4F:70	802.1x:EAP LEAP/response
31		5/5 20:53:51.459016	3.549756	1	10	75	11		Netgear:66:E5:F1	802.11 acknowledgement
32		5/5 20:53:51.471192	3.561932	1	78	76	11	00:0D:ED:A5:4F:70	Netgear:66:E5:F1	802.1x:EAPOL-EAP success
33		5/5 20:53:51.471277	3.562017	1	10	78	11		00:0D:ED:A5:4F:70	802.11 acknowledgement
34		5/5 20:53:51.471624	3.562364	1	57	50	11	Netgear:66:E5:F1	00:0D:ED:A5:4F:70	802.1x:EAP LEAP/request
35		5/5 20:53:51.471631	3.562371	1	10	73	11		Netgear:66:E5:F1	802.11 acknowledgement
36		5/5 20:53:51.484159	3.574899	1	78	73	11	00:0D:ED:A5:4F:70	Netgear:66:E5:F1	802.1x:EAP LEAP/response
37		5/5 20:53:51.484249	3.574989	1	10	78	11		00:0D:ED:A5:4F:70	802.11 acknowledgement
38		5/5 20:53:51.484563	3.575303	1	93	75	11	00:0D:ED:A5:4F:70	Netgear:66:E5:F1	802.1x: EAPOL-key
39		5/5 20:53:51.484670	3.575410	1	10	78	11		00:0D:ED:A5:4F:70	802.11 acknowledgement
40		5/5 20:53:51.484945	3.575685	1	80	76	11	00:0D:ED:A5:4F:70	Netgear:66:E5:F1	802.1x: EAPOL-key
41		5/5 20:53:51.485057	3.575797	1	10	78	11		00:0D:ED:A5:4F:70	802.11 acknowledgement
42		[illegible]	[illegible]	1	90	75	11	00:0D:ED:A5:4F:70	Netgear:66:E5:F1	802.11 encrypted data

Roaming

As a user roams about an area, the mobile station may need to reassociate with a second access point having the same SSID as the first in order to maintain a stable connection to the ESS. This may happen as the user roams out of range of one access point and within better range of another access point. Although the decision to reassociate is not specifically defined in the 802.11 standard, mobile stations typically will reassociate when either frame retransmissions exceed a specific value or when the received signal strength drops below a certain value. It can be far more complicated than only taking into consideration these two variables, but accounting for all of the proprietary decision-making algorithms used by various manufacturers is beyond the scope of this book.

When a station reassociates, it first authenticates with the new access point, and then starts the association process by sending it a reassociation request frame as shown in Figure 3.5. The reassociation request frame contains the current (old) access point's MAC address in a fixed field in the frame body as shown in Figure 3.6.

FIGURE 3.5 Reassociation Process

1	30	83	1	00:0C:30:52:93:6A	Aironet:3A:6A:C7	802.11 authentication
1	10	81	1		00:0C:30:52:93:6A	802.11 acknowledgement
1	30	83	11	Aironet:3A:6A:C7	00:0C:30:52:93:6A	802.11 authentication
1	10	80	11		Aironet:3A:6A:C7	802.11 acknowledgement
1	88	78	1	00:0C:30:52:93:6A	Aironet:3A:6A:C7	802.11 reassociation request
1	10	85	1		00:0C:30:52:93:6A	802.11 acknowledgement
1	92	85	11	Aironet:3A:6A:C7	00:0C:30:52:93:6A	802.11 reassociation response

FIGURE 3.6 Reassociation Request frame

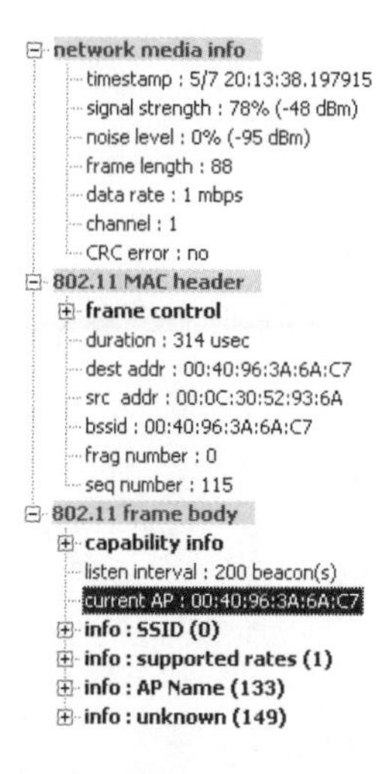

If the station successfully reassociates, the new access point then responds with a reassociation response frame which will contain the Status Code subfield indicating "successful" and with an AID value specific to that station on the new access point. The 802.11 standard indicates that it is the responsibility of the new access point to notify the DS of the reassociation.[1]

---

[1] IEEE 802.11 - 1999 (R2003) – Section 11.3.4.c

FIGURE 3.7 Reassociation Response frame

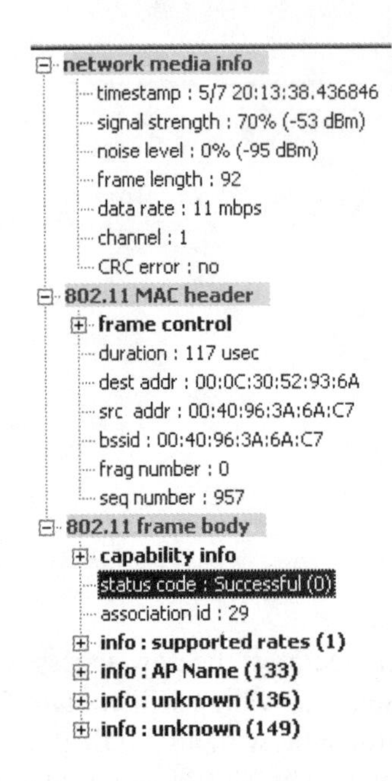

There are proprietary methods of managing this "handoff" process (such as Cisco's Fast Secure Roaming), and there are industry-standard methods such as the IEEE 802.11f Recommended Practice. 802.11f calls for a protocol called the Inter Access Point Protocol (IAPP). Functionality of this protocol is beyond the scope of this book. Additionally, wireless LAN switches are becoming prevalent in the industry. Handoff mechanisms are proprietary and handled internally in the switch's firmware.

A station may only be associated with one access point at a time. The process of authenticating to multiple access points simultaneously is common and is called preauthentication. Although preauthentication is specifically allowed by the standard, it has less relevance when used with the relatively fast Open System authentication.

Consider that, during the roaming process, a situation could arise where the new access point does not properly notify the old access point of a mobile station's reassociation. This lack of notification could be due to dissimilar vendors' equipment using incompatible handoff mechanisms. This event would result in a situation where a mobile station's MAC address would be found in two access points' association tables simultaneously. The 802.11 standard indicates that when the frames are forwarded to the station by two access points within range, the station should reply to access point to which it is not associated using deauthentication or disassociation frames only.[1]

In order for roaming to work effectively, set the SSID in all of the access points in an area to the same value. Because the stations generally refine their scanning to only access points having the same SSID as what's configured in the station, most stations will not pay attention to access points set to a different SSID. If access points don't have the same SSID, a station may first need to completely lose the signal from the associated access point. The client operating systems, such as Windows XP, can then offer new wireless networks (having the different SSID) that the user can choose and associate with.

Reassociation in a wireless 802.1X/EAP network can be a relatively slow process in comparison to using Open System or Shared Key authentication in a scenario in which "fat" or "intelligent" access points are used. Vendors that support 802.1X/EAP have realized that time-sensitive connections such as Voice over IP over WLAN (VoWLAN) may often be momentarily disrupted during a roam due to the lengthy 802.1X authentication process. For this reason, some vendors have developed proprietary roaming mechanisms (e.g. Cisco's Fast Secure Roaming) to aid in speeding the 802.1X authentication process along. Wireless LAN switches do not generally have this problem because handoffs between "thin" (limited functionality) access points (which may often amount to "smart antennas") are handled internally within the switch's firmware.

Data Protection

Confidentiality on a wireless LAN is certainly critical because of the potential exposure of data frame contents to eavesdroppers. Without solid authentication and encryption mechanisms in place, a company runs a significantly high risk of information security attacks.

---

[1] IEEE 802.11 - 1999 (R2003) – Section 5.5.c

Wired Equivalent Privacy (WEP)

The 802.11 standard defines an optional WEP security mechanism, which makes use of a secret shared key that alters the frame body of data and management frames to avoid disclosure to eavesdroppers. This process is also known as symmetric encryption: the keys must match at both the source and destination station (or access point). Figure 3.8 illustrates the WEP encryption process.

FIGURE 3.8 WEP Encryption Process

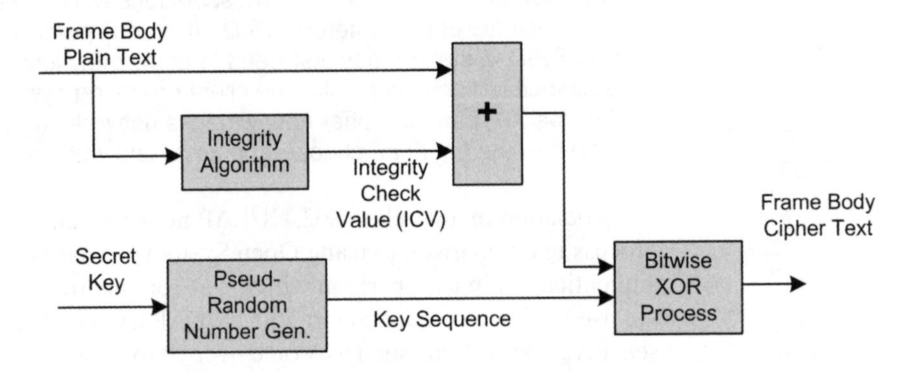

The WEP encryption process occurs as follows.

1. At the sending station, the WEP process first runs the plaintext through an integrity algorithm. This algorithm generates a four-octet integrity check value (ICV) that is sent with the data and checked at the receiving station to guard against unauthorized data modification en route.
2. The WEP process inputs the secret shared encryption key into a pseudo-random number generator (PRNG) to create a key sequence with length equal to the plaintext and ICV. The process uses a 24-bit initialization vector (IV) as part of the encryption key. The IV is sent unencrypted as part of the data frame payload.
3. WEP encrypts the data by bitwise XORing the plaintext and ICV with the key sequence to create ciphertext.

4. At the receiving station, the WEP process deciphers the ciphertext using the shared key that generates the same key sequence used initially to encrypt the frame.
5. The receiving station calculates an ICV and ensures that it matches the one sent with the frame. If the integrity check fails, the station will not process the data frame contents, and a failure indication is sent to the MAC management entity.

802.11 WEP is vulnerable to hackers, who can use freely available tools, such as Airsnort and WEPCrack, to decode WEP-encrypted data frames. These tools exploit the short IV that it sent in clear text within the data frame. Since the IV is only 24 bits, this means that data frames using the same IV must be repeated after only a short time. This cryptographic weakness makes WEP unsuitable for enterprise wireless LANs. Organizations should use stronger encryption mechanisms than WEP.

Figure 3.9 shows when the MAC header's frame control field has the WEP subfield set to 1, the frame body is encrypted. In early implementations of 802.11, there was only one choice: WEP. The "WEP" subfield has been renamed "Protected Frame" by 802.11i. Notice in the graphic that the IV and ICV are both in clear text – easily read by any protocol analyzer.

FIGURE 3.9 An Encrypted Data Frame Decode

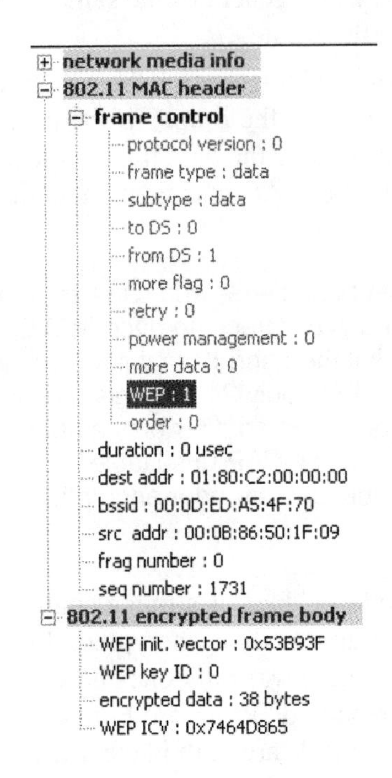

Figure 3.10 shows where the IV (not encrypted), data (encrypted), and ICV (encrypted) are positioned within the MAC frame's body.[1] The MPDU frame body has a maximum size of 2312 bytes, 2304 bytes of which may be used to carry upper layer protocols.[2] When WEP is in use, the MPDU frame body is expanded by 8 bytes to 2312 bytes.[3] These extra 8 bytes are used for the IV (4 octets) and ICV (4 octets). The encrypted data is exactly the same length as the plain text data was because that is how stream ciphers such as RC4 work. Compare the position of each item in Figure 3.9 with those shown in Figure 3.10 and notice that wireless protocol analyzers show frame contents in order from first to last.

[1] IEEE 802.11 – 1999 (R2003) – Section 8.2.3
[2] IEEE 802.11 – 1999 (R2003) – Section 6.2.1.1.2
[3] IEEE 802.11 – 1999 (R2003) – Section 7.1.3.1.9 & 8.2.5

FIGURE 3.10 WEP Data Frame Body Structure Using WEP

IEEE 802.11 MAC
FC – Frame Control
D/I – Duration/Identification
SC – Sequence Control
IV – Initialization Vector
ICV – Integrity Check Value
FCS – Frame Check Sequence

IEEE 802.11 MAC – FC field
PV – Protocol Version
TD – To Distribution System
FD – From Distribution System
MF – More Fragments
R – Retry
PM – Power Management
MD – More Data
PF – Protected Frame
O - Order

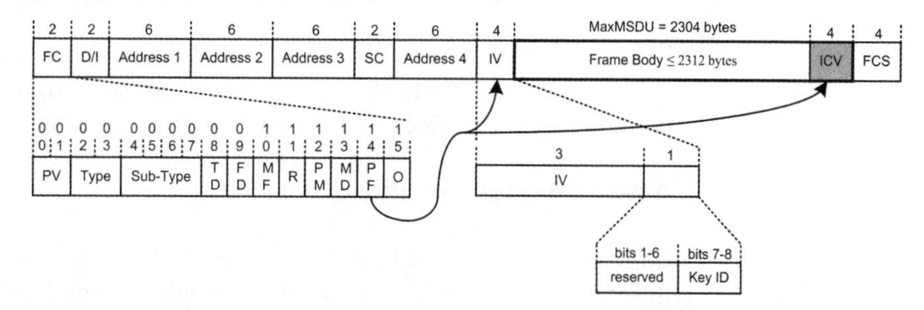

Wi-Fi® Protected Access (WPA®), 802.1X/EAP, & 802.11i

Because 802.11 WEP was easy to crack, the Wi-Fi Alliance took a bold step forward to expedite the availability of effective, standardized wireless LAN security by defining Wi-Fi Protected Access (WPA). Most wireless LAN vendors now support WPA. The initial WPA (version 1) is actually a snapshot of the 802.11i standard, which includes Temporal Key Integrity Protocol (TKIP) and the IEEE 802.1X/EAP authentication framework. The combination of these two mechanisms provides dynamic key distribution and mutual authentication, something much needed in wireless LANs. WPA2 (version 2) is the Wi-Fi Alliance's® security interoperability certification that complies with the 802.11i standard.

TKIP

As with WEP, TKIP uses the RC4 stream cipher provided by RSA Security to encrypt the frame body of each 802.11 frame before transmission. The cryptographic weakness issues with WEP have little to do with the RC4 encryption algorithm. Instead, the problems primarily relate to key generation and how encryption is implemented.

TKIP adds the following strengths:

48-bit initialization vectors

WPA with TKIP uses 48-bit IVs (as compared to 24-bit IVs with WEP) that significantly reduce IV reuse and the possibility that a hacker will collect a sufficient number of 802.11 frames to crack the encryption.

Per-packet key construction and distribution

WPA relies on IV, albeit a larger IV, for per-packet key uniqueness, same as WEP. WPA introduces the master key concept from which it periodically regenerates and redistributes new session keys.

A new message integrity check (MIC)

Called, "Michael", the new MIC is 8 bytes long and used in addition to the existing WEP ICV.[1] This new MIC adds strength to the existing ICV for prevention of in-transit bit-flipping attacks.

Figure 3.11 shows the construction of a TKIP-enhanced data frame. Note that TKIP adds 4 additional octets to extend the IV field, and the Michael MIC adds an additional 8 octets over and above the WEP ICV. The maximum MSDU remains 2304 bytes, but the frame is extended appropriately for each encryption type used (WEP (8), TKIP (20), or CCMP (16)).

A point of interest when using TKIP that the analyst should be aware of is when fragmentation is also in use. TKIP appends the MIC at the end of the MSDU payload. The Michael MIC is 8 octets in size, and the 802.11 MAC then applies its normal processing to transmit this MSDU-with-MIC as a sequence of one or more MPDUs. This means the MSDU plus MIC can be partitioned into one or more MPDUs, the WEP ICV is calculated over each MPDU, and the MIC can even be partitioned to lie in two MPDUs after fragmentation. So, an unfragmented MPDU is increased by 20 octets as we have mentioned, first by adding 8 octets to the MSDU then by adding 12 octets to the MPDU. But each MPDU of a series of fragments is only increased by 12 octets! The extra 8 octets are

---

[1] The terms ICV, MIC, and FCS are essentially interchangeable in this context.

inserted once for the series as though it were part of the original "raw" MSDU.

FIGURE 3.11 Data Frame Body Format Using TKIP

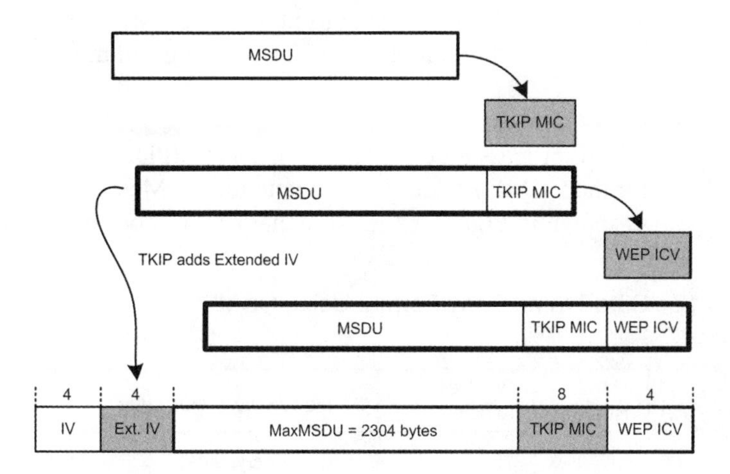

It is trivial to modify a set of bits in a MSDU such that the WEP ICV will not be able to detect the change, but the Michael MIC is much stronger, and it will detect the change. However, Michael is not invulnerable, and it is still important to protect the BSS from active attackers. If a TKIP device detects a Michael failure, it will start a clock, and if there is a second Michael failure within 60 seconds, then the BSS shuts itself down to limit the damage that an attacker can do. Before coming back online after two MIC failures in a 60-second period, the access point (or wireless LAN switch) will re-key all of the associated stations. This countermeasure limits the rate at which an attacker can forge frames, but also creates a situation where attackers could perform Denial of Service (DoS) attacks. To prevent a DoS attack against Michael, some manufacturers give you the configuration option of adjusting or disabling this countermeasure.

CCMP

The 802.11i standard calls for use of the Counter-Mode/CBC-MAC Protocol (CCMP), which provides confidentiality, authentication, integrity, and replay protection. CCMP is mandatory for RSN compliance. CCMP is based on the CCM mode of operation of the AES encryption algorithm. The CCM mode combines Counter Mode (CTR) for confidentiality and Cipher Block Chaining Message Authentication Code (CBC-MAC) for authentication and integrity. CCM protects the integrity of both the MSDU and selected portions of the 802.11 MPDU header, although only the MSDU is encrypted. All Advanced Encryption Standard (AES) processing used within CCMP uses AES with a 128 bit key and a 128 bit block size. The AES algorithm is defined in FIPS PUB 197.

CCMP processing expands the original MPDU size by 16 octets, 8 octets for the CCMP Header and 8 octets for the Message Integrity Code (MIC) as shown in Figure 3.12. Note that CCMP does not use the WEP ICV.

FIGURE 3.12 Data Frame Body Format Using CCMP

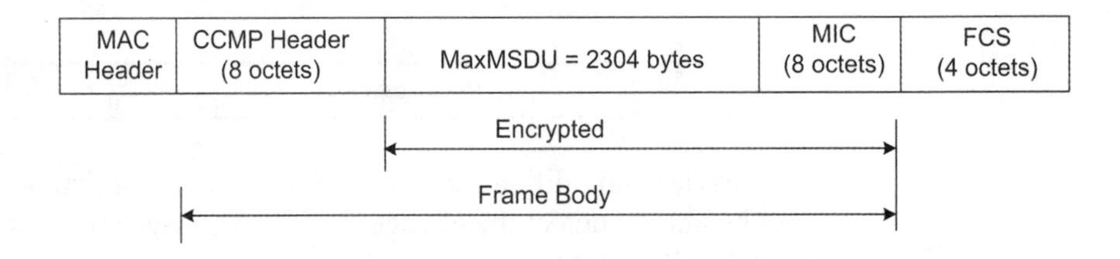

802.1X/EAP

For authentication, WPA uses a combination of 802.11 Open System authentication and 802.1X/EAP just as with most 802.1X/EAP implementations. Initially, the station authenticates with the access point using Open System authentication, then associates which authorizes the client to send frames to the access point. Next, WPA performs user-level authentication with 802.1X/EAP as we have shown in Figure 3.2.

Since WPA uses 802.1X/EAP, it must interface with an authentication server, such as RADIUS, in an enterprise environment. WPA is also capable of operating in what's known as "pre-shared key (WPA-PSK)" mode if no external authentication server is available, such as in homes and small offices. WPA-PSK requires only a shared passphrase (shared key) on each end of the link. Operationally, WPA-PSK is not much different than configuring WEP, though it uses TKIP and other more secure functions than WEP did. It is theoretically possible to upgrade most current 802.11 access points and radio card components to use WPA through relatively simple firmware upgrades. As a result, with vendor cooperation, WPA is a good solution for providing enhanced security for the existing installed base of WLAN hardware.

The use of IEEE 802.1X offers an effective framework for authenticating and controlling user traffic to a protected network, as well as dynamically varying encryption keys. 802.1X ties the Extensible Authentication Protocol (EAP) to both the wired and wireless LAN media and may support many authentication methods, such as token cards, Kerberos, one-time passwords, certificates, and public key authentication.

Initial 802.1X communication begins with an unauthenticated supplicant (client station) attempting to connect with an authenticator (access point). The access point responds by enabling a port for passing only EAP packets from the client to an authentication server located on the wired side of the access point. The access point blocks all other traffic, such as HTTP, DHCP, and POP3 packets, until the access point can verify the client's identity using an authentication server, such as RADIUS. Once the client is successfully authenticated, the access point opens the client's port for other types of traffic. This state is called "EAP Associated" and is illustrated below. Being associated and being EAP associated are different as you can see in Figure 3.13.

FIGURE 3.13 Access Point's Association Table

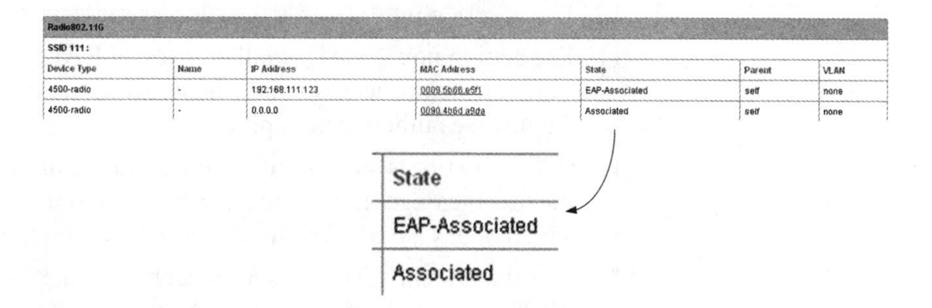

Radio802.11G

SSID 111:

Device Type	Name	IP Address	MAC Address	State	Parent	VLAN
4500-radio	-	192.168.111.123	0009.5b66.e5f1	EAP-Associated	self	none
4500-radio	-	0.0.0.0	0090.4b6d.a9da	Associated	self	none

The following describes interactions that take place within a general 802.1X/EAP framework.

1. The station sends an *EAP - Start* message to the access point. This initiates the process of EAP authentication.
2. The access point sends an access request on behalf of the client to the RADIUS server.
3. The access point replies with an *EAP – Request/Identity* message.
4. The station sends an *EAP – Response/Identity* message containing its credentials (such as username) to the access point. This message will contain ID based on the EAP type in use, such as EAP-TLS, EAP-TTLS, EAP-PEAP, EAP-LEAP, or EAP-FAST. In a password-based EAP, the user's password is NOT part of this message.
5. The access point forwards the user's ID to the RADIUS server.
6. The RADIUS server responds with a challenge message, which the access point forwards to the station as an EAP message.
7. The station encrypts the challenge message using its password (or other credential) as a secret key and sends the resulting value back to the access point.
8. The access point forwards the encrypted challenge to the RADIUS server.
9. The RADIUS server uses the password (or other credential) that it has stored for the user to encrypt the same challenge message it sent to the station. If the resultant value and the value returned by the station match, the RADIUS server sends a success message to the access point.
10. The access point forwards the success message to the station.
11. The station now sends a challenge to the RADIUS server to authenticate the access point (the network), and proceeds through the reverse authentication process.
12. If the network is successfully authenticated, the station passes a success message through the access point to the RADIUS server, which opens a port. The user is now LIVE on the network.
13. The station and RADIUS server each generate a dynamic unicast WEP key (which will match) from key material exchanged during the mutual authentication phase.
14. The RADIUS server sends the unicast WEP key to the access point in a RADIUS attribute. The attribute is encrypted using the shared key used between the access point and RADIUS server.

15. The access point generates a broadcast WEP key, encrypts it using the unicast WEP key received from the RADIUS server, and sends it to the station.
16. The station and the access point now both have the unicast and broadcast WEP keys.
17. The access point sends an EAPoL-Key message to the station indicating that they should both activate encryption.

This book is not meant to be a comprehensive wireless security text, but rather we intend for it to give the reader a good understanding of how the authentication/association framework of 802.1X/EAP and WPA should operate. This understanding will aid the analyst in troubleshooting various authentication protocols using a wireless protocol analyzer.

Because of the critical threat that unsecured wireless LANs pose to information security, organizations are quickly implementing strong wireless security mechanisms. For this reason, it is very common to use a wireless protocol analyzer in an environment where data is encrypted. Since WEP and TKIP are layer 2 security protocols, all protocols for layers above the MAC layer are encrypted. Additionally, 802.1X/EAP uses unique unicast keys for each station, which are then rotated periodically. For these reasons, most troubleshooting must be done based only on information found at layer 2. For example, troubleshooting a broken FTP session would not be possible above the Data-Link because all data from the L2/LLC to the L7/FTP would be encrypted using WEP or TKIP. The protocol analyzer will display only "WEP Data" as shown in Figure 3.14.

FIGURE 3.14 WEP Data

Packet	Source Physical	Dest. Physical	BSSID	Channel	Data Rate	Size	Protocol
98	00:09:5B:66:E5:F1	00:01:96:A4:96:6E	00:0B:86:80:EB:98	56	54.0	116	802.11 WEP Data
99		00:09:5B:66:E5:F1		56	24.0	14	802.11 Ack
100	6B:65:6B:74:99:6E	30:0E:5B:66:E5:F1	00:0B:86:80:AB:FD	56	54.0	116	802.11 WEP Data
101		00:0B:86:80:EB:98		56	24.0	14	802.11 Ack
102	00:09:5B:66:E5:F1	00:01:96:A4:96:6E	00:0B:86:80:EB:98	56	54.0	116	802.11 WEP Data
103	00:0B:86:80:EB:98	00:09:5B:66:E5:F1		56	24.0	14	802.11 Ack
104	00:01:96:A4:96:6E	00:09:5B:66:E5:F1	00:0B:86:80:EB:98	56	54.0	116	802.11 WEP Data
105		00:0B:86:80:EB:98		56	24.0	14	802.11 Ack
106	00:09:5B:66:E5:F1	00:01:96:A4:96:6E	00:0B:86:80:EB:98	56	54.0	116	802.11 WEP Data
107	00:0B:86:80:EB:98	00:09:5B:66:E5:F1		56	24.0	14	802.11 Ack
108	6B:85:6F:35:9B:6E	00:09:5B:66:E5:F1	00:0B:86:80:AB:27	56	54.0	116	802.11 WEP Data
109		00:0B:86:80:EB:98		56	24.0	14	802.11 Ack
110	00:09:5B:66:E5:F1	00:01:96:A4:96:6E	00:0B:86:80:EB:98	56	54.0	116	802.11 WEP Data
111		00:09:5B:66:E5:F1		56	24.0	14	802.11 Ack
112	00:01:96:A4:96:6E	00:09:5B:66:E5:F1	00:0B:86:80:EB:98	56	54.0	116	802.11 WEP Data
113	00:09:5B:66:E5:F1	00:0B:86:80:EB:98		56	24.0	14	802.11 Ack
114	00:09:5B:66:E5:F1	00:01:96:A4:96:6E	00:0B:86:80:EB:98	56	54.0	116	802.11 WEP Data

Summary

In an 802.11 network, authentication proceeds association. With more sophisticated security mechanisms in place, such as 802.1X/EAP, there may be a succession of authentication and association steps. The client may be required to Open System authenticate and associate before starting the process of 802.1X/EAP authentication. WEP is no longer deemed acceptable security for a corporate WLAN. Stronger layer 2 security such as WPA and WPA2/802.11i are now available.

Wireless networks' main attraction is the ability to roam freely. If roaming does not happen seamlessly, then most of the value of wireless is lost. An analyst should understand, in depth, the details of the roaming process including reassociation, security, and the amount of time expected between leaving one AP and connecting with another.

Key Terms

Before taking the exam, you should be familiar with the following terms:

IEEE 802.1X/EAP

IEEE 802.11i

protection mechanisms

Service Set Identifier (SSID)

Wi-Fi Protected Access (WPA)

Wired Equivalent Privacy (WEP)

Review Questions

1. When using Shared Key authentication, how many authentication frames are exchanged between a station and an access point?

2. A reassociation frame contains what unique identifier that allows the new access point to obtained queued frames from the old AP?

3. When static WEP is used, the MPDU frame body is extended by how many bits?

4. When TKIP is used, the MPDU frame body is extended by how many bits?

5. How is an 802.1X/EAP conversation initiated?

6. The MSDU may consist of how many bytes (maximum) of upper layer protocol?

7. Which Shared Key authentication frame contains a clear text challenge phrase?

CHAPTER 4

Configuration Options and Protection Mechanisms

CWAP Exam Objectives Covered:

❖ Explain and describe the frame exchange processes involved in:

- Power Management mode operation
- Fragmentation
- 802.11b/g mixed mode environments and protection mechanisms

In This Chapter

Power Management

Fragmentation

RTS / CTS

CTS-to-Self

Protection Mechanisms

Power Management

802.11 network interface cards consume significant amounts of energy and drain batteries fast, especially in smaller handheld devices. To prolong battery life, the 802.11 standard defines an optional "power-save mode." End users can activate power-save mode via the radio card's vendor-supplied configuration tool (client utilities) or operating system interface. With power-save mode disabled, the 802.11 network card is generally in receive mode listening for packets and occasionally in transmit mode when sending packets. These modes require the client station to keep most circuits powered-up and ready for operation.

General Operation

Stations that have their client utilities configured for power-save mode will send all of their frames to the access point with the power management bit in the frame control field of each 802.11 MAC frame header set to 1. This indicates the station's desire to remain in power-save mode, and it informs the access point that it should buffer unicast data frames for the station until polled by the station. This continues to be the case until such time that the station's client utility is reconfigured for fully awake mode. At this time, the station will send its frames to the access point with the power management bit set to 0 to indicate that it is fully awake and the access point should not buffer frames on its behalf.

When dozing, the station consumes much less power than normal by shutting off power to nearly everything except for a timing circuit. This enables the station to consume very little power and still be able to wake up periodically (at a predetermined time) to receive regular beacon transmissions coming from the access point. Each beacon frame contains a traffic indication map (TIM) that identifies which dozing stations have unicast frames buffered at the access point. These buffered frames are awaiting delivery to their respective destinations. The dozing station will wake up to view the TIM in the first beacon it hears. A station may doze at its leisure once in power-save mode. When the station discovers frames are buffered at the access point the station will send PS-Poll frames to the access point until the access point's buffer is empty. Upon receiving a PS-Poll frame, the access point may respond with a single queued data frame or it may send an ACK frame. If the access point responds with an ACK frame, it may then send the queued data frame at its leisure. Each queued data frame is sent in response to an additional PS-Poll frame from the station. As long as there are more queued data frames at the access point, each data frame sent to the station will have the More Data bit in the Frame Control field of the MAC header set to 1. The last queued data frame will have a More Data bit of 0.

FIGURE 4.1 More Data bit

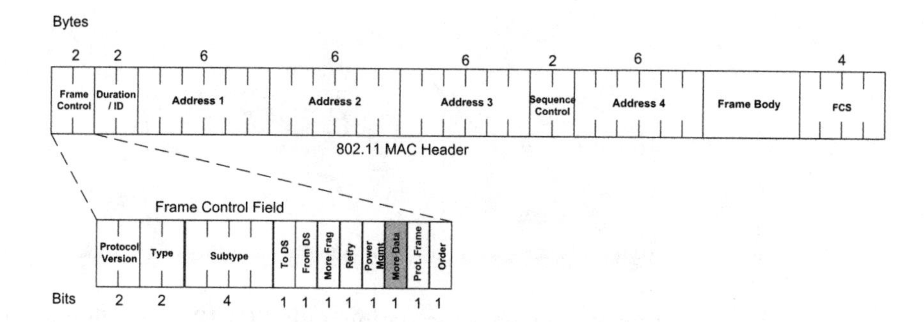

The More Data bit is the method that the 802.11 standard specifies to ensure that stations empty the access point's buffer before dozing again. After emptying the access point's buffer using the PS-Poll mechanism, the beacon will no longer show that station's AID in the TIM. The station may return to doze mode at its convenience.[1] Figure 4.2 shows a beacon management frame with the TIM indicating that AID 12 has queued data.

[1] IEEE 802.11 – 1999 (R2003) – Section 11.2.1.6

FIGURE 4.2 Beacon Showing Queued Data

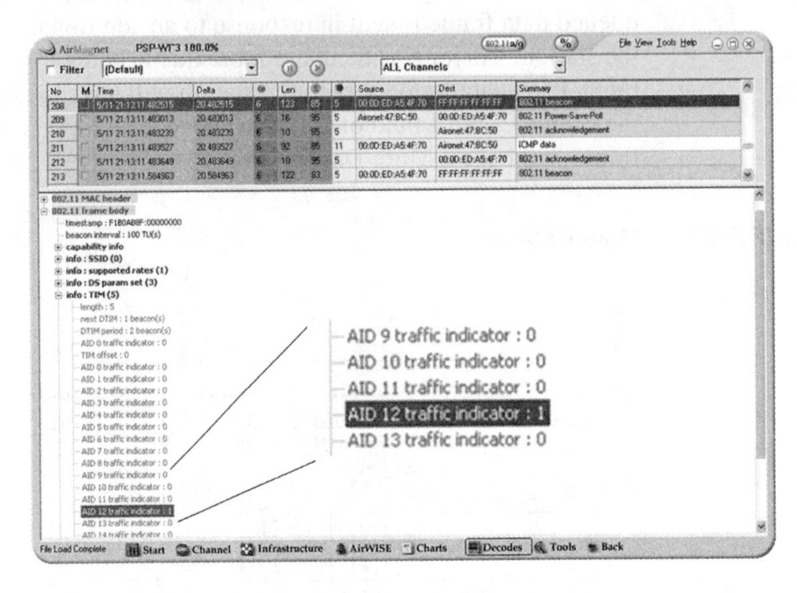

Figure 4.3 shows the station with AID 12 requesting its queued data immediately after receiving the beacon.

FIGURE 4.3 PS-Poll Frame Requesting Its Queued Data

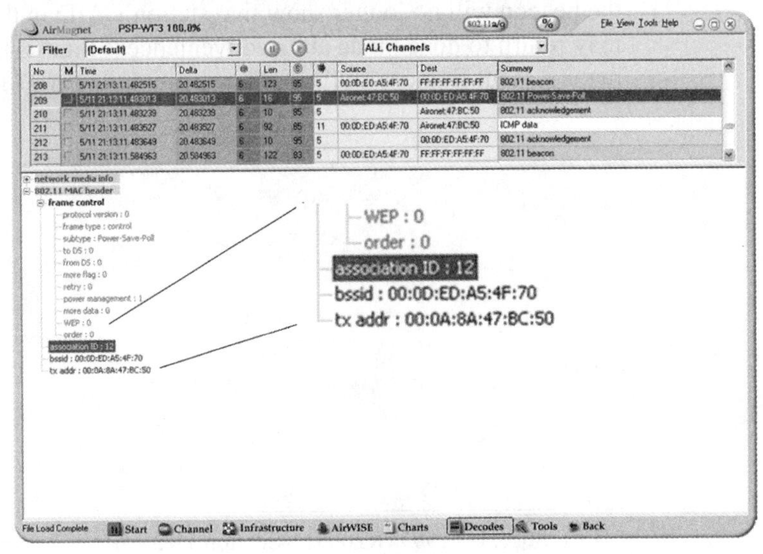

Figure 4.4 shows the requested data frame being sent to the station with the More Data bit set to 0 (meaning that this is the only queued data frame at the access point for this station).

FIGURE 4.4 Queued Data Frame

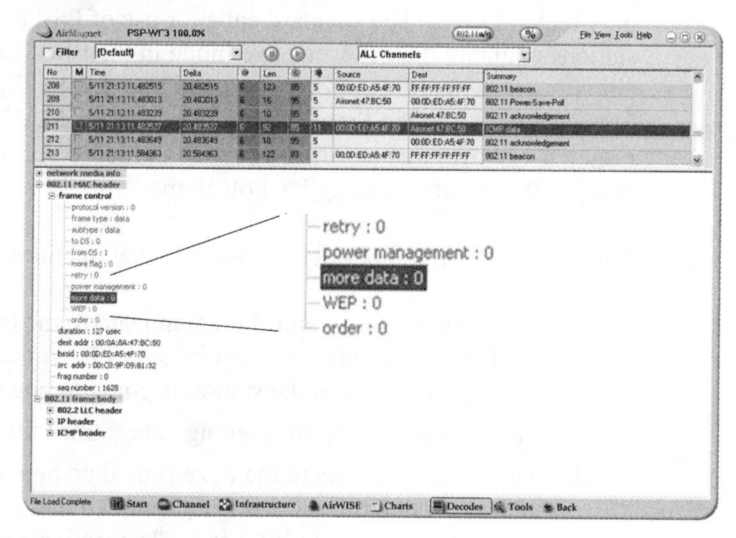

Figure 4.5 shows that after the queued data has been retrieved from the access point by the station, the access removes the station's AID from the next TIM.

FIGURE 4.5 Beacon Showing No Queued Data

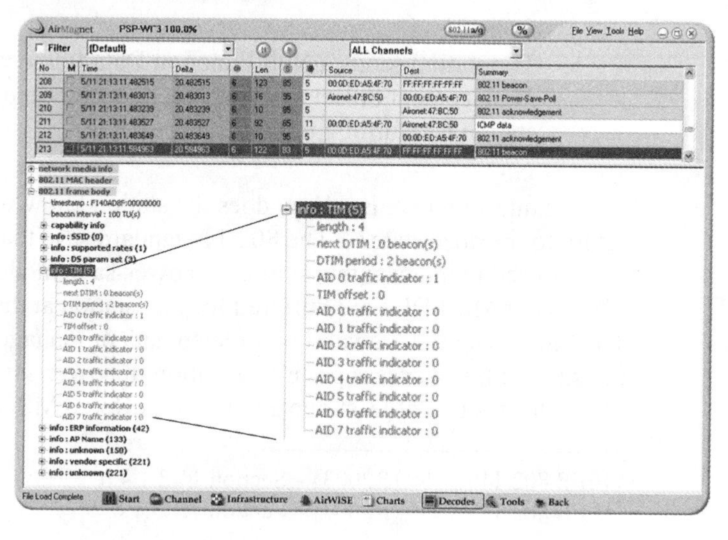

Power Management Bit Flipping[1]

The 802.11 standard method of performing queuing and retrieval of data frames at the access point for the benefit of Power-Save stations is not the only method used. The standard calls for use of PS-Poll frames while stations maintain their Power-Save mode in the BSS. Some of the new chipsets on the market perform the same function by flipping the Power Management bit in the Frame Control field of the MAC header (see Figure 4.1 above) on and off as needed in order to accomplish the same thing as those stations using PS-Poll frames.

This alternate method of queued data retrieval operates as follows.

1. The station sends a Null Function Data frame to the access point with the Power Management bit set to 1. This setting indicates to the access point that the station is going to power-save mode.
2. The access point starts queuing data frames for the station.
3. The station changes to the doze state then periodically powers up and sends a Null Function Data frame to the access point with the Power Management bit set to 0. The station sends this frame without regard to what it might have heard in the beacon.
4. The access point stops queuing data for the station, and immediately sends any queued frames to the station as fast as possible. If there are no queued frames for this station, nothing happens.
5. The station sends a Null Function Data frame with the Power Management bit set to 1 to the access point indicating that the station is returning to power-save mode.

A valid question here is, "How does the station know when it is ok to return to the doze state?" The 802.11 standard says that the More Data field is used to indicate to a station in power-save mode that more MSDUs or MMPDUs are buffered for that station at the access point.[2] The More Data field is valid in directed data or management type frames transmitted by an access point to a station in power-save mode. A value of 1 indicates that at least one additional buffered MSDU, or MMPDU, is

---

[1] IEEE 802.11 - 1999 (R2003) – Section 11.2.1.4.h
[2] IEEE 802.11 - 1999 (R2003) – Section 7.1.3.1.8

present for the same station. The standard does not specify that the More Data bit is used to indicate additional buffered frames for stations that are not in Power-Save mode. This functionality is not expressly specified in the standard, so the decision on when to return to doze state is up to the implementer.

FIGURE 4.6 Null Function Data Frame with Power Management = 1

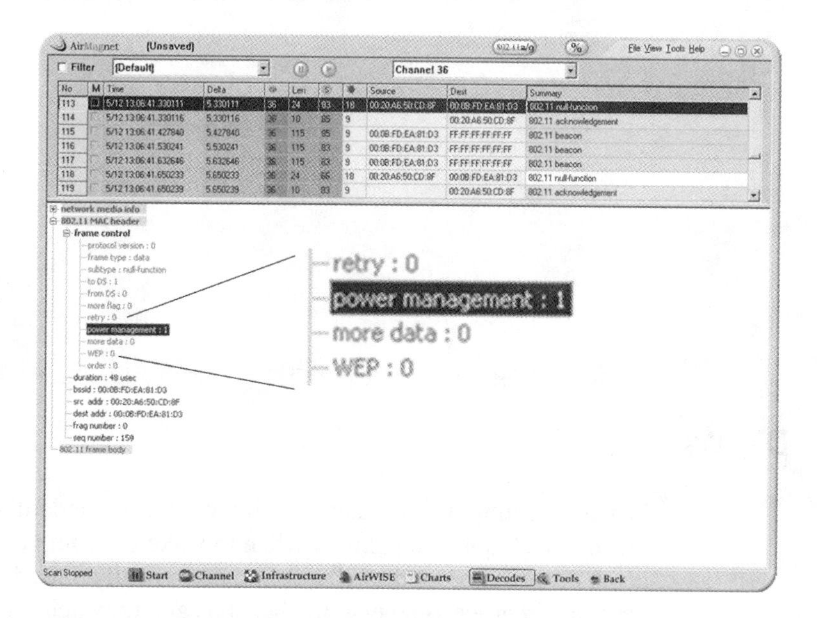

FIGURE 4.7 Null Function Data Frame with Power Management = 0

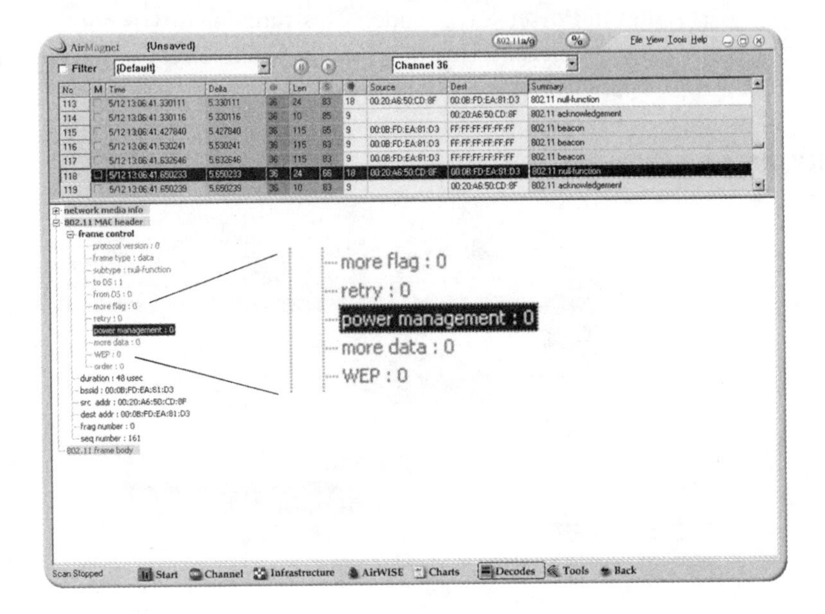

DTIMs

When operating in accordance with the 802.11 standard's power-saving mode, the stations must know when to wake up from dozing. Stations using power-save mode will awaken periodically, based on a number of beacons, in order to receive the beacons and to watch for their own AID in the beacon's TIMs. There are two types of TIMs with which a wireless network analyst should be familiar: TIMs & DTIMs. We have discussed that TIMs are used to notify power-save mode stations that they have unicast traffic queued at the access point. This section will discuss DTIMs.

A DTIM is a TIM with particular settings in its fields used to indicate (to the BSS) the presence of queued broadcast/multicast traffic at the access point. The TIM information element in the beacon management frame is structured as shown in Figure 4.8.

FIGURE 4.8 TIM Information Element

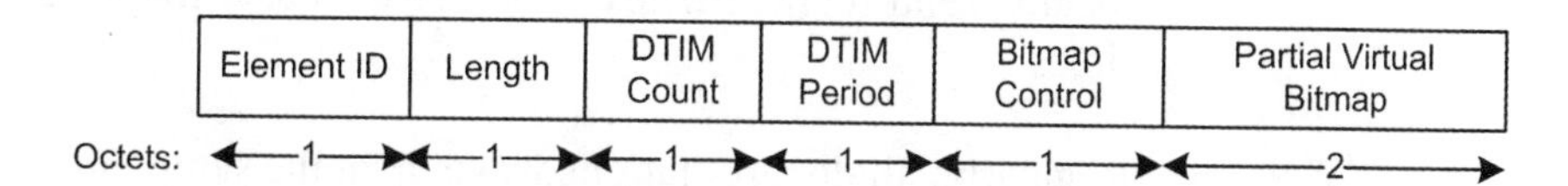

The DTIM Count field indicates how many more beacons (including the current frame) appear before the next DTIM. A DTIM count of 0 indicates that the current TIM is a DTIM. The DTIM Period field indicates the number of beacon intervals between successive DTIMs. If all TIMs are DTIMs, the DTIM Period field has the value 1. The DTIM Period value 0 is reserved. The Bitmap Control field is a single octet. Bit 0 of the Bitmap Control field contains the Traffic Indicator bit associated with Association ID 0. This bit is set to 1 in TIM elements with a value of 0 in the DTIM Count field when one or more broadcast or multicast frames are buffered at the access point. The remaining 7 bits of the Bitmap Control field form the Bitmap Offset. Each bit in the Virtual Bitmap corresponds to traffic buffered for a specific station within the BSS that the access point is prepared to deliver at the time the beacon frame is transmitted.

The DTIM interval is the interval between TIMs that are DTIMs and is configurable in the access point (or wireless LAN switch). Stations do not request broadcast/multicast traffic, but rather the traffic is delivered automatically following beacons that contain DTIMs. The bit for AID 0 (zero) is set to 1 whenever broadcast or multicast traffic is buffered.

The More Data field of each broadcast/multicast frame is set to indicate the presence of further buffered broadcast/multicast data frames. If the access point is unable to transmit all of the buffered broadcast/multicast data frames before the TBTT following the DTIM, the access point will indicate that it will continue to deliver the broadcast/multicast data frames by setting the bit for AID 0 of the TIM element of every beacon frame, until all buffered broadcast/multicast frames have been transmitted.

Ad Hoc

As with infrastructure networks, Ad Hoc stations indicate that they are entering power-save mode by setting the power management bit to 1 in all of their frames. This bit indicates to other stations in the IBSS that they may not transmit data to this station at will, but have to buffer frames locally, send ATIM frames, and then send data frames at appropriate times.

Regularly, all dozing stations wake up at the same time for what is called the announcement traffic indication message (ATIM) window, which corresponds with each beacon transmission. If a station is holding frames for a station operating in power-save mode, the station will send an ATIM frame to the power-save mode station indicating that frames are awaiting transmission. The power-save mode station that typically spends its time dozing then knows to stay awake through the next beacon interval, which is hopefully long enough for the station buffering the frame to send the frame successfully. After receiving and acknowledging receipt of the frame, the station can return to a doze state.

ATIM frames are messages that contain no frame body. Receiving stations know what ATIM frames are by frame type and subtype. Unicast ATIM frames are acknowledged, but broadcast/multicast ATIM frames are not acknowledged. The ATIM window's length is specified in the beacon's IBSS Parameter Set information element and is measured in Time Units (TUs).

Summary

The actual savings in battery life using 802.11 power-save mode is difficult to determine, and there are situations where power-save mode might not provide any benefit at all. When transmitting or receiving, the client station will consume an average of 250 milliamps, whereas current draw while dozing could be as low as 30 milliamps. Because the dozing station will wake up periodically, the aggregate current draw will vary somewhere between 30 and 250 milliamps, depending on the listen interval and doze policy set in the clients.

If the client stays awake longer to accommodate higher traffic levels, then the aggregate current will be closer to the receive/transmit values, possibly 230 milliamps, or so. As a result, the savings in battery life will not be appreciable. Also, keep in mind that to achieve significant battery savings using power-save mode, lower throughput will likely prevail for the power-save stations. In fact, some applications that require frequent communications with the clients will not operate well with power-save mode enabled.

Fragmentation

Fragmentation is sometimes essential in a wireless environment and is built into the MAC layer. Wireless LANs have a much higher frame corruption rate than wired LANs which means that it may often be more efficient to send many small frames and to retransmit a few than to send a few large frames (and to retransmit each several times). Sending smaller frames can minimize the negative effects of a corrupted frame which requires retransmission in its entirety.

FIGURE 4.9 Fragmentation May Reduce Retransmission Overhead

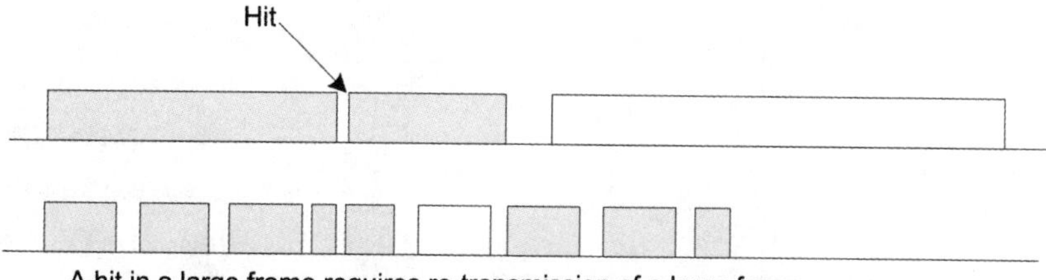

The 802.11 standard includes the ability for stations and access points to fragment directed MSDUs and MAC Management Protocol Data Units (MMPDUs) with the goal of improving performance in the presence of RF interference and marginal coverage areas. MMPDUs are management frames within the BSS. This means that only the payload of a directed data frame (called an MSDU) or a directed management frame (called an MMPDU) may be fragmented according to the 802.11 standard. [1]

Configuration

Figure 4.10 illustrates how an administrator or user can manually configure the fragmentation threshold in some wireless client utilities. If the unencrypted MPDU is equal to or less than the threshold, the station

[1] IEEE 802.11 - 1999 (R2003) – Section 9.4

will not use fragmentation. Setting the threshold to a large MPDU value, typically 2346, effectively disables fragmentation.

A good method to find out if fragmentation is worthwhile is to monitor the wireless LAN for frame retransmissions. If a relatively large number of retries are found, then try using fragmentation. This may improve throughput if the fragmentation threshold is set just right. If very few collisions (less than 5 percent) are occurring, then fragmentation will likely not improve performance. The additional headers applied to each fragment, interframe spaces, and acknowledgments will dramatically increase the overhead on the network, reducing throughput.

FIGURE 4.10 Wireless Client Utilities - Fragmentation

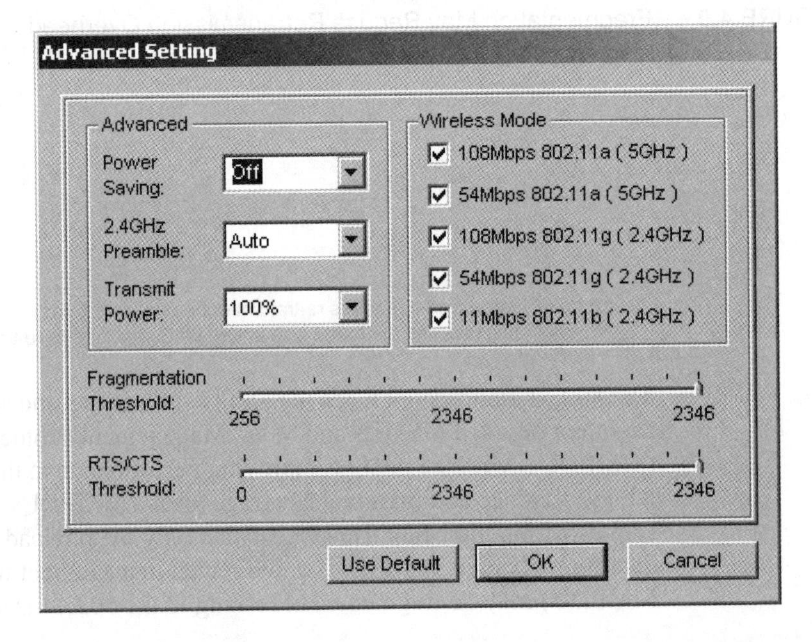

The administrator or user should configure this threshold for optimal performance. A threshold that is too small results in unnecessary fragmentation, while a threshold that is too large results in too much data being retransmitted.

When enabled, fragmentation divides 802.11 frames into smaller pieces (fragments) that are sent separately to the receiver. Each fragment consists of a separate MAC Layer header, frame check sequence (FCS), and a fragment number indicating its ordered position within the series of fragments.

FIGURE 4.11 MSDU Fragmentation

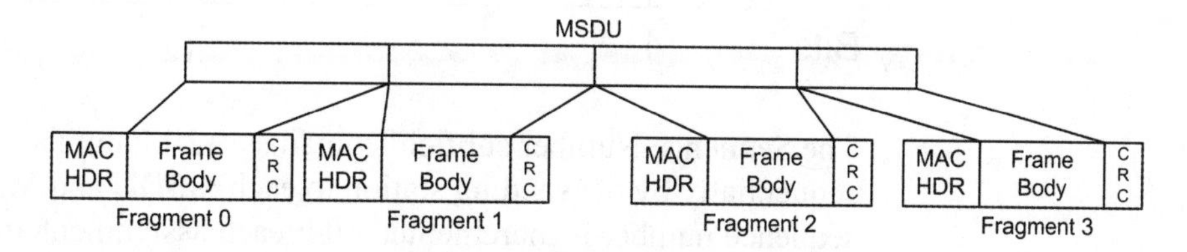

Remember that the probability of a collision is increased when fragmenting because the overall number of data bits being transmitting is greater due to the added headers and trailers.

The Bits

Figure 4.12 shows where the fragment ordering process bits reside in the MAC frame: the Sequence Control field.

FIGURE 4.12 Sequence Control Field

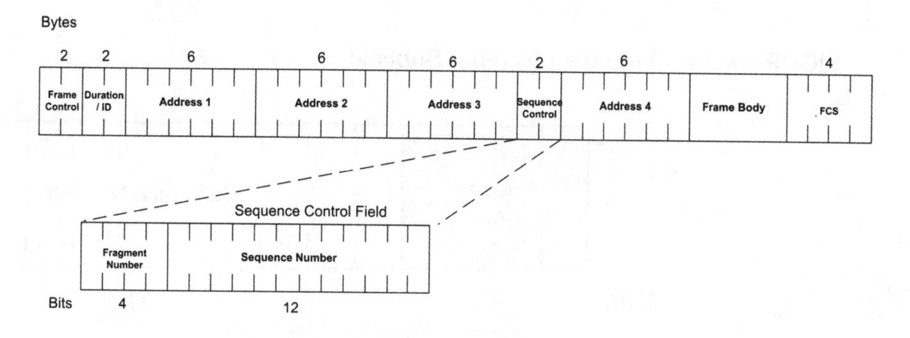

The sequence control field is a 16-bit field comprising two subfields. The subfields are a 4-bit fragment number and a 12-bit sequence number. As a whole, this field is used by a receiving station to eliminate duplicate received frames and to reassemble fragments.

FIGURE 4.13 Sequence Number Subfield

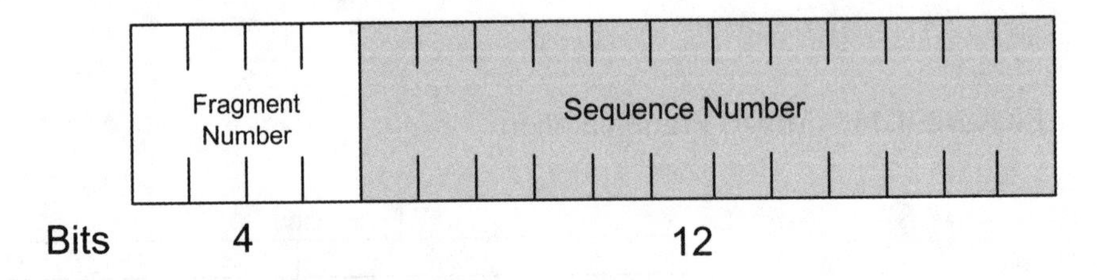

The Sequence Number subfield contains a 12-bit number assigned sequentially by the sending station to each MSDU and MMPDU. This sequence number is incremented after each assignment and returns back to zero when incremented from 4095. The sequence number for a particular MSDU is transmitted in every data frame associated with the MSDU. It is constant over all transmissions and retransmissions of the MSDU. If the MSDU is fragmented, the sequence number of the MSDU is sent with each frame containing a fragment of the MSDU.

The Fragment Number subfield contains a 4-bit number assigned to each fragment of an MSDU. The first, or only, fragment of an MSDU is assigned a fragment number of zero. Each successive fragment is assigned a sequentially incremented fragment number. The fragment number is the same in a transmission or any retransmission of a particular frame or fragment.

FIGURE 4.14 Fragment Number Subfield

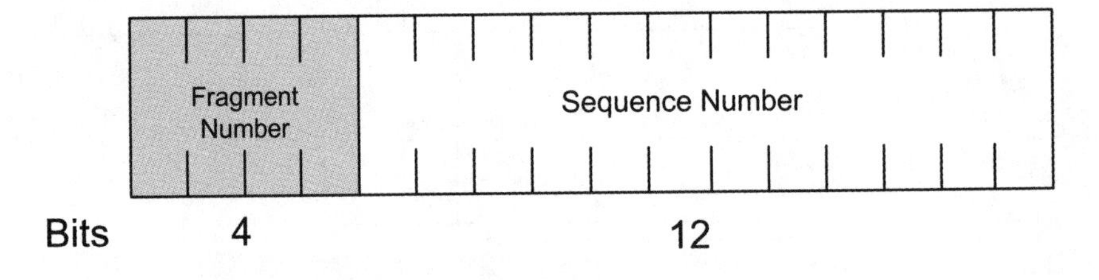

The immediate receiver reassembles the fragments back into the original MSDU or MMPDU using sequence number and fragment numbers found in the header of each frame. After ensuring the data unit is complete, the station hands it up to higher layers for processing. If the immediate receiver is an access point, and the information in the frame body is being transmitted between two client stations, the access point assembles a new frame and transmits it according to its own fragmentation settings. Even though fragmentation involves more overhead, its use may result in better performance if tuned properly.

Sequencing

Let's take a look at how a station using a 300 byte fragmentation threshold (the minimum is 256 bytes) will fragment and sequence a 1200 byte MSDU. Figure 4.15 shows the 1200 byte MSDU with a sequence number of 542 and a fragmentation threshold of 300 bytes.

FIGURE 4.15 Fragment Sequencing & Sizes

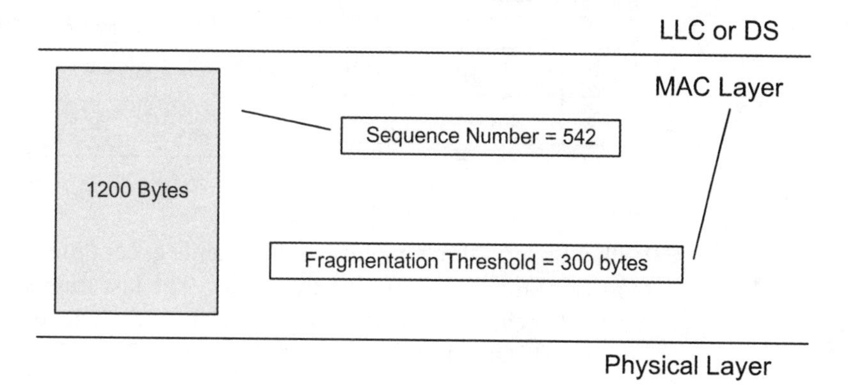

Figure 4.16 is displaying the fragmentation from the bottom upward because the fragments move down the OSI stack. The first fragment to be transmitted will be fragment #0, and the "More Fragments" bit in the Frame Control field will be set to 1. The More Fragments bit will stay set to 1 until the last fragment, at which time it will change to 0, indicating that there are no more fragments. The fragment number will increment by 1, starting at 0, until the last frame. The sequence number will remain unchanged throughout the fragment burst. Notice that the unencrypted MPDU, which includes the MAC header, frame body, and CRC, must not exceed the fragmentation threshold.[1] Fragmentation does not take into account frame body expansion due to encryption, thus encrypted

[1] IEEE 802.11 - 1999 (R2003) – Page 484, dot11FragmentationThreshold

fragments may exceed the fragmentation threshold size. Since the unencrypted MPDU cannot exceed the fragmentation threshold, the size of the header and CRC must be taken into account when deciding how much of the original MSDU can be placed into the frame body of a fragment. In this case, we're using a 24 byte header (all header fields except address 4) and a 4 byte CRC. 300 – 28 = 272, so the first four data fragments will carry 272 bytes of the original 1200 byte MSDU.

FIGURE 4.16 Fragment Sequencing & Sizes

LLC or DS

300 Bytes

CRC | 112 Bytes | MAC — Seq. #= 542; Fragment # = 4; More Frag. = 0

CRC | 272 Bytes | MAC — Seq. #= 542; Fragment # = 3; More Frag. = 1

CRC | 272 Bytes | MAC — Seq. #= 542; Fragment # = 2; More Frag. = 1

CRC | 272 Bytes | MAC — Seq. #= 542; Fragment # = 1; More Frag. = 1

CRC | 272 Bytes | MAC — Seq. #= 542; Fragment # = 0; More Frag. = 1

Physical Layer

Without encryption enabled, all fragments except the last fragment will be the size of the fragmentation threshold. The last fragment, not counting the MAC header and CRC, will be the size of whatever is left of the original MSDU. In this case, 1200 – 272 – 272 – 272 – 272 = 112.

FIGURE 4.17 Transmission Order of Fragments

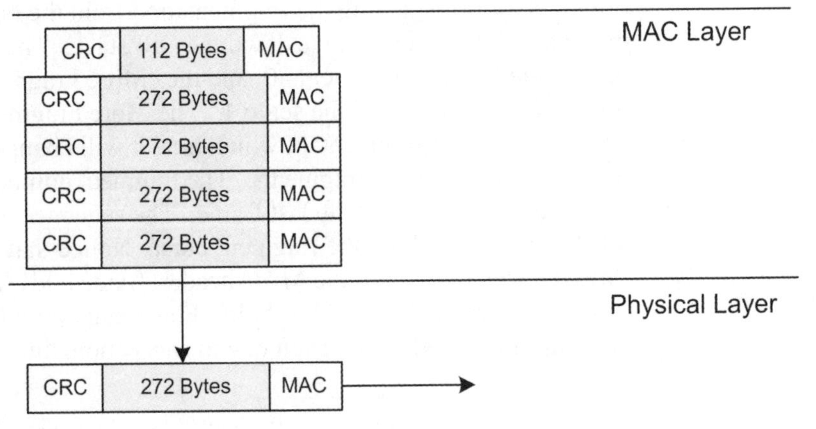

Fragment Bursting

Fragments are always sent in bursts. Once the transmitting station gains control of the medium, it maintains control through two mechanisms: duration values (which set other stations' NAVs) and SIFS.

First, the value of the duration field in the MAC header of data fragments and ACK frames is used to reserve the medium for the next fragment. Second, as a sort of "backup" mechanism, SIFS is used between data fragments and ACK frames in order to preempt those stations that are trying to gain control of the medium using DIFS.

FIGURE 4.18 Fragment Burst

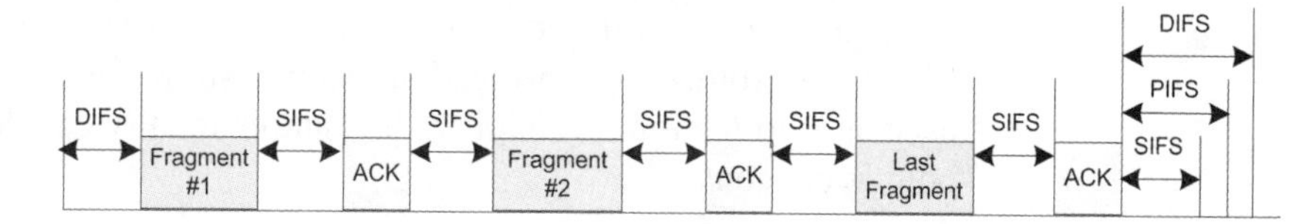

Because the transmitting station sends each fragment independently, the receiving station replies with a separate acknowledgement for each fragment. If a fragment is not acknowledged, then retries begin at the unacknowledged fragment (using DIFS), not at the beginning of the original MSDU.[1] This functionality is the reason that using fragmentation in the presence of RF interference may increase throughput.

RTS/CTS

As an optional feature, the 802.11 standard includes the Request-to-Send/Clear-to-Send (RTS/CTS) function to reserve medium access. With RTS/CTS enabled, a station may transmit a data frame after it completes an RTS/CTS handshake with the immediate receiver of the data frame.

A station (or access point) initiates the four step frame exchange sequence by sending an RTS frame to the intended receiver of the subsequent data frame. The immediate receiver of the RTS responds with a CTS frame.

[1] IEEE 802.11 - 1999 (R2003) – Section 9.2.5.5

The station that sent the RTS frame must receive the CTS frame before sending the data frame. The RTS and CTS frames each contain values in their duration fields that signifies the amount of time needed to complete the transfer of the subsequent data frame and acknowledgement. This duration field value alerts nearby stations to hold off from transmitting for the duration of the four step frame exchange sequence.

The RTS/CTS protocol provides positive control over the use of the shared medium. The purpose of the RTS/CTS protocol is to reserve the wireless medium in order to minimize collisions among hidden stations. This "hidden node" problem can occur when users and access points are spread out throughout a facility or when 802.1b and 802.11g stations coexist in the same BSS or BSA. Using the RTS/CTS protocol to alleviate collisions in this kind of scenario is a "protection mechanism" as described earlier. The main difference between use of the RTS/CTS protocol as a manually-configured medium reservation tool and use of the RTS/CTS protocol as a protection mechanism is that when RTS/CTS is used as a protection mechanism, it is automatically enabled by the access point's beacons.

Hidden Node Example

Imagine there are two 802.11b end users (Station A and Station B) and one access point. Station A and Station B cannot hear each other because of high attenuation (e.g., substantial range), but they can both communicate with the same access point. Because of this situation, Station A may begin sending a frame without noticing that Station B is currently transmitting (or vice versa). This situation will very likely cause a collision between Station A and Station B to occur at the access point. As a result, both Station A and Station B would need to retransmit their respective packets, which results in higher overhead and lower throughput.

If either Station A or Station B activates RTS/CTS, however, the collision is less likely to happen. Before transmitting a data frame, Station-B would send an RTS frame to the access point. The RTS frame is very small, and would finish arriving at the access point quickly. There is still a chance that the RTS could collide with Station-A's data frame, but the chance is much smaller than when Station-B sends a large data frame. The RTS frame would be answered by the access point with a CTS frame.

Station A would hear the CTS frame (because Station A is associated with the access point and must be within range), and set its NAV to the value of the CTS frame's duration field. This means that Station A would not transmit for a period of time sufficient for Station B to transmit its data to the access point and receive an acknowledgement.

Stations near Station B would set their NAV upon hearing the RTS frame, and then set it again upon hearing the CTS frame from the access point. Stations that could not hear Station B (like Station A) would set their NAV upon hearing the CTS only. Both the RTS and CTS frames would set the NAVs of other stations to expire at exactly the same time – after the data transmission and ACK. Thus, the use of RTS/CTS reduces collisions and increases the performance of the network if hidden stations are present. Figure 4.19 shows a time table on when stations set their NAVs upon hearing the RTS and/or CTS frames.

FIGURE 4.19 Network Allocation Vectors

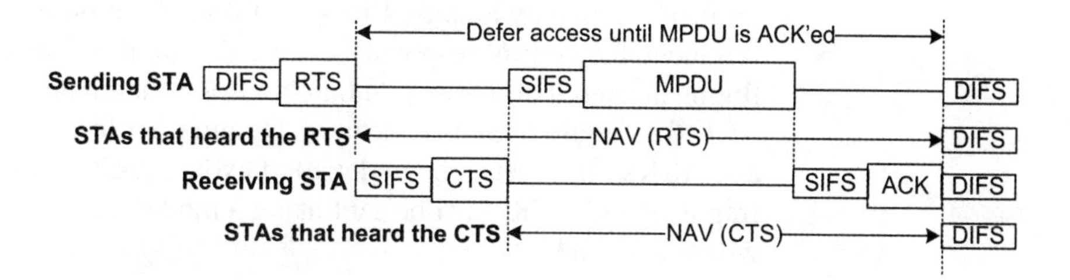

Duration Values & Modulation

There are a complex set of rules regarding duration values specified by the 802.11g standard[1], which will be translated into layman's terms in this text.

In a fragment burst, modulation of frames is as follows.

[1] IEEE 802.11g – (2003) – Sections 9.2.5.6, 9.6, & 9.10

NonERP modulation is used with:

- RTS & CTS frames
- All ACK frames except the last ACK frame in the fragment burst
- All data fragments except the last fragment in a burst

ERP modulation is used with:

- The last ACK frame in a fragment burst
- The last data fragment in a burst

ACK frames should be sent at the same rate and modulation as the data frame which preceded it. If they are not, then the station that transmitted the data frame may not understand the ACK and may begin retransmissions.

If a protection mechanism, such as RTS/CTS, is being used, a fragment sequence may only employ ERP-OFDM modulation for the final fragment and control response because the duration values of data fragments and their corresponding NonERP-modulated data fragment and ACK frames are used as a virtual RTS and CTS for subsequent fragments and ACKs. In order to be understood by NonERP stations, all but the last fragment and ACK must be sent using a modulation that NonERP stations will understand.

Each ACK frame sets the NAV of NonERP (and ERP) stations in the BSS and BSA to a value equal to the subsequent SIFS+DATA+SIFS+ACK (ACKs that are not the last ACK) or to a value of 0 (the last ACK). The data fragments (except the last fragment) also set the NAV of ERP and NonERP stations in the BSS by having a duration value equal to that of the subsequent SIFS+ACK. Therefore, these data fragments must also be sent using a modulation type that is understood by NonERP stations. The last data fragment and ACK are covered by the duration value of the immediately preceding ACK frame, so they can be transmitted using ERP-OFDM without any problems.

Notice in Figures 4.20 and 4.21 that each frame shown contains a duration value equal to subsequent interframe spaces and frames in accordance with the rules listed above. A frame's duration value *never* takes into account its own length, but rather a certain number of interframe spaces and frames that come after it in a frame exchange sequence. In Figure 4.20, the data frame and ACK may be transmitted using ERP-OFDM modulation because there is no fragmentation in use.

FIGURE 4.20 Duration – no Fragmentation

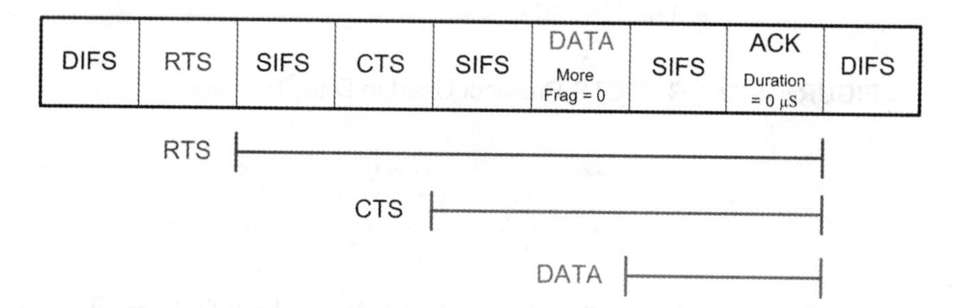

RTS and CTS frame duration values only provide for the first data fragment and its corresponding acknowledgement as shown in Figure 4.21. The duration value found in subsequent data fragments and their ACK frames reserve the medium for enough time for the next fragment and ACK. In Figure 4.21, the first data fragment must use NonERP modulation (such as BPSK, QPSK, or CCK), and the second data fragment will use OFDM modulation. There is no need for the last data fragment and ACK to be understood (for NAV-setting purposes) by NonERP stations in the BSS because the previous ACK reserved the medium using NonERP modulation.

FIGURE 4.21 Duration - Fragmentation

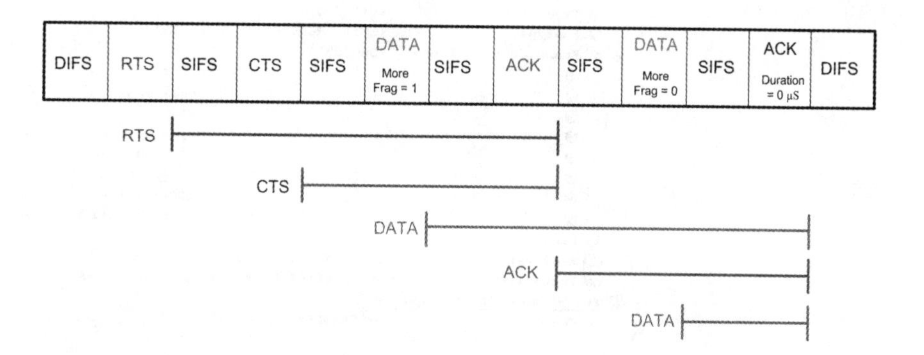

Duration Values Without Fragmentation

In Figure 4.22 a wireless station is announcing its intention to transmit a 1500 byte FTP Data frame using the RTS/CTS protocol. The details of this announcement are shown and include both using and not using fragmentation.

FIGURE 4.22 RTS/CTS Protocol Used in Data Transfer

Packet	Source Physical	Dest. Physical	BSSID	Channel	Data Rate	Size	Protocol
11	00:09:5B:69:FC:98	00:0D:ED:A5:4F:70		1	24.0	20	802.11 RTS
12	00:0D:ED:A5:4F:70	00:09:5B:69:FC:98		1	24.0	14	802.11 CTS
13	00:09:5B:69:FC:98	00:C0:9F:09:81:32	00:0D:ED:A5:4F:70	1	54.0	1536	FTP Data
14	00:0D:ED:A5:4F:70	00:09:5B:69:FC:98		1	24.0	14	802.11 Ack

Notice that the duration value in the RTS frame decode in Figure 4.23 shows 352 µs. This is the time it will take to send the subsequent SIFS+CTS+SIFS+DATA+SIFS+ACK.

FIGURE 4.23 WLAN Protocol Analyzer Displaying an RTS Frame's Duration Value

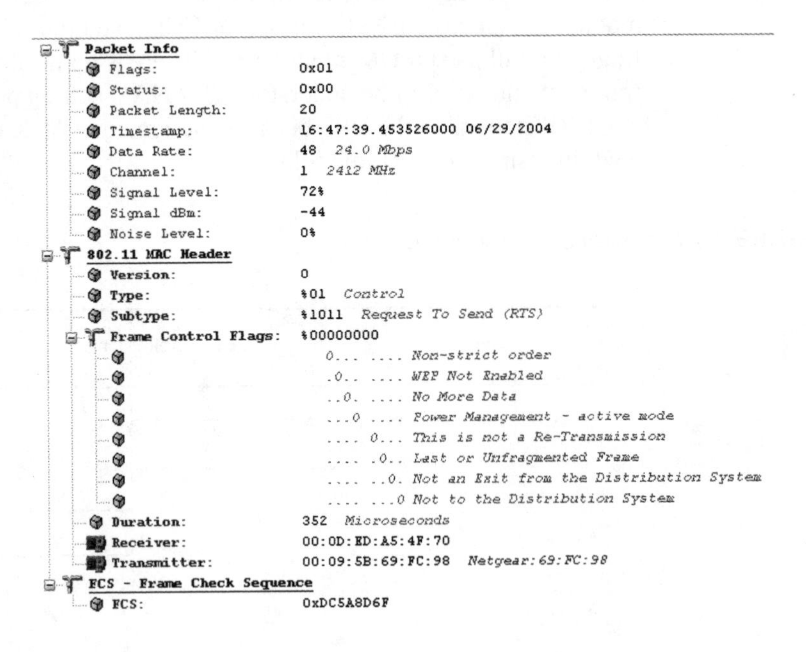

The CTS frame's duration value is calculated from the RTS frame's duration value. It should equal the RTS duration –SIFS–CTS. This equates to SIFS+DATA+SIFS+ACK. Notice in Figure 4.24, that the CTS frame has a value of 308 µs. This means that the previous SIFS and CTS frames had a combined duration of 44 µs.

FIGURE 4.24 WLAN Protocol Analyzer Displaying a CTS Frame's Duration Value

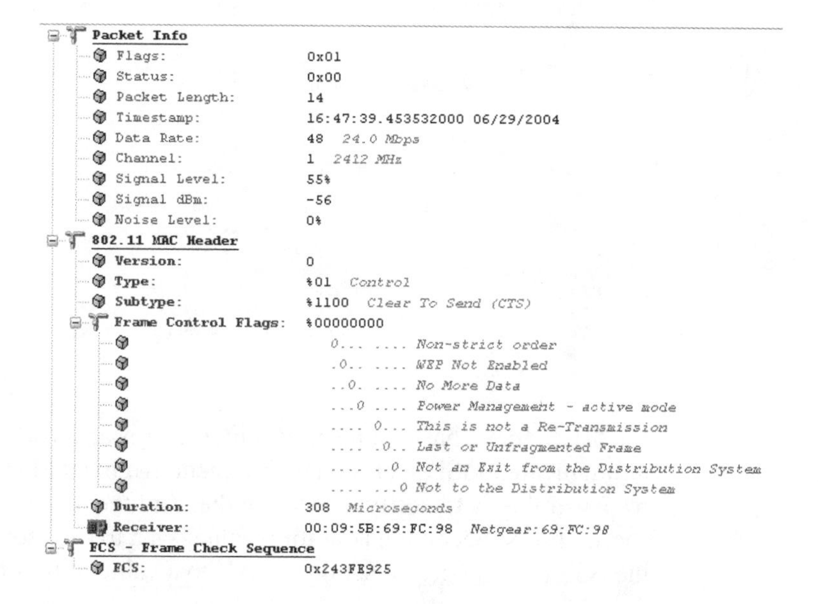

The data frame also reserves the medium for the sake of the SIFS+ACK to follow. In Figure 4.25, we can see that the data frame is reserving the medium for 44 µs. Note that an ACK frame and a CTS frame are constructed in exactly the same fashion, and have the same length. Therefore, SIFS+CTS = SIFS+ACK.

FIGURE 4.25 WLAN Protocol Analyzer Displaying a Data Frame's Duration Value

```
Packet Info
    Flags:                    0x00
    Status:                   0x00
    Packet Length:            1536
    Timestamp:                16:47:39.452459000 06/29/2004
    Data Rate:                108  54.0 Mbps
    Channel:                  1  2412 MHz
    Signal Level:             67%
    Signal dBm:               -48
    Noise Level:              0%
802.11 MAC Header
    Version:                  0
    Type:                     %10  Data
    Subtype:                  %0000  Data Only
    Frame Control Flags:      %00000001
                                 0... .... Non-strict order
                                 .0.. .... WEP Not Enabled
                                 ..0. .... No More Data
                                 ...0 .... Power Management - active mode
                                 .... 0... This is not a Re-Transmission
                                 .... .0.. Last or Unfragmented Frame
                                 .... ..0. Not an Exit from the Distribution System
                                 .... ...1 To the Distribution System
    Duration:                 44  Microseconds
    BSSID:                    00:0D:ED:A5:4F:70
    Source:                   00:09:5B:69:FC:98  Netgear:69:FC:98
    Destination:              00:C0:9F:09:81:32  Quanta Comp:09:81:32
    Seq. Number:              2962
    Frag. Number:             0
```

The ACK frame has two responsibilities if it is acknowledging an unfragmented MSDU or the last fragment in a burst. First, its primary responsibility is to notify the station that sent the data frame that it arrived successfully. Second, its duration value is set to 0 μs so that all stations in the BSS that can hear it set their NAV to a value of 0, if not there already.

FIGURE 4.26 WLAN Protocol Analyzer Displaying an ACK Frame's Duration Value

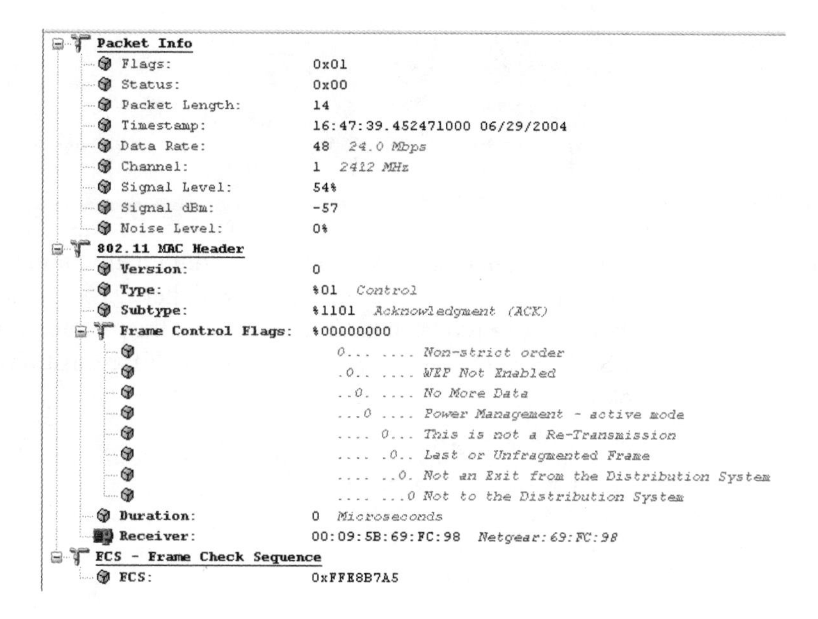

Duration Values Using Fragmentation

The next three Figures are of wireless LAN analyzer screenshots, used here so that you will recognize fragmentation when you see it in real world applications (although shown here without encryption). Figures 4.27 – 4.29 illustrate a 1500 byte MSDU that has been transmitted using a 600 byte threshold. Notice that all fragments except the last are cut at the size of the threshold, and each fragment is acknowledged individually. Some analyzers interpret frame fragments better than others, so you may have to take a close look at the frame types to understand that they are only fragments of an MSDU. RTS and CTS frames work exactly the same in a fragmented environment as they do in a non-fragmented environment. Instead of reserving the medium for the entire frame exchange, they simply reserve the medium for the first data fragment exchange with its subsequent ACK.

FIGURE 4.27 WLAN Protocol Analyzer – Fragment Burst

No	M	Time	Delta		Len			Source	Dest	Summary
147		6/29 17:05:36.688181	12.688181	1	600	88	54	Netgear:69:FC:98	00:0D:ED:A5:4F:70	UDP port:1043 -> port:1034
148		6/29 17:05:36.688185	12.688185	1	10	95	24		Netgear:69:FC:98	802.11 acknowledgement
149		6/29 17:05:36.688485	12.688485	1	600	88	54	Netgear:69:FC:98	00:0D:ED:A5:4F:70	LLC SSAP:20 DSAP:65
150		6/29 17:05:36.688489	12.688489	1	10	95	24		Netgear:69:FC:98	802.11 acknowledgement
151		6/29 17:05:36.688756	12.688756	1	380	88	54	Netgear:69:FC:98	00:0D:ED:A5:4F:70	LLC SSAP:65 DSAP:6D
152		6/29 17:05:36.688761	12.688761	1	10	100	24		Netgear:69:FC:98	802.11 acknowledgement

The first fragment shows that the entire MSDU was 1500 bytes, though this first fragment is only 600 bytes. Additionally, notice that the first frame's duration value is significantly higher (216 µs) than shown previously (44 µs) because it must reserve the medium for the subsequent SIFS+ACK+DATA+SIFS+ACK. The math for this transmission would look like:

```
44 µs (SIFS+ACK) + 128 µs (DATA FRAG #2) + 44 µs
(SIFS+ACK) = 216 µs
```

FIGURE 4.28 WLAN Protocol Analyzer – Fragment #1

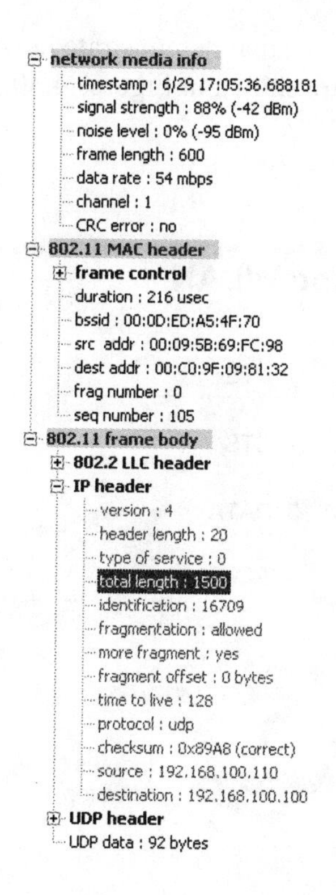

In Figure 4.29 the first ACK's duration value is 172 μs. Since this ACK is not the last ACK in the burst, this ACK is responsible for reserving the medium for the subsequent SIFS+DATA+SIFS+ACK.

FIGURE 4.29 WLAN Protocol Analyzer – ACK duration

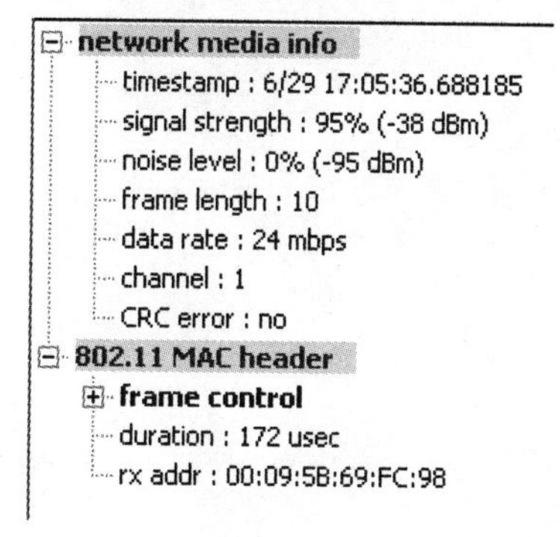

Scenarios

There are three specific scenarios that demonstrate how RTS/CTS is most often used. These scenarios are shown in Figures 4.30 – 4.32.

FIGURE 4.30 RTS/CTS – Ad Hoc WLAN

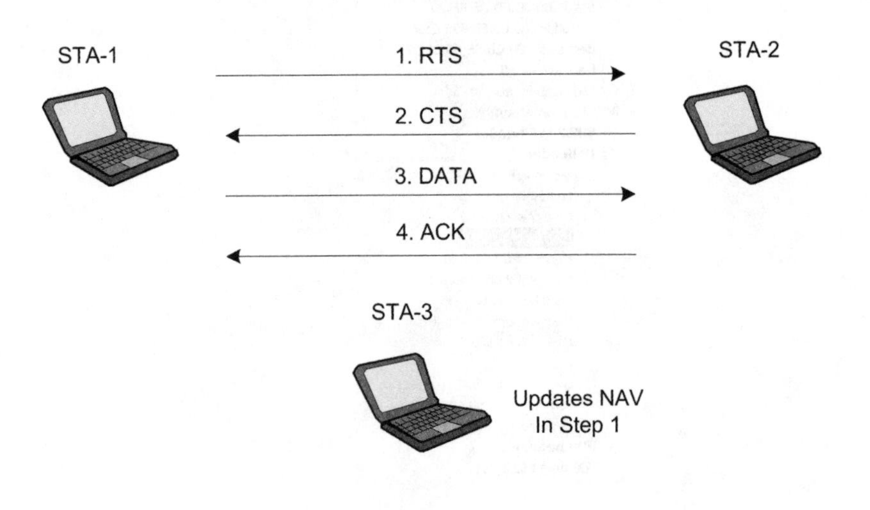

FIGURE 4.31 RTS/CTS – Infrastructure Example #1

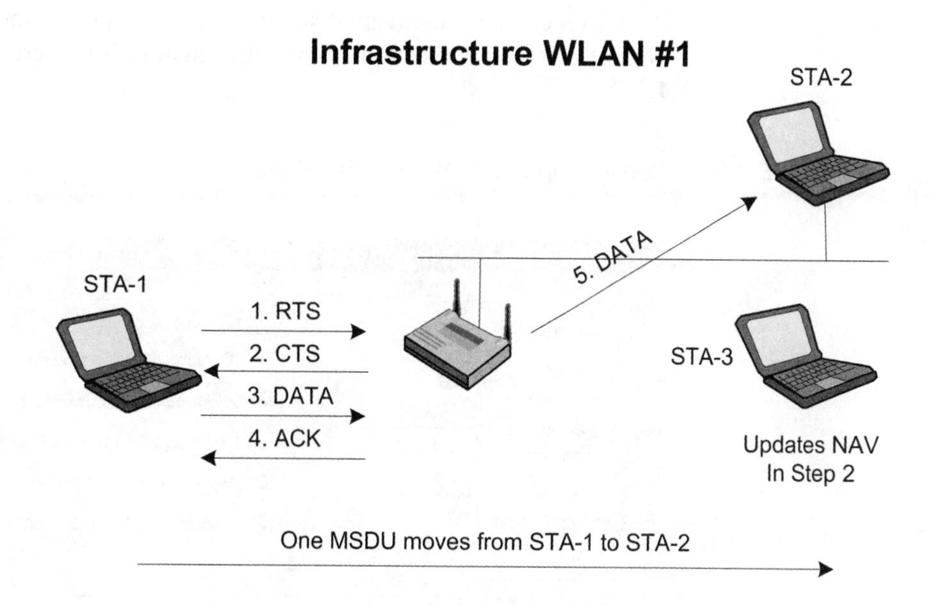

FIGURE 4.32 RTS/CTS - Infrastructure Example #2

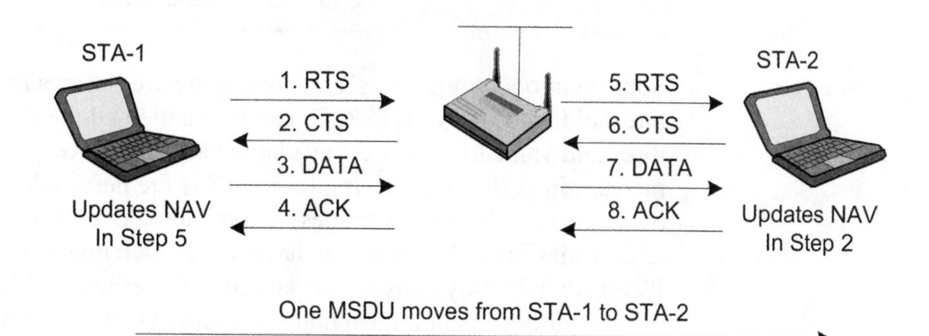

Configuration

RTS/CTS can be effectively disabled by setting the threshold value to the highest available value as shown in the client utilities screenshot in Figure 4.33.

FIGURE 4.33 Configuring the RTS/CTS Threshold

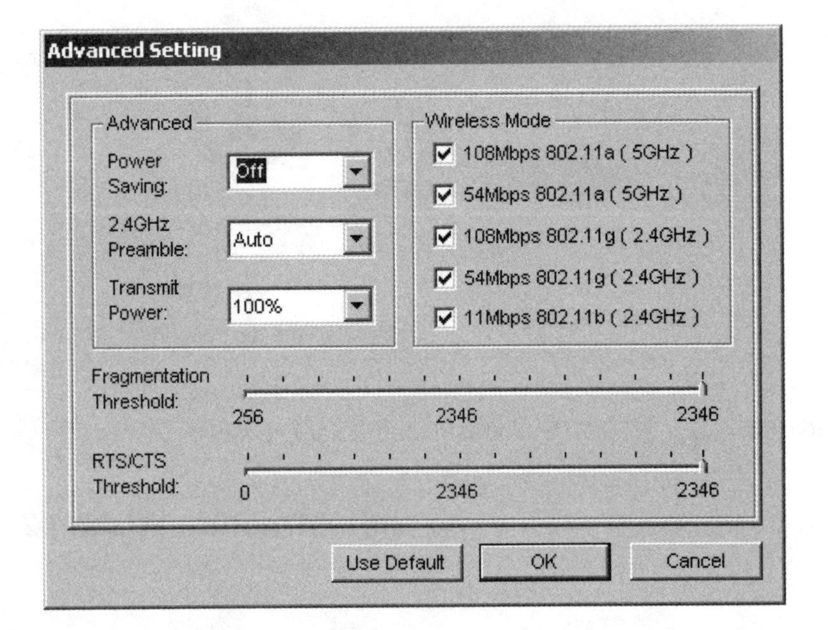

Some vendors provide "on" and "off" software settings in addition to the threshold value only. RTS/CTS can be enabled all the time by setting the threshold value to the lowest available value of 0. Keep in mind that an increase in performance using RTS/CTS is the net result of introducing overhead (i.e., RTS/CTS frames) and reducing overhead (i.e., fewer retransmissions). If you do not have any hidden nodes, then the use of RTS/CTS will only increase the amount of overhead, which reduces throughput. A slight hidden node problem may also result in performance degradation if you implement RTS/CTS. In this case, the additional RTS/CTS frames cost more in terms of overhead than what you gain by reducing retransmissions.

As with fragmentation, one of the best ways to determine if you should activate RTS/CTS is to monitor the wireless LAN for retransmissions. If you find a large number of retransmissions and the users are relatively far apart and likely out of range, then try enabling RTS/CTS on the applicable user wireless NICs. Keep in mind that user mobility can change the results. A highly mobile user may be hidden for a short period of time, perhaps when you perform the testing, then be closer to other stations most of the time. If collisions are occurring between users within range of each other, the problem may be the result of high network utilization or possibly RF interference.

After activating RTS/CTS, test to determine if the number of retransmissions is less and the resulting throughput is better. Because RTS/CTS introduces overhead, you should shut it off if you find a drop in throughput, even if you have fewer retransmissions. After all, the goal is to improve performance.

Except in the case of access points contending for the same channel in the same BSA, initiating RTS/CTS in the access point is not useful because the hidden station problem does not exist from the perspective of the access point. All stations having valid associations are within range and not hidden from the access point. Forcing the access point to implement the RTS/CTS handshake will significantly increase the overhead and reduce throughput.

CTS-to-Self[1]

This optional mechanism is designed to advise NonERP stations that a transmission is pending, so that they will properly update their NAVs and not transmit during an ERP-OFDM transmission. This mechanism allows ERP stations to exchange frames using the ERP-OFDM modulation that is undetectable by the DSSS or HR/DSSS stations. In a small BSS, without hidden nodes, this mechanism would suffice to alert NonERP stations to defer for a frame exchange sequence even though the data frame will be undetectable to the NonERP stations. CTS-to-Self is a standard CTS frame transmitted using a NonERP modulation with a destination address of the transmitting station. Obviously the transmitting

[1] IEEE 802.11g – (2003) – Sections 7.2.1.2, 9.2.1.1, and 9.7

station cannot hear its own transmission in a half-duplex medium, so the transmission of this frame is the human vocal equivalent of shouting "Be Quiet!" All nearby stations are alerted that a frame exchange sequence is pending.

FIGURE 4.34 CTS-to-Self Operation

The CTS-to-Self MAC Control frame is only generated by ERP stations if the access point has signaled *Use_Protection* in its beacon or Probe Response MMPDUs, which indicates to ERP stations that NonERP stations are present. Unless the NonERP stations are associated to this access point, the access point is allowed to turn protected mode off (even though it knows that some NonERP stations are present) if it senses that the NonERP stations are not sending much traffic. The access point has real-time discretion over this decision.

For every associated station, the access point must be within range of that station. From the station's perspective, it is in range of the access point, and other station may be in range of that station as well. If an ERP station has sent a CTS-to-Self MAC control frame, we can assume that the access point has received it, and knows what it means. We are assuming that the access point is capable of understanding both HR and ERP transmissions. All stations, including ERP stations and NonERP stations, within range of a station's radio will understand the meaning of the CTS-to-Self frame and update their NAVs accordingly. All stations will see this CTS frame as a normal CTS; only the originating station knows that it sent a CTS-to-

Self frame. Stations do not need to keep state on whether they have seen an RTS Control frame in order to know if a CTS Control frame is a response to an RTS or a CTS-to-Self message. They must simply process the CTS frame statelessly. All STAs within range of the originating station, regardless of whether they are NonERP or ERP stations, will update their NAVs.

CTS-to-Self is automatically invoked on the access point, so there is no manual configuration that might be accomplished by the administrator. Only stations near the transmitting station may be able to hear the CTS frame.

FIGURE 4.35 Access Point Invoke Protection Mode

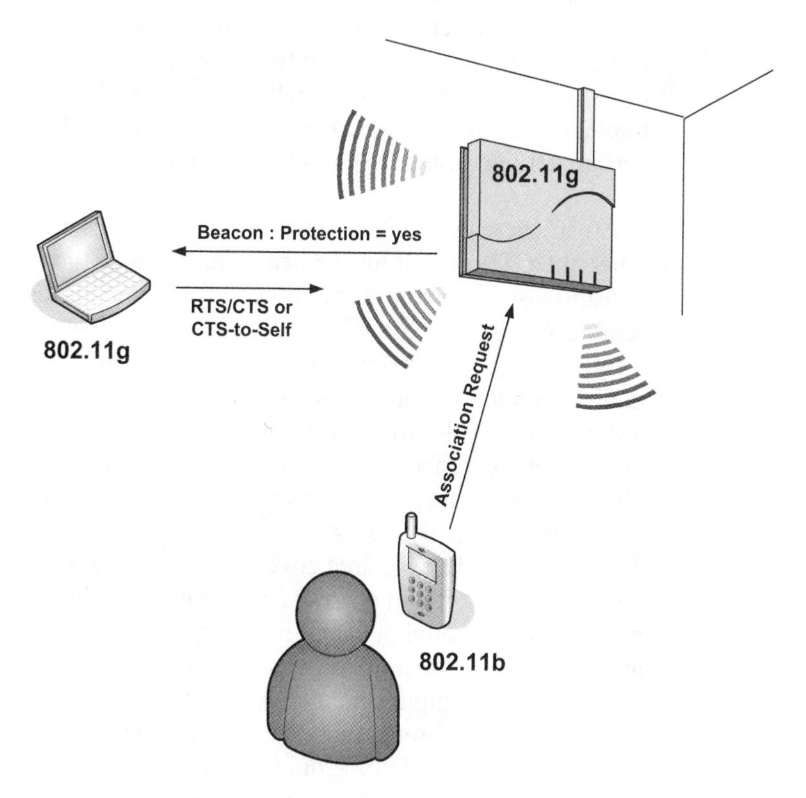

The CTS-to-Self procedure is not perfect, but since faster modulations tie up the medium for as much as 80 percent less time (for a given packet size), it might be reasonable to suppose that the mechanism need not be perfect. A 1500-octet IP packet takes about as much time to transmit at 54 Mbps as a 320-octet IP packet takes at 11 Mbps. Moreover, the CTS-to-Self mechanism is not mandatory, and the IEEE 802.11g standard does still permit the use of RTS/CTS if desired. However, while the RTS/CTS exchange would seem to be sufficient to alert the BSS to the impending transmission (provided that the RTS and CTS are transmitted using a modulation that all HR- and ERP-STAs can understand), the RTS/CTS exchange incurs the cost of additional latency before an ERP station can begin to send data. This latency significantly reduces the maximum throughput that can be achieved.

Given that the CTS-to-Self mechanism can only be used by ERP stations, another issue is, "When should the station generate such a frame?" It is a given that the access point must have enabled protected mode, but the station need not send an RTS before its frame. There does not appear to be any standards-based way to force the station to use either RTS/CTS or CTS-to-Self as a protection mechanism, so this looks like a choice that implementers are free to make; this freedom extends all the way to legitimately ignoring the possibility of implementing CTS-to-Self.

In cases where the BSS is small and most stations are in range of each other, the CTS-to-Self mechanism, coupled with the fact that the frame transmission times are shorter (due to the faster speeds), means that stations using CTS-to-Self may see higher throughput. If a station fails to receive an ACK after sending a CTS-to-Self frame and a Data frame, the implementation might choose to infer that there are hidden nodes and it is not safe to use CTS-to-Self for a while. This behavior – using collisions to infer the existence of hidden nodes – is not specified in the IEEE 802.11g-2003 standard, either as a mandate or a recommendation. However, anyone who tries to implement CTS-to-Self should probably define his own decision tree governing the usage of this feature, since blindly enabling it will not always give the best throughput. CTS-to-Self does not help throughput at all in a busy BSS that is large enough such that a significant fraction of the nodes are hidden relative to each other. The goal of any implementation should be to provide the end user with the best possible throughput. Again, CTS-to-Self is not mandatory, and it is perfectly reasonable to avoid implementing it.

Protection Mechanisms

Protection mechanisms are so prevalent in today's wireless LANs that they are worth discussing a little deeper. Both RTS/CTS and CTS-to-Self have detrimental impacts on throughput. RTS/CTS has more negative impact on throughput, but a more positive impact on hidden nodes in most wireless environments. It is typical to see that half of a BSS's throughput is lost due to protection mechanisms alone (when they are enabled). Additionally, one ERP access point's decision to enable protection may affect the entire wireless LAN infrastructure as a whole in a negative fashion due to vendor-specific implementations.[1]

Each ERP access point's beacons have an ERP Information Element. This ERP IE has three active fields as shown in Figure 4.36.

FIGURE 4.36 ERP Access Point's ERP Information Element

```
ERP Information
  Element ID:        42  ERP Information
  Length:            1
  ERP Flags:         %00000000
                       x... .... Reserved
                       .x.. .... Reserved
                       ..x. .... Reserved
                       ...x .... Reserved
                       .... x... Reserved
                       .... .0.. Not Barker Preamble Mode
                       .... ..0. Disable Use of Protection
                       .... ...0 Non-ERP Not Present
```

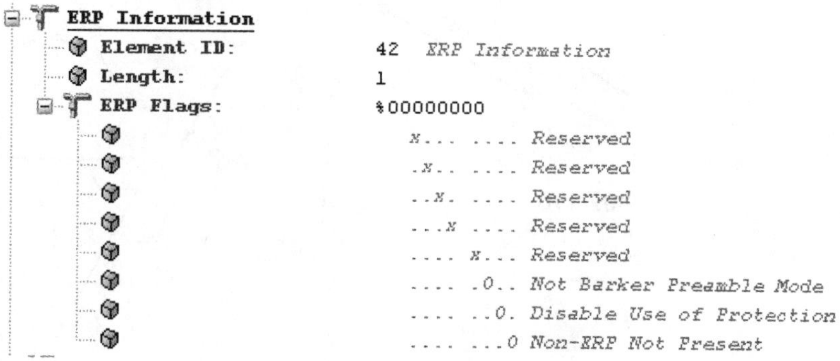

The Barker_Preamble_Mode bit is used to specify whether long or short preambles are to be used when transmitting frames modulated with BPSK, QPSK, or CCK. These frames include RTS, CTS, and data fragments (except the last fragment) that are transmitted by an ERP station that has been notified by the access point, using the NonERP_Present field, that a NonERP station is associated with the access point. The Use_Protection bit is used by the access point in

---

[1] For more information on Protection Mechanism use in ERP wireless LANs, see Protection Ripple in 802.11 WLANs (Devin Akin, June 2004) at www.cwnp.com

beacons to alert ERP stations in the BSS that they should use protection mechanisms such as RTS/CTS and CTS-to-Self before transmitting data using OFDM modulation.

Depending on the vendor's implementation of the standard, co-channel and adjacent channel interference between access points may also lead to a situation in which an access point will enable protection in its own beacons if it hears another access point enable protection in its beacons. This reaction can lead to a situation in which a single NonERP station causes protection mechanisms to be enabled throughout part or all of a wireless LAN infrastructure. Figures 4.37 and 4.38 show the concepts of co-channel and adjacent channel protection proliferation.

FIGURE 4.37 Co-channel Protection Proliferation

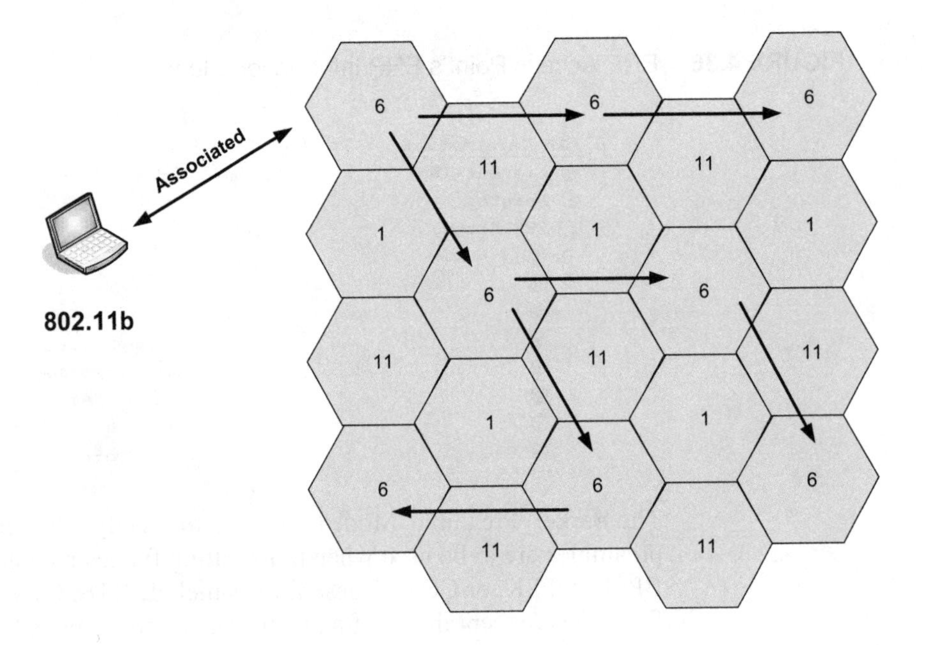

FIGURE 4.38 Adjacent channel Protection Proliferation

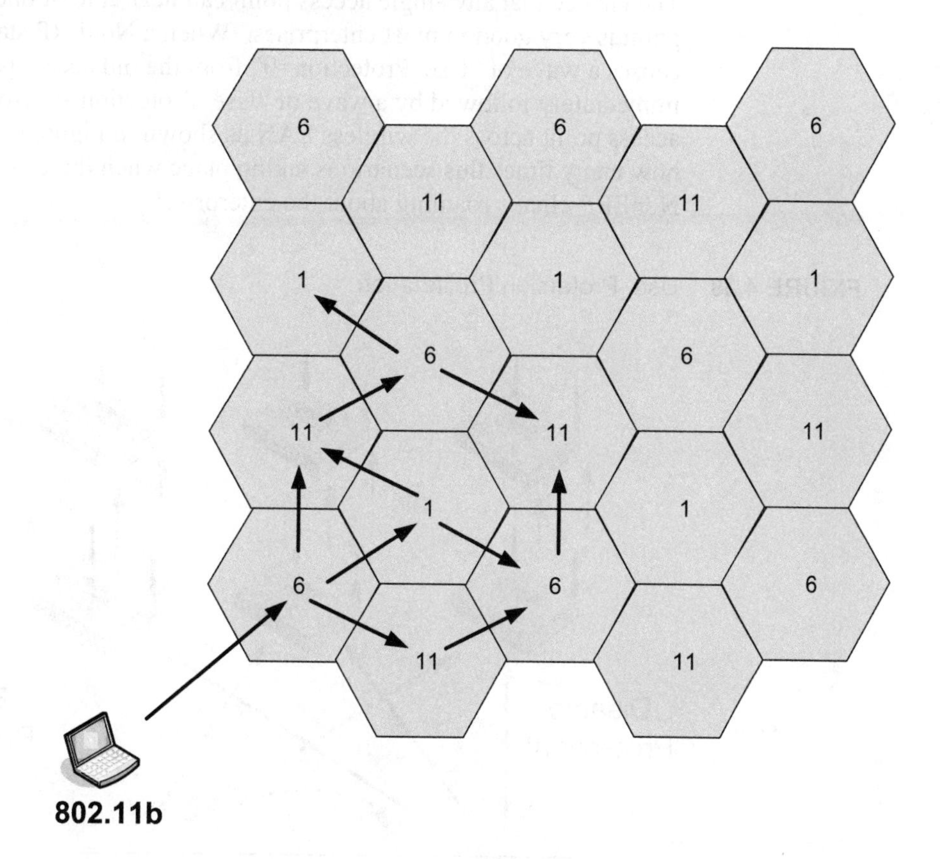

Suppose that a NonERP station successfully roams to another access point in the ESS. This roaming will cause the new access point to enable protection and notify the old access point that it has now reassociated this particular mobile station. This notification should cause the old access point to drop the association with the NonERP station immediately. When the new access point enables protection, its beacons may then cause the old access point to either keep protection enabled or to immediately re-enable it. The old access point may make this change even though the NonERP station has left its BSS.

This cause and effect scenario demonstrates that everywhere a NonERP station goes in an ERP wireless LAN, protection mechanisms are not only enabled on the local access point, but are also triggered elsewhere depending on which access points can hear which other access points. The chance that any single access point can hear at least one other access point is very good in most enterprises. When a NonERP station roams, it causes a wave of "Use_Protection=0" from the old access point immediately followed by a wave of "Use_Protection=1" from the new access point across the wireless LAN as shown in Figure 4.39. Imagine how many times this scenario is taking place when there are many NonERP clients roaming about the enterprise!

FIGURE 4.39 Use_Protection Proliferation

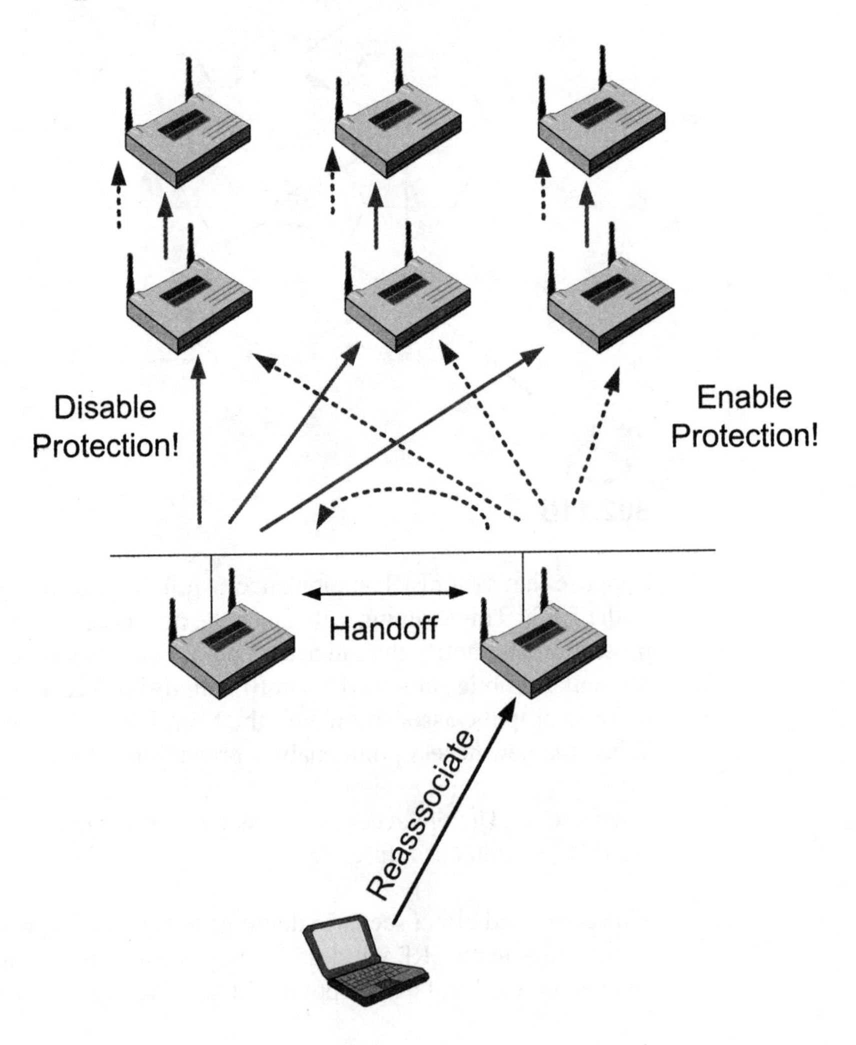

Throughput degradation when using protection in a BSS is severe. Figures 4.40 and 4.41 show how much protection mechanism affect throughput in a single BSS. The loss in throughput is approximately half for the entire BSS even if the 802.11b station just associates. The 802.11b station does not have to transmit any traffic onto the wireless medium for this to happen, but doing so makes the situation far worse.

FIGURE 4.40 Throughput Decline Due to Protection

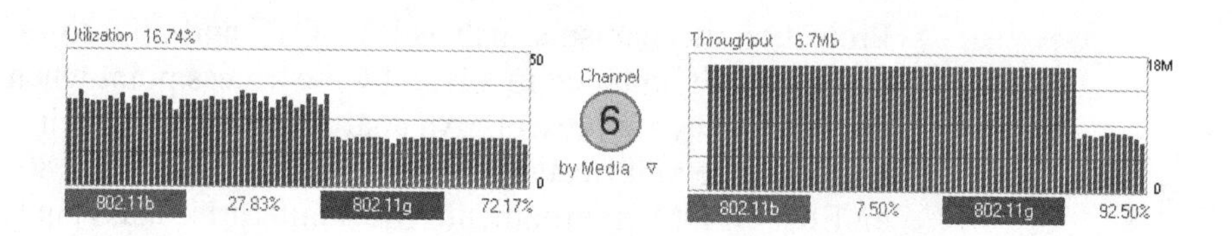

FIGURE 4.41 802.11b/g Mixed Mode Environment Throughput Scale

Number of 802.11b clients \ Number of 802.11g clients	0	1	2	3	4	5	6	7	8	9	10
10	5.9	6.2	6.5	6.8	7.0	7.2	7.4	7.6	7.8	8.0	8.2
9	5.9	6.2	6.5	6.8	7.1	7.4	7.6	7.8	8.0	8.2	8.3
8	5.9	6.3	6.6	6.9	7.2	7.5	7.7	8.0	8.2	8.4	8.5
7	5.9	6.3	6.7	7.1	7.4	7.7	7.9	8.2	8.4	8.6	8.8
6	5.9	6.4	6.8	7.2	7.6	7.9	8.2	8.4	8.7	8.9	9.1
5	5.9	6.5	7.0	7.4	7.8	8.2	8.5	8.7	9.0	9.2	9.4
4	5.9	6.6	7.2	7.7	8.2	8.5	8.9	9.2	9.4	9.6	9.8
3	5.9	6.8	7.6	8.2	8.7	9.1	9.4	9.7	9.9	10.2	10.4
2	5.9	7.2	8.2	8.9	9.4	9.8	10.2	10.4	10.7	10.9	11.1
1	5.9	8.2	9.4	10.2	10.7	11.1	11.3	11.6	11.7	11.9	12.0
0	0.0	22.1	22.1	22.1	22.1	22.1	22.1	22.1	22.1	22.1	22.1

Number of 802.11g clients

Summary

The information covered by this chapter is often treated as "mystical" by WLAN administrators and engineers alike because it is a bit subjective. Determining the best fragmentation and RTS/CTS threshold can be a daunting task since it must be set on every transmitter individually. Some agree that it is best left alone, and others think that the added throughput realized by optimization of settings is worth the trouble.

Protection mechanisms such as RTS/CTS and CTS-to-Self are automatically initiated in a BSS by the access point when certain circumstances are present. An analyst should thoroughly understand the performance ramifications of using 802.11 DSSS and 802.11b stations in an ERP-OFDM environment. Most enterprise class analyzers are more than adequate for realizing the performance affects due to 802.11b/g co-location.

Key Terms

Before taking the exam, you should be familiar with the following terms:

Basic Service Set (BSS)

Distributed Coordination Function (DCF)

Distribution System (DS)

Extended Service Set (ESS)

fragmentation

Independent Basic Service Set (IBSS)

Network Allocation Vector (NAV)

portal

power-save mode

Request-to-Send / Clear-to-Send (RTS/CTS)

Review Questions

1. In which frame types is the association ID (AID) found?

2. Which information element contains the *Barker_Preamble_Mode* bit and what is its intended purpose?

3. Acknowledgement frames (ACKs) must be sent using which modulation and at what speed?

4. When are queued multicast and broadcast frames transmitted onto the wireless medium by the access point?

5. If a transmitting station has a fragmentation threshold of 428 bytes, and it is transmitting a 1200 byte MSDU to an access point, how many frames will be needed to complete the transfer?

6. How may an access point respond to a PS-Poll frame from a station?

7. What is the purpose of the More Data bit found in the Frame Control field of a MAC frame?

8. When is the CTS-to-Self frame required to be used?

9. What is one of the best ways to determine whether RTS/CTS and/or fragmentation thresholds should be modified?

10. Which station types may use CTS-to-Self frames?

CHAPTER

5

802.11 MAC Frame Format

CWAP Exam Objectives Covered:

- ❖ Explain the structure of each 802.11 MAC layer frame type
  - MAC Layer terminology used in the 802.11 series of standards
  - Header fields and subfields
  - Frames sizes
  - MAC layer addressing

In This Chapter

MAC Frame Format

There are three IEEE 802.11 MAC frames types: management, control, and data frames. The combination of these frame types establishes an overall means for carrying data between 802.11 stations.

Nomenclature

Figure 5.1 illustrates the proper nomenclature for frames at all layers of the 802.11 architecture. The MAC layer takes protocol data from LLC called a MAC Service Data Unit (MSDU). This MSDU is put into a MAC Protocol Data Unit (MPDU), which is essentially adding a header and a 32-bit CRC to the MSDU.

The MPDU complies with a specific format that the 802.11 standard defines, which includes a series of header fields. There are cases when an MPDU will have a null (empty) frame body instead of an MSDU. The MAC layer in turn hands the MPDU down to the 802.11 physical (PHY) layer as a Physical Layer Convergence Protocol (PLCP) Service Data Unit (PSDU). The PSDU and MPDU are the same data unit. It is appropriate to refer to this data unit as an MPDU when you are discussing MAC layer operations, and a PSDU when discussing PHY layer operations.

When the PSDU is accepted by the PLCP layer (the upper half of Layer 1), it receives a preamble and header. The combination of the PLCP preamble, the PLCP header, and the PSDU form the PLCP Protocol Data Unit (PPDU). The PPDU is passed down to the PMD to be transmitted onto the RF medium as symbols representing the 0s and 1s of the PPDU.

FIGURE 5.1 Frame Nomenclature

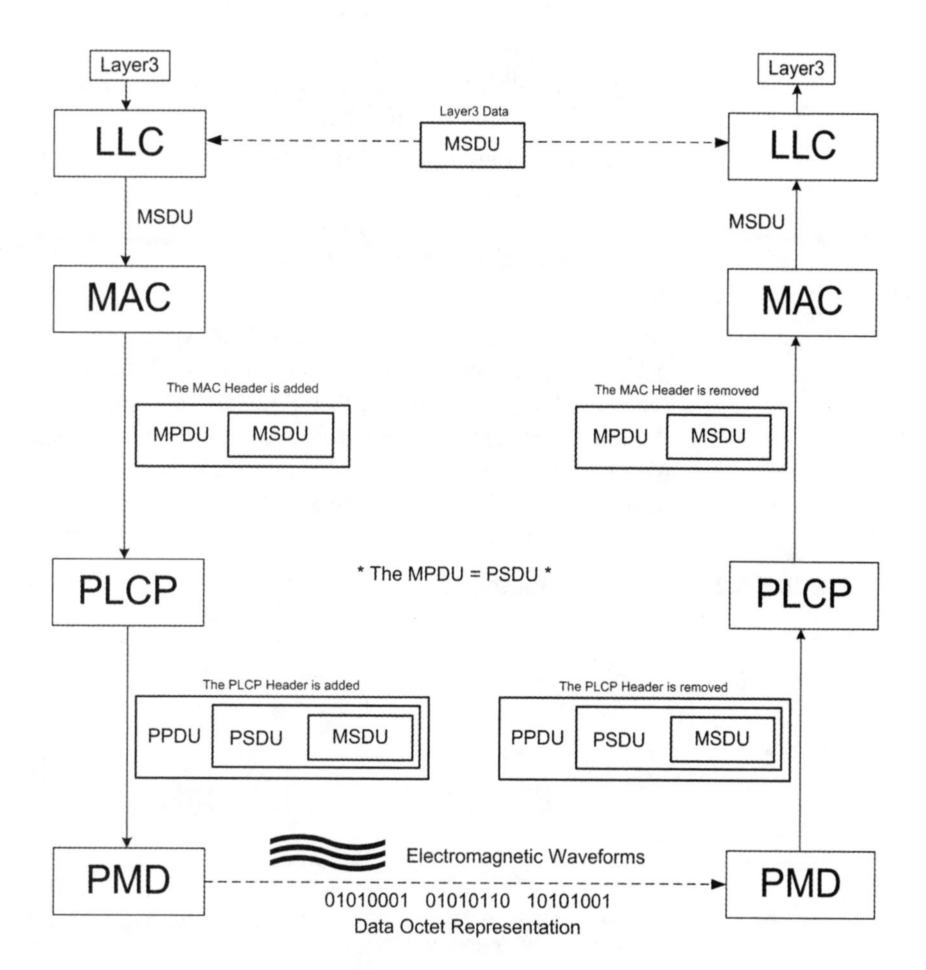

The text above described logical data units at each layer and their purposes. The concept of a successively increasing frame being passed from MAC to PLCP to PMD is simpler to contemplate than the actual process. The reality is more like the following.

1. The MAC signals the PLCP that an MPDU is ready for transmission, the number of octets in the MPDU, and the intended transmission rate.

2. The PLCP calculates the PLCP header and signals the PMD entity to transmit a certain preamble and the octets of the PLCP header.
3. The PMD transmits the corresponding symbols and signals the PLCP that it is finished.
4. The moment the PMD is finished transmitting the PLCP preamble and header, the PLCP begins requesting, one at a time, the octets of the MPDU from the MAC and passing them, one at a time, to the PMD for transmission.

At the receiving end, the receiver's PMD synchronizes using the preamble, and then begins receiving the PLCP header. The PLCP layer reads the length field of the PLCP header, and informs the MAC entity of the idle/busy status of the wireless medium and to prepare to receive an MPDU with certain attributes. The MPDU arrives as a series of octet transfers from the PMD to the PLCP, and the PLCP to the MAC.

FIGURE 5.2 Signaling Between Layers

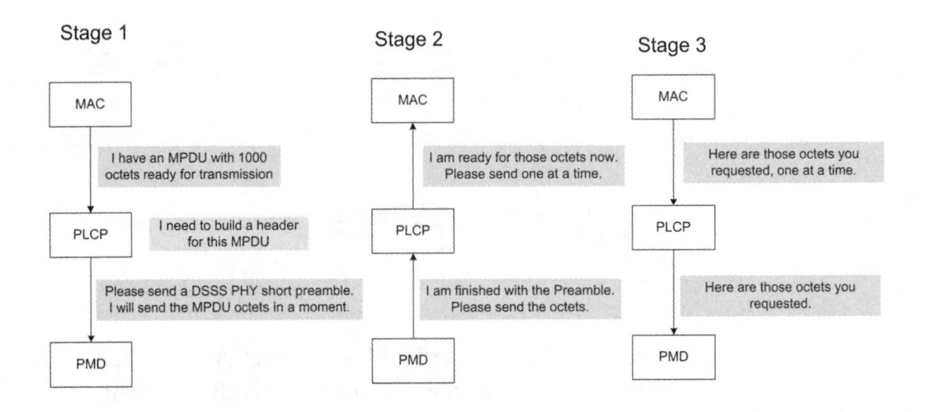

Frame Structure

The 802.11 standard specifies an overall MAC frame (MPDU) format, as shown in Figure 5.3. This frame structure is found in all frames that stations send, regardless of frame type. Not all parts of the frame are used by all frame types, but the overall frame structure remains the same across all frames.

FIGURE 5.3 802.11 Overall Frame Structure

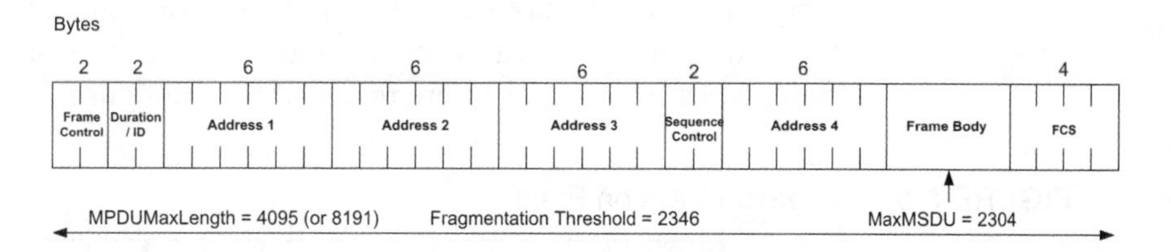

In the following sections, we will discuss each part of the 802.11 frame structure down to the bits and bytes. Additionally, we will cover the structure of each frame type and what fields and subfields are used.

Frame Control Field

The 16-bit Frame Control field includes information about the frame being sent and specific protocol operations. Figure 5.4 illustrates specific subfields within the frame control field.

FIGURE 5.4 Frame Control Field

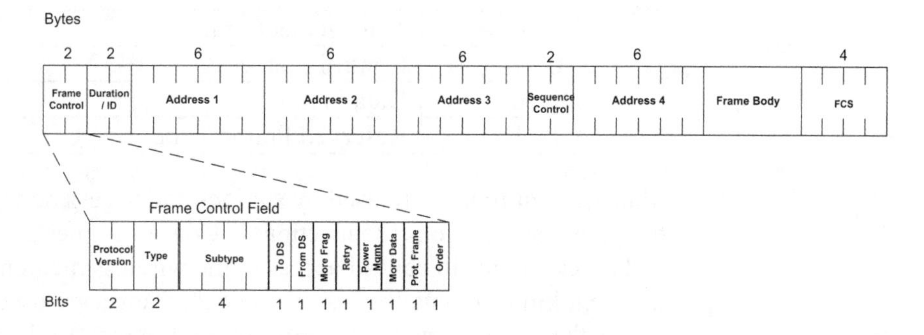

Each of the subfields within the frame control field is defined below.

Protocol Version Field

For the current standard, the 2-bit Protocol Version field is 00. IEEE will add version numbers in the future if a newer version of the standard is fundamentally incompatible with an earlier version. If the Protocol Version field is equal to a value that the station cannot interoperate with, the station must discard the entire frame. Since there are 2 bits for this field, types 01, 10, and 11 are reserved for future use.

FIGURE 5.5 Protocol Version Field

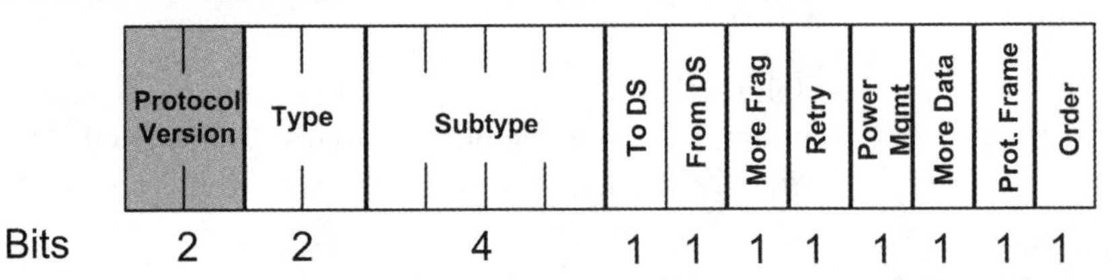

Type Field

The Type field defines whether the frame is a management, control, or data frame as indicated by the bits in Figure 5.6.

FIGURE 5.6 Frame Types

0 , 0	Management Frame
0 , 1	Control Frame
1 , 0	Data Frame
1 , 1	Reserved Frame Type

Management frames are used by stations and access points to build and tear down associations (connections). Control frames are used by stations and access points to control access to the wireless medium. There are different kinds of data frames. Some data frames are used to move data across the wireless medium, some are used to request data, and still others are used to inform other stations that there is no data to transmit.

FIGURE 5.7 Type Field

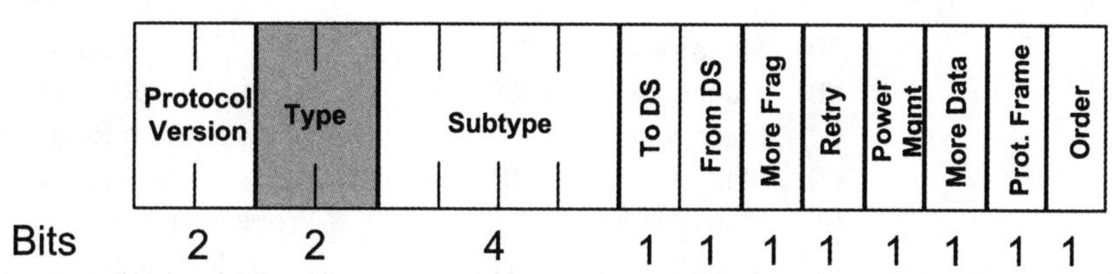

Subtype Field

The Subtype field defines the specific function of the frame, as shown in Figure 5.8. Figure 5.9 shows where the Subtype field is positioned in the Frame Control field.

FIGURE 5.8 Frame Subtypes

00	Management	0000	Association Request
00	Management	0001	Association Response
00	Management	0010	Reassociation Request
00	Management	0011	Reassociation Response
00	Management	0100	Probe Request
00	Management	0101	Probe Response
00	Management	0110-0111	Reserved
00	Management	1000	Beacon
00	Management	1001	Announcement Traffic Indication Message (ATIM)
00	Management	1010	Disassociation
00	Management	1011	Authentication
00	Management	1100	Deauthentication
00	Management	1101-1111	Reserved
01	Control	0000-1001	Reserved
01	Control	1010	Power Save (PS)-Poll
01	Control	1011	Request To Send (RTS)
01	Control	1100	Clear To Send (CTS)
01	Control	1101	Acknowledgement (ACK)
01	Control	1110	Contention-Free (CF)-End
01	Control	1111	CF-End + CF-Ack
10	Data	0000	Data
10	Data	0001	Data + CF-Ack
10	Data	0010	Data + CF-Poll
10	Data	0011	Data + CF-Ack + CF-Poll
10	Data	0100	Null function (no data)
10	Data	0101	CF-Ack (no data)
10	Data	0110	CF-Poll (no data)
10	Data	0111	CF-Ack + CF-Poll (no data)
10	Data	0000-1111	Reserved
11	Reserved	0000-1111	Reserved

FIGURE 5.9 Subtype Field

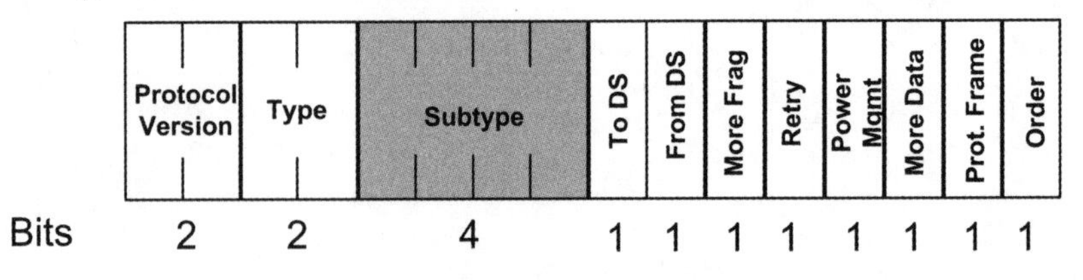

ToDS Field

The MAC coordination sets the single-bit ToDS field to 1 in a data frame destined for the distribution system of an access point. For example, the ToDS field is always set to 1 when a data frame is transmitted by a client station associated to an access point. ToDS is always set to 0 for all control and management frames, and data frames not destined for the distribution system.

FIGURE 5.10 ToDS Field

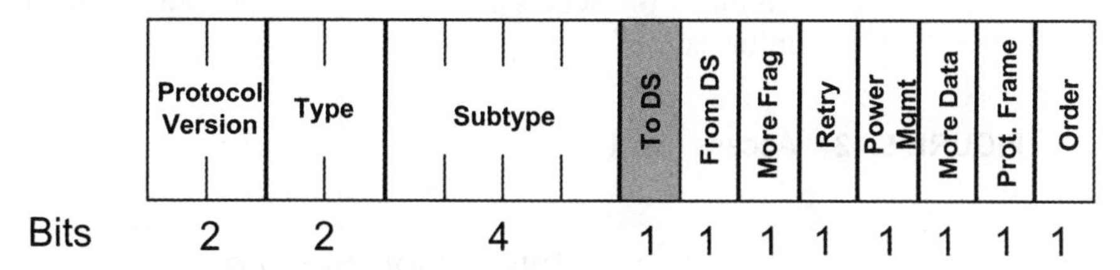

FromDS Field

The MAC layer sets the single-bit FromDS field to 1 in data frame leaving the distribution system of an access point. For example, the FromDS field is set to a 1 in data frames being sent from an access point to a mobile station. The FromDS field is set to 0 for all control and management frames and data frames not coming from the distribution system. Both the ToDS and FromDS fields are set to 1 if the frame is being sent directly from one access point to another access point as part of a wireless distribution system (WDS). They are both set to 0 if the data frame is being sent directly from one mobile station to another, as in the case of Ad Hoc mode.

FIGURE 5.11 FromDS Field

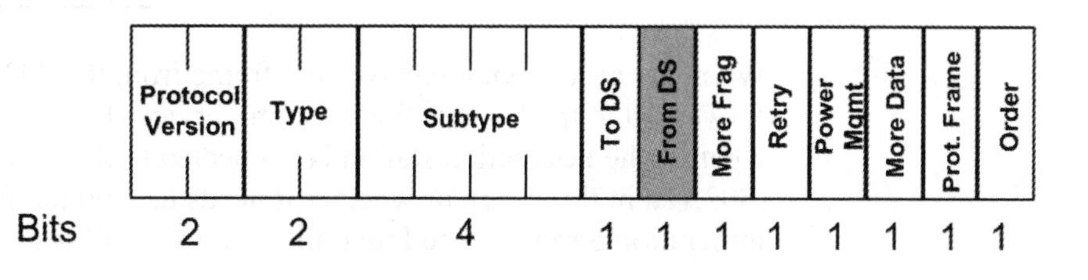

ToDS / FromDS Example

Figure 5.12 illustrates the internal workings of a dual-radio access point. The 802.11a station is shown sending a data frame to an 802.11g station which is connected to the same access point. The 802.11a station's data frame has settings of (ToDS=1, From DS=0) because it is the access point that makes the forwarding decision on where to send the data frame. All data frames sourced from a mobile station destined to either the access point itself or another node connected to the access point (wired or wireless) will have ToDS=1. This setting is much like the mobile station saying to the access point, "Please figure out where to send this information."

FIGURE 5.12 Access Point

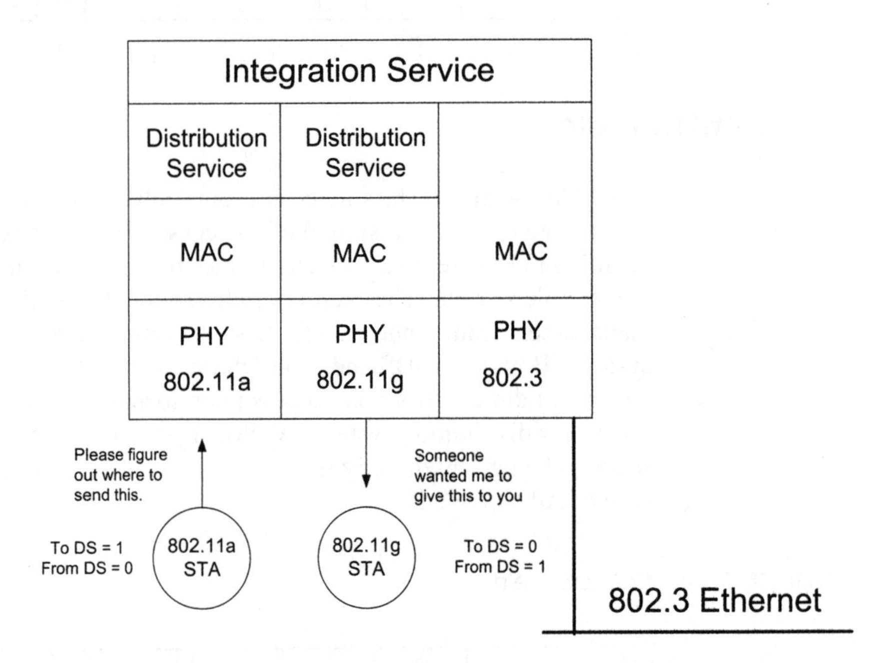

When the access point receives the frame from the 802.11a station, it is forwarded to the Distribution System Service (DSS) which decides whether the destination station is connected to the same access point radio (802.11a in this case), or whether it needs to send the frame up to the Integration Service. The Integration Service will decide whether the frame should be forwarded to the 802.11g radio or onto the 802.3 Ethernet link based on known MAC addresses.

More Frag Field

The single-bit More Frag field is set to 1 if another fragment of the same MSDU follows in a subsequent frame. MAC layer services provide fragmentation service to support the division of MSDUs and MMPDUs into smaller elements for transmission. Fragmentation can increase the efficiency of transmission because it decreases the overhead of retransmitting corrupted frames rather than corrupted frame fragments.

The MAC layer fragments only frames having a unicast receiver address (Address 1 in the MAC frame). It never fragments broadcast and multicast frames because frames sent to a multicast Address 1 are never acknowledged and retransmitted in any event. If the length of the resulting unencrypted MPDU needing transmission exceeds the aFragmentationThreshold parameter located in the MAC's management information base (MIB), then the MAC protocol will fragment the MSDU. Each fragmented frame consists of a MAC header, frame body, and FCS, which together comprise an MPDU. The unencrypted MPDU must not exceed the fragmentation threshold value, so the header and FCS must be taken into account when calculating the MSDU fragment size in each MPDU. Each fragment gets a fragment number (to be discussed later) indicating its ordered position in the sequence of fragments that carry the original MSDU. Each of the fragments is sent independently and requires separate acknowledgements (ACK frames) from the receiving station. The More Fragment field in the Frame Control field indicates whether or not a frame is the last of a series of fragments and is always set to 0 for unfragmented MSDUs.

The destination station will combine all fragments of the same sequence number in the correct order to reconstruct the corresponding MSDU. Based on the fragment numbers, the destination station will discard any duplicate fragments, and then pass the MSDU up to the next highest layer for processing. If the station is an access point the next higher layer is the DSS, otherwise it is the LLC.

FIGURE 5.13 More Frag Field

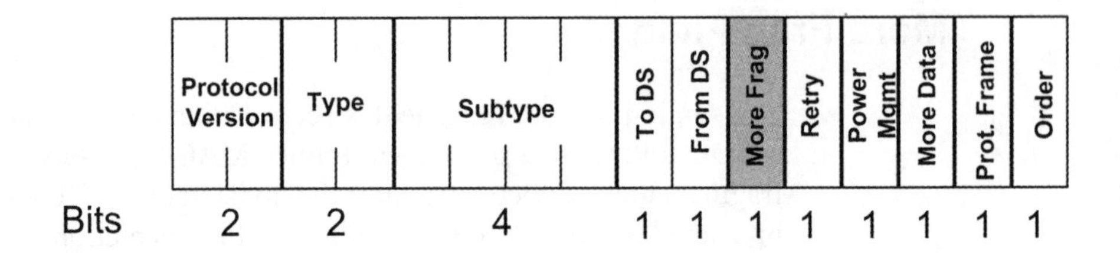

Retry Field

If the frame is a retransmission of an earlier frame, the single-bit Retry field is set to 1. The Retry field is 0 when the frame is being sent for the first time. The reason for retransmission could be that the first frame was corrupted or that no ACK frame was received by the transmitting station after sending the data frame.

FIGURE 5.14 Retry Field

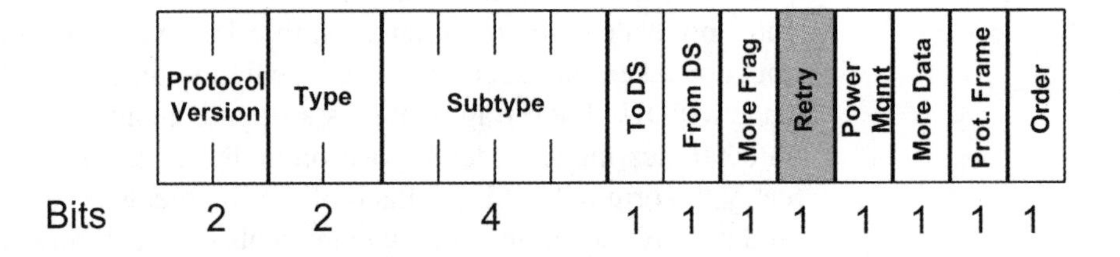

Power Management Field

The single-bit Power Management field indicates the power management mode that the transmitting station will be in after the current frame exchange sequence is complete. The MAC layer places a 1 in this field if the station will be in Power Save mode. In this mode, the access point (or other stations in an Ad Hoc network) will refrain from sending unsolicited frames to the station and buffer the frames for possible delivery at a later time when solicited. The station indicating a change of power save mode must not change power save mode until immediately after a successful frame exchange sequence. A 0 in the Power Management field indicates that the station will be in Active mode, in other words, never dozing.

FIGURE 5.15 Power Management Field

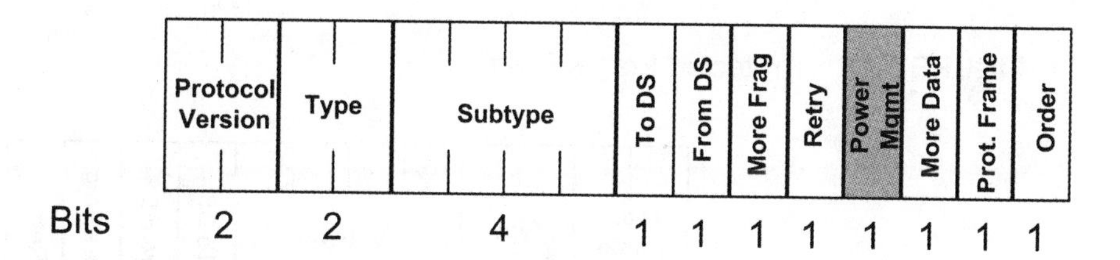

More Data Field

If a station has additional MSDUs to send to a station that is in power save mode, or to a point coordinator in response to a CF-Poll, then the transmitting station will place a 1 in this field. The More Data field is 0 for all other transmissions. The More Data feature alerts the receiving station to send more PS-Poll frames to empty the access point's frame buffer.

FIGURE 5.16 More Data Field

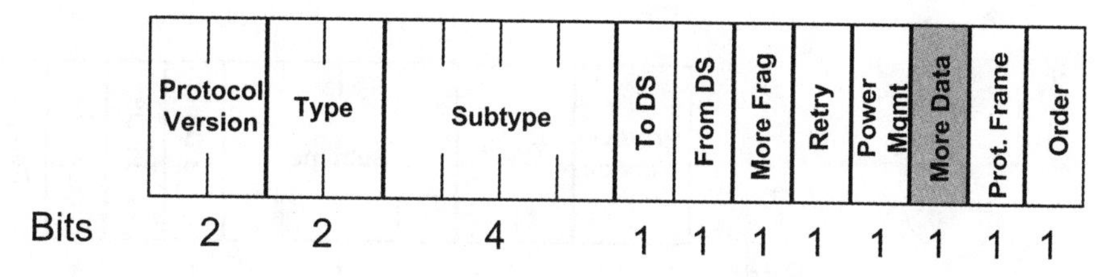

Protected Frame Field[1]

A 1 in the single-bit Protected Frame field tells the receiving station that the Frame Body field has been processed by an encryption algorithm. This field was called the "WEP" field in the 802.11 series of standards until 802.11i was ratified in 2004. A protected Frame Body field is increased by 8 octets for WEP, 20 octets for TKIP, and 16 octets for CCMP. The Protected Frame field is 0 when encryption is not in use with

---

[1] 802.11i – 2004, Section 7.1.3.1.9

data frames. The Protected Frame field is always 0 for management and control frames, except the third frame in a four frame Shared Key authentication process.

FIGURE 5.17 Protected Frame Field

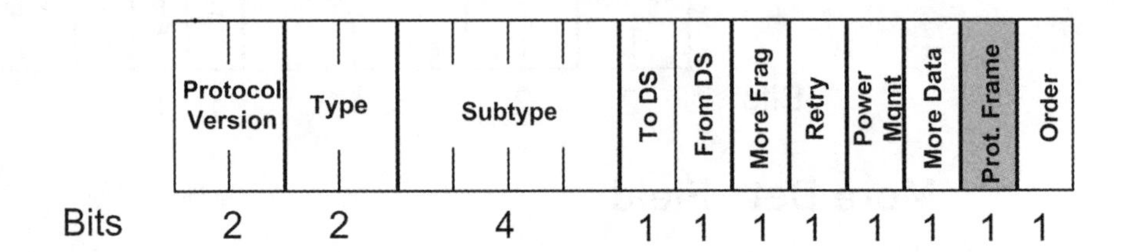

Order Field

The single-bit Order field is set to 1 in any data frame when a higher layer has requested that the data be sent using a strictly ordered class of service, which tells the receiving station that frames must be processed in order. Use of the Order bit prohibits stations from using power save mode.

FIGURE 5.18 Order Field

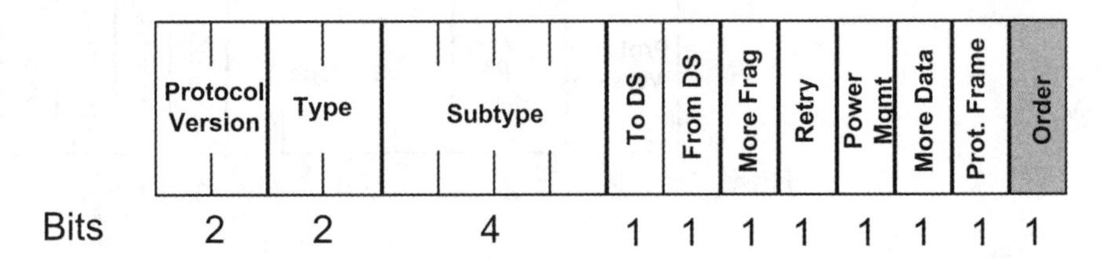

Duration/ID Field

The duration/ID field is 16 bits in length. It alternately contains duration information for updating the NAV or a short ID, called the association identifier (AID), used by a power save station to retrieve frames that are buffered for it at the access point. Figure 5.19 shows where the Duration/ID field is positioned in the MPDU.

FIGURE 5.19 Duration / ID field

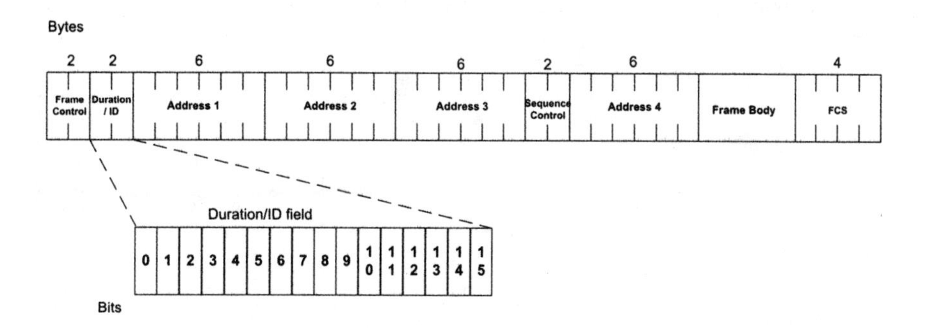

Only power-save poll (PS-Poll) frames use this field as the AID. In PS-Poll frames, the AID is aligned in the least significant 14 bits of the field, and the most significant two bits of the field are both set to 1. The maximum allowable value for the AID is 2007, and all values larger than 2007 are reserved as shown in Figure 5.20.

FIGURE 5.20 The Reserved Values

Bit 15	Bit 14	Bits 13-0	Usage
0	0-32767		Duration
1	0	0	Fixed value within frames transmitted during the CFP
1	0	1-16383	Reserved
1	1	0	Reserved
1	1	1-2007	AID in PS-Poll frames
1	1	2008-16383	Reserved

When bit 15 of the field is zero, the value in bits 14-0 represent the duration of a frame exchange sequence remaining after the frame in which the duration value is found. This value is used to update the NAVs of other stations, preventing a station receiving this field from beginning a transmission that might cause corruption of the ongoing frame exchange sequence.

The value of the duration/ID field is set to 32,768 (i.e., bit 15 is one and all other bits are zero) in all frames transmitted during a contention-free period (CFP). Other than the AID values, where bits 15 and 14 are set to one, all other values in this field are reserved.

FIGURE 5.21 Does it mean Duration or ID?

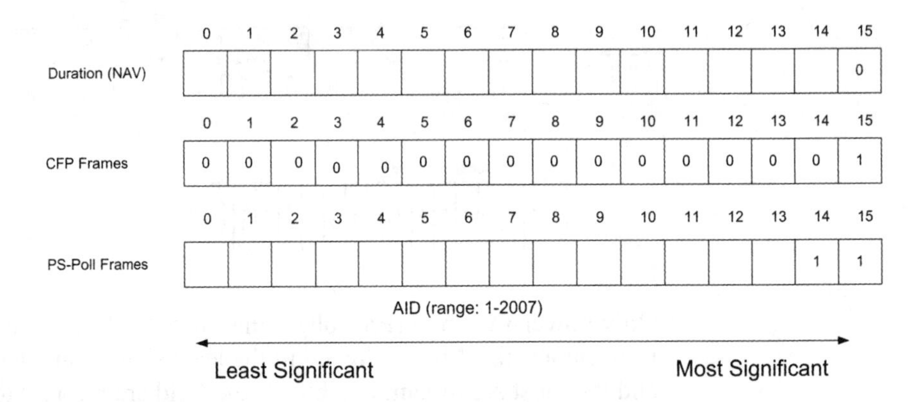

The association ID (AID) is a 16-bit field that contains an arbitrary number assigned to the station by the access point when a station associates with a BSS. The numeric value in the least significant 11 bits (0 – 10) of this field are used by the mobile station to identify which bit in a traffic information map information element indicates that the access point has frames buffered for the mobile station

Address Fields

MPDU address fields contain different types of 48-bit MAC addresses, depending on the type of frame being sent. These address types may include the basic service set identification (BSSID), source address, destination address, transmitting station address, and receiving station address. The first 24 bits (3 bytes) of the MAC address of a particular station signifies the manufacturer of the station's adapter, which is how packet analyzers are able to display manufacturer's names of specific stations' NICs. Three types of addresses may be used in address fields in an MPDU: unicast, broadcast, or multicast. Unicast addresses are sometimes referred to as individual addresses because they address a single MAC entity, and broadcast and multicast addresses are sometimes referred to as groups since they can address more than one MAC entity simultaneously. Figure 5.22 shows the position of each address within the MPDU. Figure 5.23 shows how many of these four address fields are used in a frame traversing the WM for a given situation.

FIGURE 5.22 Address Fields

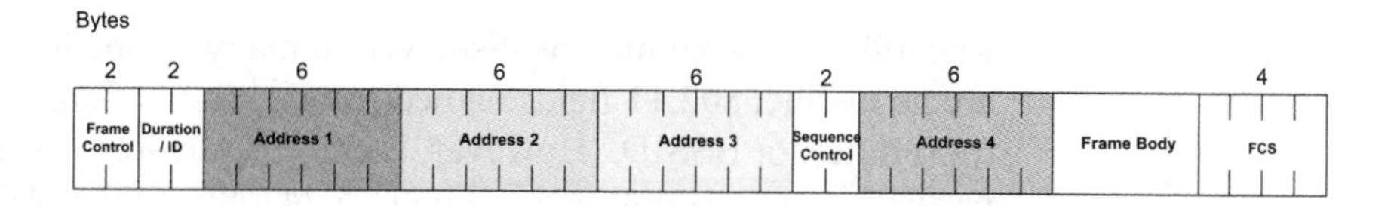

FIGURE 5.23 Address Usage

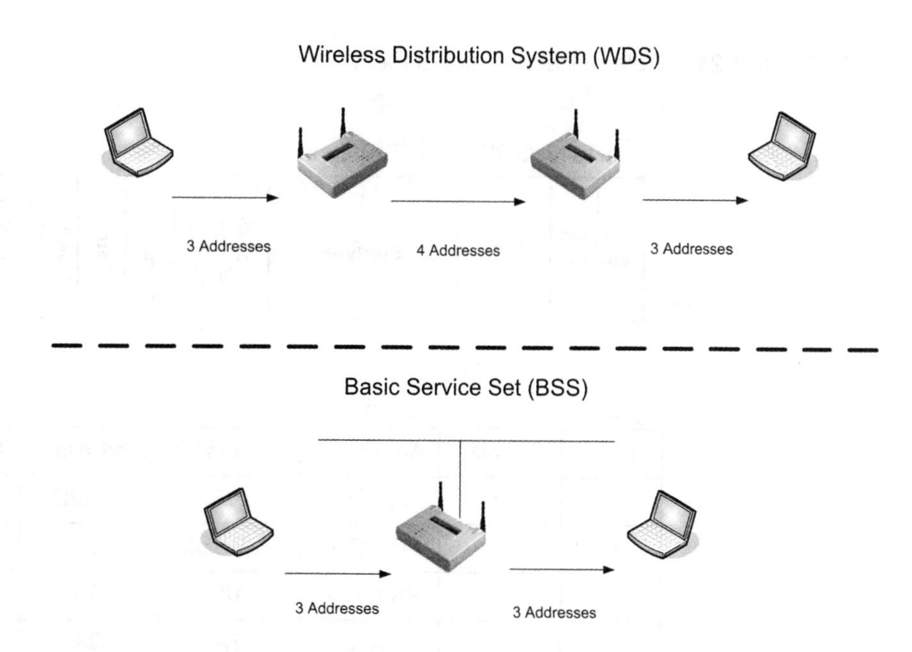

The IEEE 802.11-1999 standard defines the following address types:

- Destination Address (DA): The final recipient(s) of the MSDU contained in the frame body field (individual or group address).
- Source Address (SA): The address of the MAC entity that initiates the MSDU transfer (always an individual address).
- Receiver Address (RA): The address of the immediate station(s) for the information contained in the frame body field (individual or group address).

- Transmitter Address (TA): The address of the station transmitting the information contained in the frame body field onto the wireless medium (always an individual address).

Regardless of what may be displayed in many protocol analyzers, there are no distinct 802.11 fields named source, destination, receiver, transmitter, or BSSID. However, each of Addresses 1-4 may be used in particular ways as outlined in Figure 5.24. Notice that Address 1 is always the RA, and Address 2 is always the TA regardless of the frame type or where the frame body is sourced or destined.

FIGURE 5.24 Role of Each Address Field

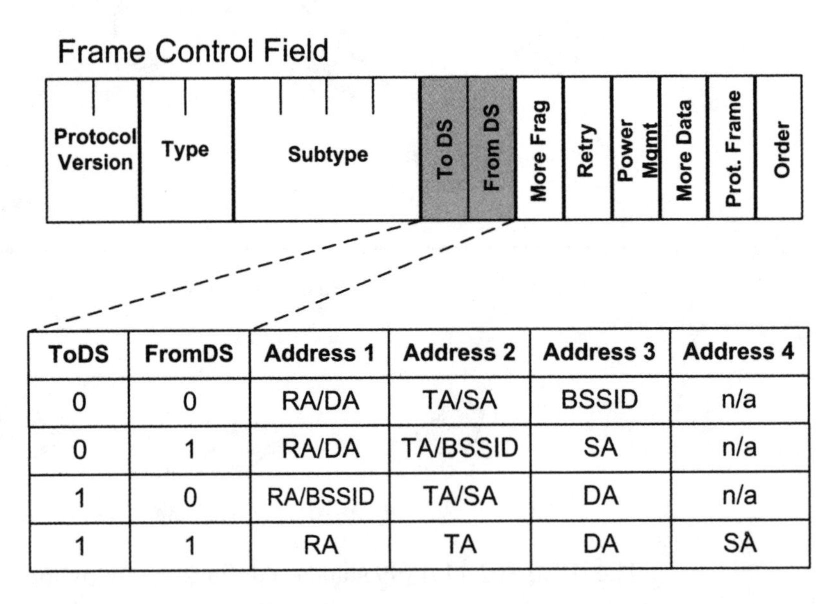

ToDS	FromDS	Address 1	Address 2	Address 3	Address 4
0	0	RA/DA	TA/SA	BSSID	n/a
0	1	RA/DA	TA/BSSID	SA	n/a
1	0	RA/BSSID	TA/SA	DA	n/a
1	1	RA	TA	DA	SA

Figure 5.25 shows how a wireless protocol analyzer may display Addresses 1-4 as column headings. The example in Figure 5.25 shows a frame capture of a Ping Request and Reply across a wireless distribution system (WDS). The frames traversing the wireless link between the two access points is the only time that all four address fields are used simultaneously.

FIGURE 5.25 Protocol Analyzer Showing Addresses 1 - 4

Packet	Address 1	Address 2	Address 3	Address 4	Data Rate	Size	Protocol
1	00:0D:65:C9:32:76	00:0A:8A:47:BF:4A	00:0A:8A:47:BC:1A		11.0	96	PING Req
2	00:0A:8A:47:BF:4A				11.0	14	802.11 Ack
3	00:0D:ED:A5:4F:70	00:0D:65:C9:32:76	00:0A:8A:47:BC:1A	00:0A:8A:47:BF:4A	11.0	102	PING Req
4	00:0D:65:C9:32:76				11.0	14	802.11 Ack
5	00:0A:8A:47:BC:1A	00:0D:ED:A5:4F:70	00:0A:8A:47:BF:4A		11.0	96	PING Req
6	00:0D:ED:A5:4F:70				11.0	14	802.11 Ack
7	00:0D:ED:A5:4F:70	00:0A:8A:47:BC:1A	00:0A:8A:47:BF:4A		11.0	96	PING Reply
8	00:0A:8A:47:BC:1A				11.0	14	802.11 Ack
9	00:0D:65:C9:32:76	00:0D:ED:A5:4F:70	00:0A:8A:47:BF:4A	00:0A:8A:47:BC:1A	11.0	102	PING Reply
10	00:0D:ED:A5:4F:70				11.0	14	802.11 Ack
11	00:0A:8A:47:BF:4A	00:0D:65:C9:32:76	00:0A:8A:47:BC:1A		11.0	96	PING Reply
12	00:0D:65:C9:32:76				11.0	14	802.11 Ack

Figure 5.26 shows the same information as Figure 5.25, but the analyzer has been configured to use the names of four address types instead of the Address 1-4 field names as column headings. Most protocol analyzers on the market today offer this feature. It is also common to see an analyzer fill in these virtual address fields with information from other frames in the same frame exchange sequence as shown in Figure 5.26. Notice that the ACK control frames, which have only one address field, are shown with four "addresses". The analyzer has made this assumption.

FIGURE 5.26 Protocol Analyzer Showing SA, DA, TA, RA

Packet	Source Physical	Dest. Physical	Transmitter	Receiver	Data Rate	Size	Protocol
1	00:0A:8A:47:BF:4A	00:0A:8A:47:BC:1A	00:0A:8A:47:BF:4A	00:0D:65:C9:32:76	11.0	96	PING Req
2	00:0D:65:C9:32:76	00:0A:8A:47:BF:4A	00:0D:65:C9:32:76	00:0A:8A:47:BF:4A	11.0	14	802.11 Ack
3	00:0A:8A:47:BF:4A	00:0A:8A:47:BC:1A	00:0D:65:C9:32:76	00:0D:ED:A5:4F:70	11.0	102	PING Req
4	00:0D:ED:A5:4F:70	00:0D:65:C9:32:76	00:0D:ED:A5:4F:70	00:0D:65:C9:32:76	11.0	14	802.11 Ack
5	00:0A:8A:47:BF:4A	00:0A:8A:47:BC:1A	00:0D:ED:A5:4F:70	00:0A:8A:47:BC:1A	11.0	96	PING Req
6	00:0A:8A:47:BC:1A	00:0D:ED:A5:4F:70	00:0A:8A:47:BC:1A	00:0D:ED:A5:4F:70	11.0	14	802.11 Ack
7	00:0A:8A:47:BC:1A	00:0A:8A:47:BF:4A	00:0A:8A:47:BC:1A	00:0D:ED:A5:4F:70	11.0	96	PING Reply
8	00:0D:ED:A5:4F:70	00:0A:8A:47:BC:1A	00:0D:ED:A5:4F:70	00:0A:8A:47:BC:1A	11.0	14	802.11 Ack
9	00:0A:8A:47:BC:1A	00:0A:8A:47:BF:4A	00:0D:ED:A5:4F:70	00:0D:65:C9:32:76	11.0	102	PING Reply
10	00:0D:65:C9:32:76	00:0D:ED:A5:4F:70	00:0D:65:C9:32:76	00:0D:ED:A5:4F:70	11.0	14	802.11 Ack
11	00:0A:8A:47:BC:1A	00:0A:8A:47:BF:4A	00:0D:65:C9:32:76	00:0A:8A:47:BF:4A	11.0	96	PING Reply
12	00:0A:8A:47:BF:4A	00:0D:65:C9:32:76	00:0A:8A:47:BF:4A	00:0D:65:C9:32:76	11.0	14	802.11 Ack

Figure 5.27 shows each of the four address fields but spread across five virtual fields, one for each of the five types. Since the analyzer "helps" you by *interpreting* and *assuming*, using a wireless analyzer can get confusing at times.

FIGURE 5.27 Protocol Analyzer Showing SA, DA, BSSID, TA, RA

Packet	Source Physical	Dest. Physical	BSSID	Transmitter	Receiver	Data Rate	Size	Protocol
1	00:0A:8A:47:BF:4A	00:0A:8A:47:BC:1A	00:0D:65:C9:32:76	00:0A:8A:47:BF:4A	00:0D:65:C9:32:76	11.0	96	PING Req
2	00:0D:65:C9:32:76	00:0A:8A:47:BF:4A		00:0D:65:C9:32:76	00:0A:8A:47:BF:4A	11.0	14	802.11 Ack
3	00:0A:8A:47:BF:4A	00:0A:8A:47:BC:1A		00:0D:65:C9:32:76	00:0D:ED:A5:4F:70	11.0	102	PING Req
4	00:0D:ED:A5:4F:70	00:0D:65:C9:32:76		00:0D:ED:A5:4F:70	00:0D:65:C9:32:76	11.0	14	802.11 Ack
5	00:0A:8A:47:BF:4A	00:0A:8A:47:BC:1A	00:0D:ED:A5:4F:70	00:0D:ED:A5:4F:70	00:0A:8A:47:BC:1A	11.0	96	PING Req
6	00:0A:8A:47:BC:1A	00:0D:ED:A5:4F:70		00:0A:8A:47:BC:1A	00:0D:ED:A5:4F:70	11.0	14	802.11 Ack
7	00:0A:8A:47:BC:1A	00:0A:8A:47:BF:4A	00:0D:ED:A5:4F:70	00:0A:8A:47:BC:1A	00:0D:ED:A5:4F:70	11.0	96	PING Reply
8	00:0D:ED:A5:4F:70	00:0A:8A:47:BC:1A		00:0D:ED:A5:4F:70	00:0A:8A:47:BC:1A	11.0	14	802.11 Ack
9	00:0A:8A:47:BC:1A	00:0A:8A:47:BF:4A		00:0D:ED:A5:4F:70	00:0D:65:C9:32:76	11.0	102	PING Reply
10	00:0D:65:C9:32:76	00:0D:ED:A5:4F:70		00:0D:65:C9:32:76	00:0D:ED:A5:4F:70	11.0	14	802.11 Ack
11	00:0A:8A:47:BC:1A	00:0A:8A:47:BF:4A	00:0D:65:C9:32:76	00:0D:65:C9:32:76	00:0A:8A:47:BF:4A	11.0	96	PING Reply
12	00:0A:8A:47:BF:4A	00:0D:65:C9:32:76		00:0A:8A:47:BF:4A	00:0D:65:C9:32:76	11.0	14	802.11 Ack

Sequence Control Field

The sequence control field is a 16-bit field comprising two fields. The fields are a 4-bit fragment number and a 12-bit sequence number. As a whole, this field is used by a receiving station to eliminate duplicate received frames and to reassemble fragments. Figure 5.28 shows the positioning of the Sequence Control field in the MPDU.

A transmitting station can have no more than one outstanding, unacknowledged frame. On reception of a frame, a station can filter duplicate frames by monitoring the sequence and fragment numbers. The station knows the frame is a duplicate if the sequence number and fragment number are equal to one recently received, or the Retry bit is set to 1.

Duplicate frames can occur when a station receives a frame without errors and sends an ACK frame back to the transmitting station, and then transmission errors corrupt the reception of the ACK frame. After not receiving the ACK over a specific time period, the transmitting station retransmits a duplicate frame. The receiving station performs an acknowledgement of the retransmitted frame even if the frame is discarded due to duplicate filtering.

FIGURE 5.28 Sequence Control Positioning

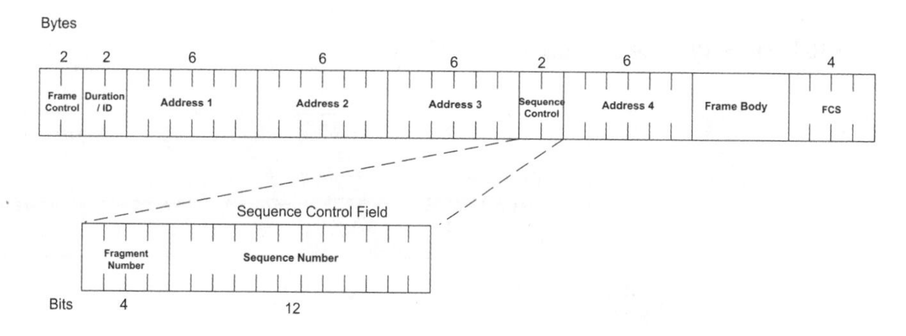

The Fragment Number field contains a 4-bit number assigned to each fragment of an MSDU or MMPDU. This number starts with 0000 for the first fragment, then increments by 1 for each successive fragment. The first, or only, fragment of an MSDU is assigned a fragment number of 0000 (or just 0). Each successive fragment is assigned a sequentially incremented fragment number. The fragment number is the same in a transmission or any retransmission of a particular frame or fragment.

FIGURE 5.29 Fragment Number Field

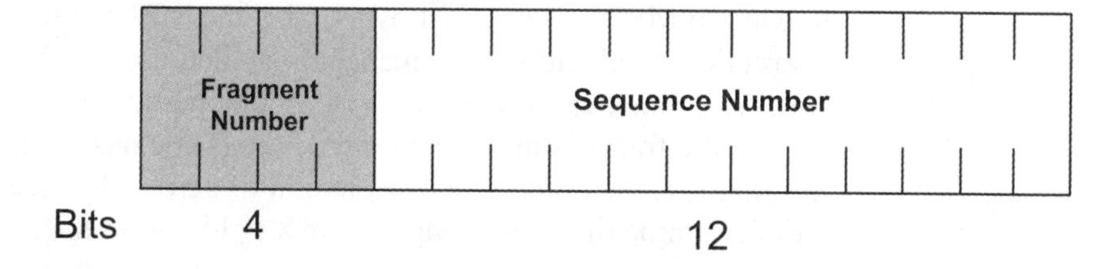

The Sequence Number field contains a 12-bit number assigned sequentially by the sending station to each MSDU or MMPDU. The sequence number field starts at 000000000000 (or just 0), increments by 1 for each subsequent MSDU transmission, and wraps back to zero when the maximum value of 4095 is reached. The sequence number for a particular MSDU is transmitted in every data frame associated with the MSDU, and it is a constant value over all transmissions and retransmissions of the MSDU. If the MSDU is fragmented, the sequence number of the MSDU is sent with each frame containing a fragment of the MSDU. Each station chooses its own set of fragment and sequence numbers independent from other stations.

FIGURE 5.30 Sequence Number Field

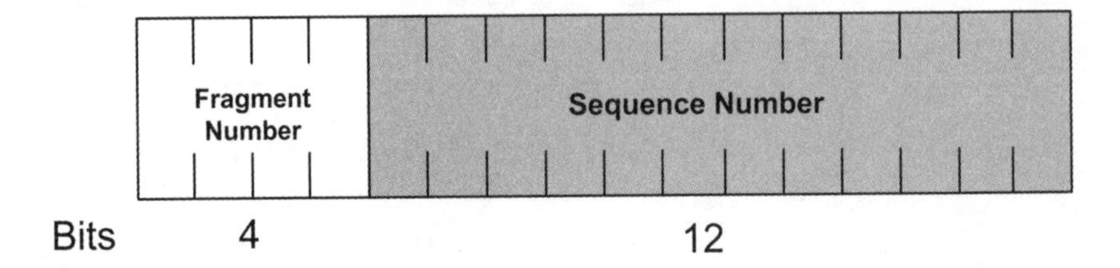

Frame Body Field

This field has a variable length and carries information that pertains to the specific frame being sent. The maximum MSDU length is 2304 bytes when encryption is not in use. When WEP is in use, the Frame Body is expanded by 8 bytes to 2312 bytes to accommodate the WEP IV and ICV. TKIP adds 20 (WEP's 8 plus an additional 12) octets, and CCMP adds 16 octets. In the case of a data frame, the frame body contains an MSDU which is data passed down to the MAC layer by higher layer protocols, LLC or DSS. The maximum size of the MPDU is determined by the maximum MSDU size (2304 bytes) plus the MPDU header, FCS, and any encryption overhead. MAC management and control frames may include specific parameters in the Frame Body that pertain to the particular service the frame is implementing. If the frame has no need to carry information, then this field has a length of zero. The receiving station will determine the frame length from a field within the PLCP header.

FIGURE 5.31 Frame Body Field

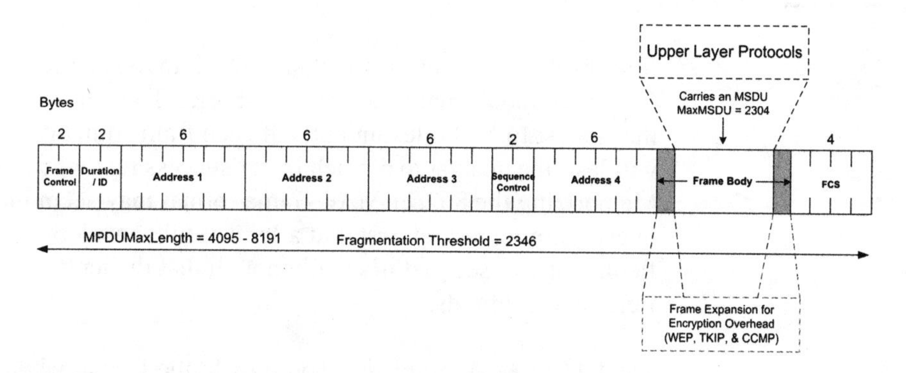

Frame Check Sequence (FCS)

The MAC layer at the transmitting station calculates a 32-bit frame check sequence (FCS) using a cyclic redundancy code (CRC) over all the fields of the MAC header and the frame body and places the result in this field.

FIGURE 5.32 32-bit Frame Check Sequence

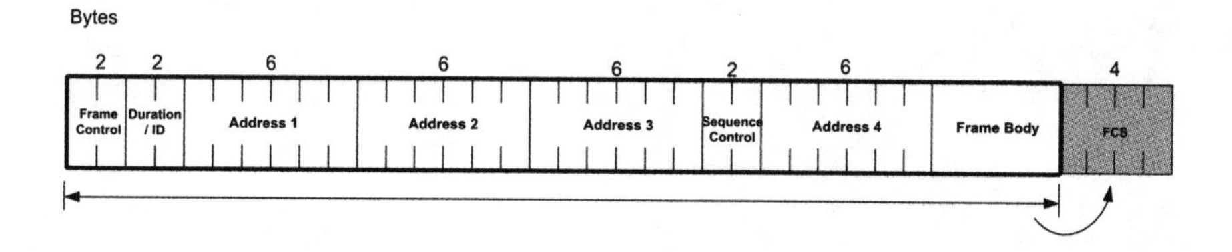

Summary

It is absolutely imperative that an 802.11 analyst understand and memorize the format of the 802.11 frame. This knowledge will enable the analyst to find relevant parts of each frame quickly in the analyzer and will lead the analyst to a much better success rate when troubleshooting. Memorizing the 802.11 MAC frame format may seem like an overwhelming task at first, but after careful consideration of each frame field, you will see that it's really not all that difficult to remember the fields and subfields.

The 802.11 MAC frame is a complex frame format when compared with an Ethernet frame. There are four address fields instead of the two most analysts are accustomed to. There are several header fields and subfields used by the transmitting station to inform the receiver of certain parameters. There are different types of frames: control, management, and data frames. Each of these frames types use the same overall frame structure, but may not use all parts of it simultaneously.

Key Terms

Before taking the exam, you should be familiar with the following terms:

ACK frame

address fields

Duration/ID field

Frame Body field

Frame Check Sequence

Frame Control field

FromDS field

MMPDU

More Data field

More Frag field

MPDU

MSDU

Order field

Power Management field

Protected Frame field

Protocol Version field

Retry field

Sequence Control field

Subtype field

ToDS field

Type field

Review Questions

1. Which MPDU field carries the More Data field, and how is the More Data field used in an 802.11 WLAN?

2. Stations operating in Power Save mode may begin dozing under what conditions?

3. A basic service set's BSSID is how many bytes long and is carried where in the 802.11 beacon frame?

4. How many bits does the 802.11 FCS consist of?

5. The maximum MSDU size in all cases is _______ bytes.

6. What is the maximum MMPDU length in bytes?

7. The Sequence Control field immediately follows which 802.11 frame field?

8. The Address-1 field always carries what information in a data frame?

9. Under what condition(s) does the Duration/ID field carry the value of 32,768?

10. If you are observing an 802.11g network with a WLAN protocol analyzer, and a frame decode shows the Protocol Version = 01, what can you assume?

11. The Protected Frame field is part of what other 802.11 MAC frame field?

12. If the ToDS and FromDS values in a frame are both set to 1, what type of frame is it and where is it being sourced and sunk?

CHAPTER

6

802.11 Management Frames

CWAP Exam Objectives Covered:

- Distinguish the intended purpose of each 802.11 MAC layer frame type
  - Management frames
- Explain the structure of each 802.11 MAC layer frame type
  - Control, Management, and Data frame payload contents and sizes

In This Chapter

Management Frames

Management frames establish communications between stations and access points. They support Power Management and Contention Free modes. MAC management frames are also called MAC Management Protocol Data Units (MMPDUs). Management frames provide such services as authentication, association, and reassociation. Management frame bodies are never relayed through an access point. Instead they are "sourced" (generated) and sunk (read and disposed of) at the MAC layer, and therefore are never passed to the distribution system service or LLC. The Address 4 field is never used in an MMPDU. Unicast MMPDUs may be acknowledged, retransmitted, and fragmented[1]. Figure 6.1 depicts the common format of all management frames.

FIGURE 6.1 MMPDU Frame Structure

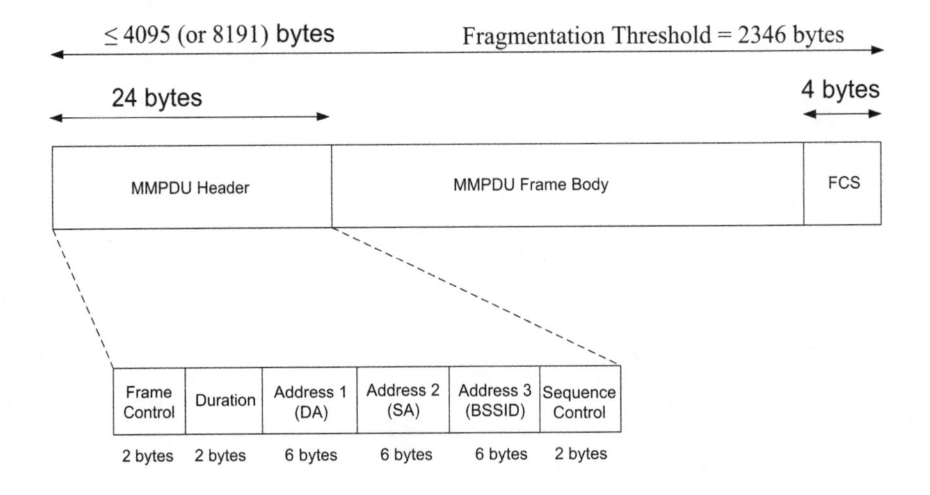

Association Request Frame

A station will send an Association Request frame to an access point if it wants to associate with that access point. This frame exchange sequence

[1] IEEE 802.11 - 1999 (R2003) – Section 7.1.3.4

ends successfully with an acknowledgement. A station becomes associated with an access point after the access point responds with an acceptance.

FIGURE 6.2 Association Request Frame Body Contents

Order	Information	Notes
1	Capability Information	
2	Listen Interval	
3	SSID	
4	Supported Rates	
5	Extended Supported Rates	The Extended Supported Rates element is present whenever there are more than eight supported rates, and it is optional otherwise.

Association Response Frame

After an access point receives an Association Request frame and acknowledges it, the access point will send an Association Response frame to indicate whether or not it is accepting the association with the requesting station. This second frame exchange sequence ends successfully with an acknowledgment. The Association Response frame provides the status (acceptance or rejection) and an AID (if the association was accepted).

FIGURE 6.3 Association Response Frame Body Contents

Order	Information	Notes
1	Capability Information	
2	Status Code	
3	Association ID (AID)	
4	Supported Rates	
5	Extended Supported Rates	The Extended Supported Rates element is present whenever there are more than eight supported rates, and it is optional otherwise.

Reassociation Request Frame

A station will send a Reassociation Request frame to an access point if it already associated to the ESS and wants to reassociate to the ESS through another access point. This frame exchange sequence is ended successfully with an acknowledgment. A reassociation may occur if a station moves out of range from one access point and within range of another access point. The station will need to reassociate (not merely associate) with the new access point so that the new access point knows that it will need to negotiate the forwarding of data frames from the old access point and update its association table. Notice in Figure 6.4 that the Current AP Address (old AP address) fixed field is part of the Reassociation Frame.

FIGURE 6.4 Reassociation Request Frame Body Contents

Order	Information	Notes
1	Capability Information	
2	Listen Internal	
3	Current AP Address	
4	SSID	
5	Supported Rates	
6	Extended Supported Rates	The Extended Supported Rates element is present whenever there are more than eight supported rates, and it is optional otherwise.

Reassociation Response Frame

After an access point receives a Reassociation Request frame, the access point will send a Reassociation Response frame to indicate whether or not it is accepting the reassociation with the sending station. This frame exchange sequence is ended successfully with an acknowledgment. Reassociation is dependent on authentication with/through the new access point. If the reassociation is successful, the new access point will indicate this status in the status code element and include a unique AID for this station to use if operating in Power Save Poll mode.

FIGURE 6.5 Reassociation Response Frame Body Contents

Order	Information	Notes
1	Capability Information	
2	Status Code	
3	Association ID (AID)	
4	Supported Rates	
5	Extended Supported Rates	The Extended Supported Rates element is present whenever there are more than eight supported rates, and it is optional otherwise.

Probe Request Frame

A station sends a Probe Request frame to obtain information from another station or access point. For example, a station may send a Probe Request frame to determine whether a certain access point is available. Mobile stations use Probe Request frames as part of the active scanning process.

FIGURE 6.6 Probe Request Frame Body Contents

Order	Information	Notes
1	SSID	
2	Supported Rates	
3	Undefined	May be assigned for proprietary use by a vendor.
4	Extended Supported Rates	The Extended Supported Rates element is present whenever there are more than eight supported rates, and it is optional otherwise.

Probe Response Frame

If a station or access point receives a Probe Request frame, the station will respond to the requesting station with a Probe Response frame containing specific parameters about itself. All access points and the station which last generated the beacon frame (if operating as an IBSS) can respond to probe requests with Probe Response frames.

FIGURE 6.7 Probe Response Frame Body Contents

Order	Information	Notes
1	Timestamp	
2	Beacon Interval	
3	Capability Information	
4	SSID	
5	Supported Rates	
6	FH Parameter Set	The FH Parameter Set information element is present with Probe Response frames generated by stations using FH PHYs.
7	DS Parameter Set	The DS Parameter Set information element is present with Probe Response frames generated by stations using Clause 15, 18, & 19 PHYs.
8	CF Parameter Set	The CF Parameter Set information element is only present with Probe Response frames generated by access points supporting a PCF.
9	IBSS Parameter Set	The IBSS Parameter Set information element is only present within Probe Response frames generated by stations in an IBSS.
10 - 17	Undefined	May be assigned for proprietary use by a vendor(s).
18	ERP Information	ERP Information element is present within Probe Response frames generated by STAs using ERP PHYs and is optionally present in other cases.
19	Extended Supported Rates	The Extended Supported Rates element is present whenever there are more than eight supported rates, and it is optional otherwise.

Beacon Frame

An access point (or mobile station in an Ad Hoc network) periodically sends a beacon frame at a rate based on the *aBeaconPeriod* parameter in the MIB. The beacon provides synchronization among stations of a BSS, and includes a timestamp that all stations within its BSS use to update what 802.11 defines as a timing synchronization function (TSF) timer. The TSF is basically a station's internal 802.11 clock. The Probe Response frame and beacon frame are identical, except that the beacon also carries the traffic indication map (TIM) information element.

If the access point supports the Point Coordination Function, then it uses a beacon frame to announce the beginning of a contention-free period. If the network is an independent BSS (that is, it has no access points), all stations periodically send beacons for synchronization purposes.

FIGURE 6.8 Beacon Frame Body Contents

Order	Information	Notes
1	Timestamp	
2	Beacon Interval	
3	Capability Information	
4	SSID	
5	Supported Rates	
6	FH Parameter Set	The FH Parameter Set information element is present with Beacon frames generated by STAs using FH PHYs.
7	DS Parameter Set	The DS Parameter Set information element is present with Beacon frames generated by STAs using Clause 15, 18, & 19 PHYs.
8	CF Parameter Set	The CF Parameter Set information element is only present with Beacon frames generated by APs supporting a PCF.
9	IBSS Parameter Set	The IBSS Parameter Set information element is only present within Beacon frames generated by STAs in an IBSS.
10	TIM	The TIM information element is only present within Beacon frames generated by APs.
11 - 18	Undefined	May be assigned for proprietary use by a vendor(s).
19	ERP Information	The ERP Information element is present within Beacon frames generated by STAs using ERP PHYs and is optionally present in other cases.
20	Extended Supported Rates	The Extended Supported Rates element is present whenever there are more than eight supported rates, and it is optional otherwise.

ATIM Frame

In an IBSS a station with frames buffered for other stations sends an Announcement Traffic Indication Message (ATIM) frame to each of these stations during the ATIM window, which immediately follows a beacon transmission in Ad Hoc mode. The station that sent the ATIM then transmits the buffered frames to the applicable recipients during a data window. The transmission of the ATIM frame alerts stations in power-save mode to stay awake long enough to solicit and receive their respective frames. The ATIM frame body is null.

Disassociation Frame

If a station or access point wants to terminate an association, it will send a Disassociation frame to the other station that is part of the association. A single Disassociation frame sent to the broadcast address by an access point can terminate associations with more than one station at a time.

FIGURE 6.9 Disassociation Frame Body

Order	Information
1	Reason Code

Authentication Frame

A station sends an Authentication frame to an access point that it wants to authenticate with. The authentication process consists of the transmission of two or four authentication frames, depending on the type of authentication being implemented, Open System or Shared Key respectively. If Shared Key, the third authentication frame of the four is WEP encrypted. Each authentication frame requires an acknowledgment.

FIGURE 6.10 Authentication Frame Body

Order	Information	Notes
1	Authentication Algorithm Number	
2	Authentication Transaction Sequence Number	
3	Status Code	The status code information is reserved and set to 0 in certain Authentication frames
4	Challenge Text	The challenge text information is only present in certain Authentication frames

Deauthentication Frame

A station sends a Deauthentication frame to a station or access point with which it wants to terminate communications.

FIGURE 6.11 Deauthentication Frame Structure

Order	Information
1	Reason Code

Management Frame Summary

The chart below provides a quick reference to the contents and sizes (in octets) of each management frame type per the 802.11 series of standards).

FIGURE 6.12 Management Frame Quick Reference Chart

FC	D	Address 1	Address 2	Address 3	SC	Frame Body	FCS
2	2	6	6	6	2		4

Beacon

Field	Length
Timestamp	8
Beacon Interval	2
Capability Information	2
SSID	2 to 34
Supported Rates	3 to 10
FH Parameter Set	7
DS Parameter Set	3
CF Parameter Set	8
IBSS Parameter Set	4
Traffic Indication Map	7 to 256
ERP Information	3
Ext. Supported Rates	3 to 257

Probe Request

Field	Length
SSID	2 to 34
Supported Rates	3 to 10
Ext. Supported Rates	3 to 257

Probe Response

Field	Length
Timestamp	8
Beacon Interval	2
Capability Information	2
SSID	2 to 34
Supported Rates	3 to 10
FH Parameter Set	7
DS Parameter Set	3
CF Parameter Set	8
IBSS Parameter Set	4
ERP Information	3
Ext. Supported Rates	3 to 257

Authentication

Field	Length
Auth. Alg. No.	2
Auth. Trans. Seq. No.	2
Status Code	2
Challenge Text	3 to 255

Deauthentication

Field	Length
Reason Code	2

Disassociation

Field	Length
Reason Code	2

Association Request

Field	Length
Capability Information	2
Listen Interval	2
SSID	2 to 34
Supported Rates	3 to 10
Ext. Supported Rates	3 to 257

Association Response

Field	Length
Capability Information	2
Status Code	2
Association ID	2
Supported Rates	3 to 10
Ext. Supported Rates	3 to 257

(IBSS) Announcement Traffic Indication Message

Field	Length
empty	0

Reassociation Request

Field	Length
Capability Information	2
Listen Interval	2
SSID	2 to 34
Current AP Address	6
Supported Rates	3 to 10
Ext. Supported Rates	3 to 257

Reassociation Response

Field	Length
Capability Information	2
Status Code	2
Association ID	2
Supported Rates	3 to 10
Ext. Supported Rates	3 to 257

Frame Body Fields and Elements

The 802.11 standards describe the frame body fixed fields and information elements of various management frames. Some vendors also add proprietary extensions to 802.11 management frames that provide functionality beyond the standard. Figure 6.13 identifies which fixed fields and information elements are found in specific frames, and the following paragraphs summarize each of the fields and elements

FIGURE 6.13 Frame Body Fields and Elements Quick Reference Chart

	Association Request	Association Response	Reassociation Request	Reassociation Response	Probe Response	Probe Request	Beacon	Disassociation	Authentication	Deauthentication
Authentication Algorithm Number									X	
Authentication Transaction Sequence #									X	
Beacon Interval					X		X			
Current AP Address			X							
Listen Interval	X		X							
Reason Code								X		X
Association ID (AID)		X		X						
Status Code		X		X					X	
Timestamp					X		X			
Service Set Identifier (SSID)	X		X		X	X	X			
Supported Rates	X	X	X	X	X	X	X			
Extended Supported Rates	X	X	X	X	X	X	X			
FH Parameter Set					X		X			
DS Parameter Set					X		X			
CF Parameter Set					X		X			
Capability Information	X	X	X	X	X		X			
Traffic Indication Map (TIM)							X			
IBSS Parameter Set					X		X			
Challenge Text									X	
ERP Information					X		X			

Figure 6.14 shows an example of fields and elements found in a typical beacon's frame body. Figure 6.16 shows an example of vendor-specific frame body elements placed into a beacon management frame.

FIGURE 6.14 Typical Beacon's Frame Body

```
802.11 frame body
  timestamp : 83398520
  beacon interval : 100 TUs
  capability info
  info : SSID (0)
  info : supported rates (1)
  info : DS param set (3)
  info : TIM (5)
  info : ERP information (42)
  info : extended supported rates (50)
  info : AP Name (133)
  info : cell power (150)
  info : vendor specific (221)
```

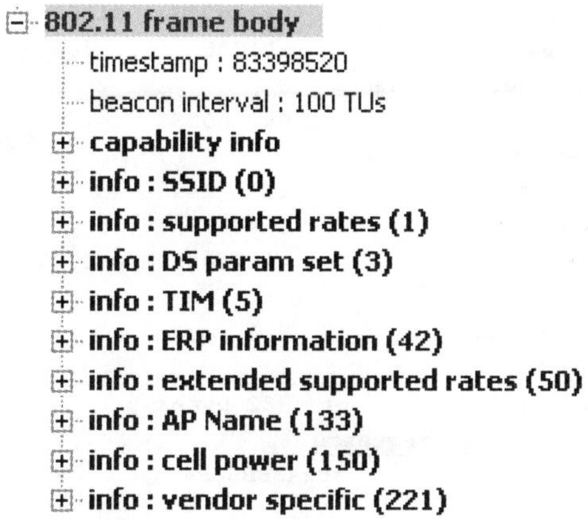

A chart outlining element identifiers is shown below. See if the elements in the screenshot above match those in the chart below.

FIGURE 6.15 Information Element Identifiers

Information Element	Element ID
SSID	0
Supported rates	1
FH Parameter Set	2
DS Parameter Set	3
CF Parameter Set	4
TIM	5
IBSS Parameter Set	6
Reserved	7-15
Challenge Text	16
Reserved for challenge text extension	17-41
ERP Information	42
Reserved	43-49
Extended Supported Rates	50
Reserved	51-255

FIGURE 6.16 Vendor Specific Frame Body Elements

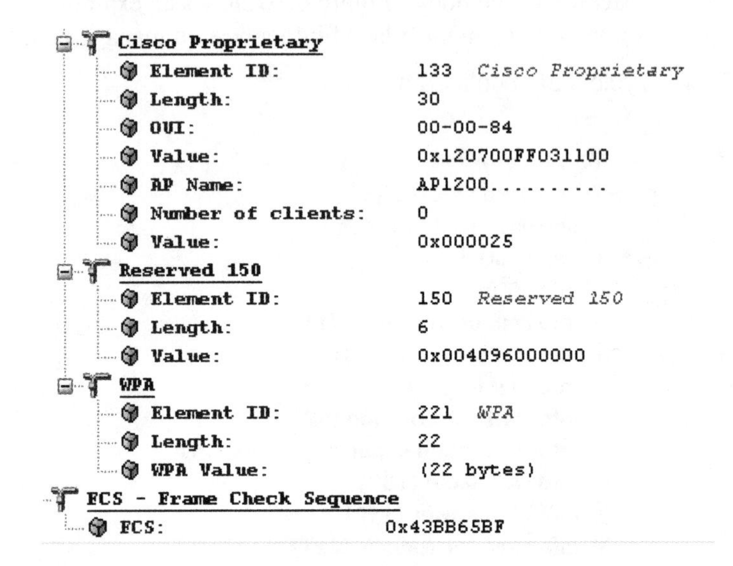

Association ID (AID) Field

The AID, which is assigned by an access point during association, is the 16-bit identification of a station corresponding to a particular association. The range of AIDs is from 1 to 2007, with the most significant two bits of the AID field set to 1. Figure 6.17 illustrates the format of the AID field.

FIGURE 6.17 Association ID Field Format

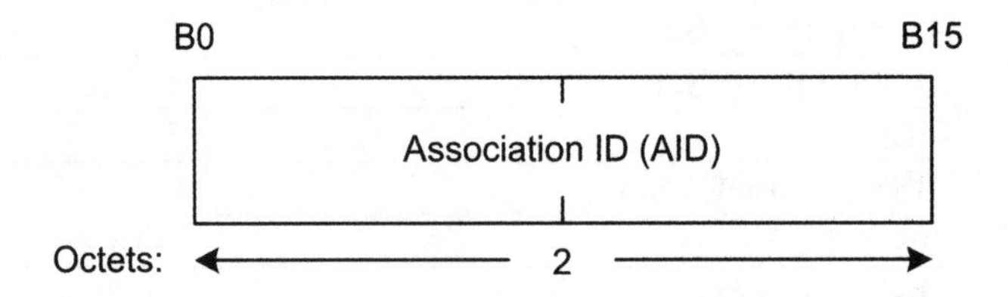

Authentication Algorithm Number Field

The Authentication Algorithm Number is a 16-bit field that specifies the authentication algorithm that the authenticated stations and access points are to use. The value is either 0 for Open System authentication or 1 for

Shared Key authentication. All other values are reserved for possible future use. Figure 6.18 illustrates the Authentication Algorithm Number field format.

FIGURE 6.18 Authentication Algorithm Number Field Format

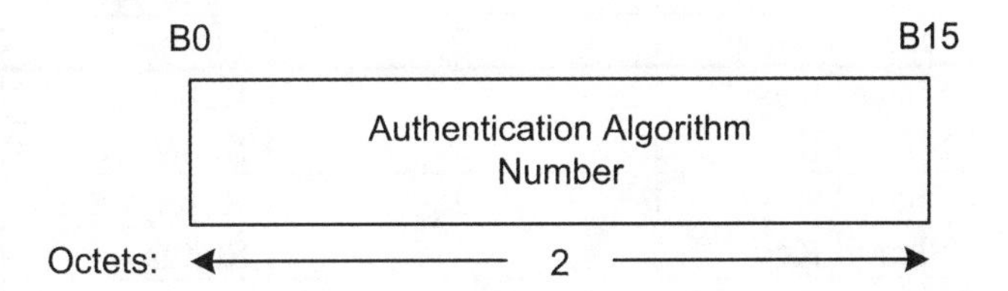

Authentication Transaction Sequence Number Field

The 16-bit Authentication Transaction Sequence Number field indicates the state of progress of the authentication process. The initial value is 1, and it increases sequentially as stations step through the authentication process. Figure 6.19 illustrates the Authentication Transaction Sequence Number field format, and Figure 6.20 illustrates the sequential increases for both Open System and Shared Key authentication.

FIGURE 6.19 Authentication Transaction Sequence Number Field Format

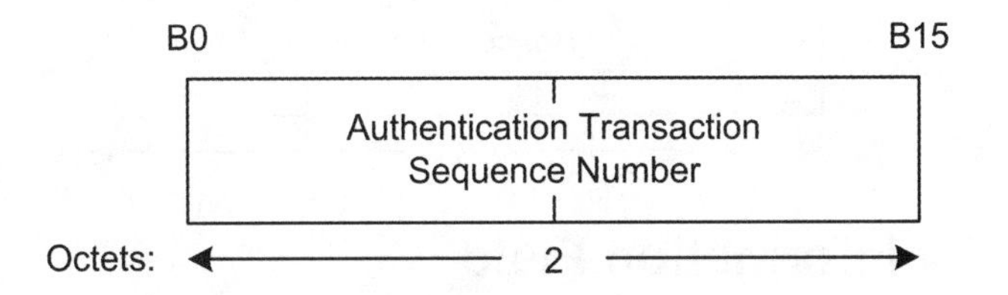

FIGURE 6.20 Authentication Transaction Sequence Numbers

Authentication Algorithm	Authentication Transaction Sequence Number	Status Code	Challenge Text
Open System	1	Reserved	Not present
Open System	2	Status	Not present
Shared Key	1	Reserved	Not present
Shared Key	2	Status	Present
Shared Key	3	Reserved	Present
Shared Key	4	Status	Not present

Beacon Interval Field

This 16-bit Beacon Interval value is the number of time units (TUs) between target beacon transmission times (TBTTs). Each TU equals 1024 microseconds (1.024 milliseconds), which is what most vendors refer to as a Kilomicrosecond (Kμs). Figure 6.21 illustrates the Beacon Interval field format.

FIGURE 6.21 Beacon Interval Field Format

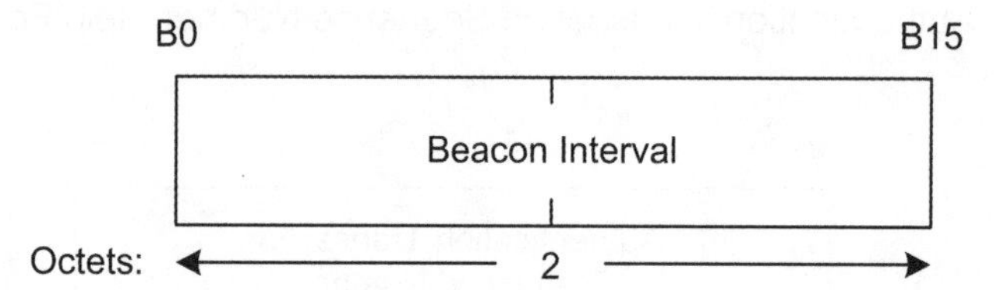

Capability Information Field

The 16-bit Capability Information field contains a number of subfields that are used to indicate requested or advertised optional capabilities. Figure 6.22 shows a Capability Information field as shown in a wireless protocol analyzer.

FIGURE 6.22 Typical Beacon's Capability Information Field

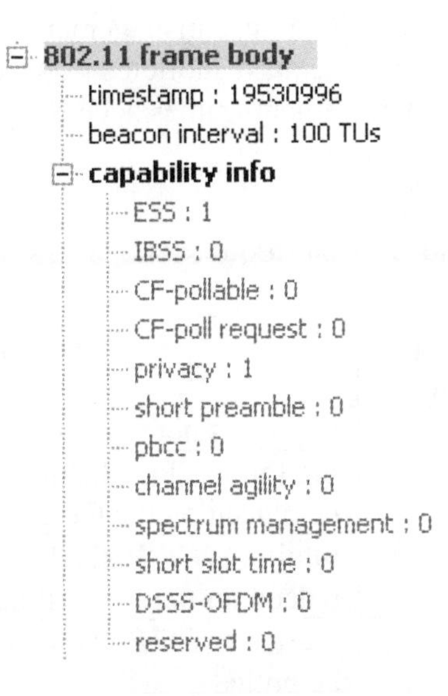

The Capability Information field consists of the following subfields: ESS, IBSS, CF-Pollable, CF-Poll Request, Privacy, Short Preamble, Packet Binary Convolutional Code (PBCC), Channel Agility, Short Slot Time, and DSSS-OFDM. The format of the Capability Information field is illustrated in Figure 6.25. Notice from a comparison of Figures 6.22 and 6.23 that some manufacturers may choose to use the reserved field bits. No subfield is supplied for ERP as a station supports ERP operation if it includes the three Clause 19 (802.11g) mandatory rates in its supported rate set.

FIGURE 6.23 Capability Information Field Format

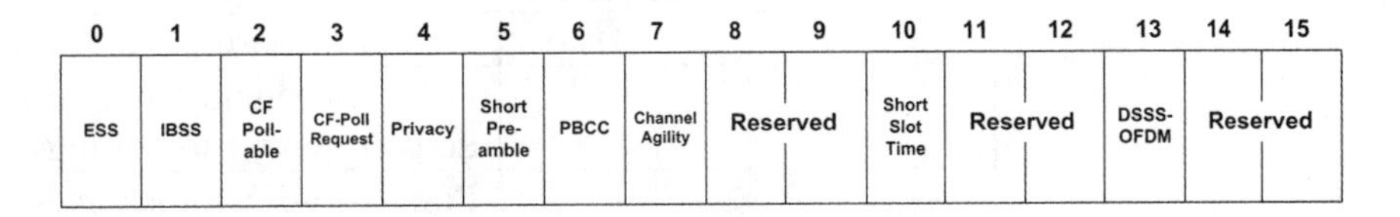

Each Capability Information subfield is interpreted only in the management frame subtypes for which the transmission rules are defined. Access points set the ESS subfield to 1 and the IBSS subfield to 0 within

transmitted Beacon or Probe Response management frames. Stations within an IBSS set the ESS subfield to 0 and the IBSS subfield to 1 in transmitted Beacon or Probe Response management frames. Stations set the CF-Pollable and CF-Poll Request subfields in Association and Reassociation Request management frames according to the values shown in Figure 6.24.

FIGURE 6.24 Station CF-Pollable and CF-Poll Request subfields

CF-Pollable	CF-Poll Request	Meaning
0	0	Station is not CF-Pollable
0	1	Station is CF-Pollable, not requesting to be placed on the CF-Polling list
1	0	Station is CF-Pollable, requesting to be placed on the CF-Polling list
1	1	Station is CF-Pollable, requesting never to be polled

Access points set the CF-Pollable and CF-Poll Request subfields in Beacon, Probe Response, Association Response, and Reassociation Response management frames according to Figure 6.25. An access point sets the CF-Pollable and CF-Poll Request subfield values in Association Response and Reassociation Response management frames equal to the values in the last Beacon or Probe Response frame that it transmitted.

FIGURE 6.25 Access Point CF-Pollable and CF-Poll Request subfields

CF-Pollable	CF-Poll Request	Meaning
0	0	No point coordinator at access point
0	1	Point coordinator at access point for delivery only (no polling)
1	0	Point coordinator at access point for delivery and polling
1	1	Reserved

Access points set the Privacy subfield to 1 within transmitted Beacon, Probe Response, Association Response, and Reassociation Response management frames if WEP encryption is required for all data frames exchanged within the BSS. If WEP encryption is not required, the Privacy field is set to 0. Stations within an IBSS set the Privacy field to 1 in transmitted Beacon or Probe Response management frames if WEP encryption is required for all data type frames exchanged within the IBSS. If WEP encryption is not required, the Privacy field is set to 0.

Access points (as well as stations in IBSSs) should set the Short Preamble subfield to 1 in transmitted Beacon, Probe Response, Association Response, and Reassociation Response management MMPDUs to indicate that the use of the Short Preamble option, as described in section 18.2.2.2 of the 802.11g standard, is allowed within this BSS. To indicate that the use of the Short Preamble option is not allowed, the Short Preamble subfield should be set to 0 in Beacon, Probe Response, Association Response, and Reassociation Response management MMPDUs transmitted within the BSS.

Stations should set the Short Preamble subfield to 1 in transmitted Association Request and Reassociation Request MMPDUs when the MIB attribute *dot11ShortPreambleOptionImplemented* is true. Otherwise, stations should set the Short Preamble subfield to 0 in transmitted Association Request and Reassociation Request MMPDUs.

Access points (as well as stations in IBSSs) should set the Short Preamble subfield to 1 in transmitted Beacon, Probe Response, Association Response, and Reassociation Response management MMPDUs to indicate that the use of the Short Preamble, as described in 18.2.2.2 of the 802.11g standard, is allowed within this BSS. To indicate that the use of the Short Preamble is not allowed, the Short Preamble subfield should be set to 0 in Beacon, Probe Response, Association Response, and Reassociation Response MMPDUs transmitted within the BSS. ERP stations should set the MIB variable *dot11ShortPreambleOptionImplemented* to true since all ERP stations support both long and short preamble formats.

Access points (as well as stations in IBSSs) should set the PBCC subfield to 1 in transmitted Beacon, Probe Response, Association Response, and Reassociation Response management MMPDUs to indicate that the

PBCC Modulation option, as described in sections 18.4.6.6 and 19.6 of the 802.11g standard, is allowed within the BSS. To indicate that the PBCC Modulation option is not allowed, the PBCC subfield should be set to 0 in Beacon, Probe Response, Association Response, and Reassociation Response management MMPDUs transmitted within the BSS. Stations that support PBCC should set the PBCC subfield to 1 in transmitted Association Request and Reassociation Request MMPDUs. Otherwise, stations should set the PBCC subfield to 0 in transmitted Association Request and Reassociation Request MMPDUs.

Bit 7 of the Capability Information field should be used to indicate Channel Agility capability by the High Rate direct sequence spread spectrum (HR/DSSS) or ERP PHYs. Stations should set the Channel Agility bit to 1 when Channel Agility is in use and shall set it to 0 otherwise.

Stations should set the Short Slot Time subfield to 1 in transmitted Association Request and Reassociation Request MMPDUs when the MIB attribute *dot11ShortSlotTimeOptionImplemented* and *dot11ShortSlotTimeOptionEnabled* are true. Otherwise, the station should set the Short Slot Time subfield to 0 in transmitted Association Request and Reassociation Request MMPDUs.

If a station that does not support Short Slot Time associates, the access point should use the long slot time beginning at the first beacon subsequent to the association of the station using long slot times. Access points should set the Short Slot Time subfield in transmitted Beacon, Probe Response, Association Response, and Reassociation Response MMPDUs to indicate the slot time value currently in use within this BSS.

Stations should set the MAC variable *aSlotTime* to the short slot value upon transmission or reception of Beacon, Probe Response, Association Response, and Reassociation Response MMPDUs from the BSS that the station has joined or started and that have the Short Slot Time subfield set to 1 when the MIB attribute *dot11ShortSlotTimeOptionImplemented* is true. Stations should set the MAC variable *aSlotTime* to the long slot value upon transmission or reception of Beacon, Probe Response, Association Response, and Reassociation Response MMPDUs from the BSS that the station has joined or started and that have the Short Slot Time subfield set to 0 when the MIB attribute *dot11ShortSlotTimeOptionImplemented* is true. Stations should set the MAC variable *aSlotTime* to the long slot value at all times when the MIB attribute *dot11ShortSlotTimeOptionImplemented* is false. When the *dot11ShortSlotTimeOptionImplemented* MIB attribute is not present, or when the PHY supports only a single slot time value, then the station should set the MAC variable *aSlotTime* to the slot value appropriate for the attached PHY. For IBSS, the Short Slot Time subfield shall be set to 0.

Access points (as well as STAs in IBSSs) should set the DSSS-OFDM subfield to 1 in transmitted Beacon, Probe Response, Association Response, and Reassociation Response management MMPDUs to indicate that the use of DSSS-OFDM, as described in section 19.7 of the 802.11g standard, is allowed within this BSS or by stations that want to use DSSS-OFDM within an IBSS. To indicate that the use of DSSS-OFDM is not allowed, the DSSS-OFDM subfield should be set to 0 in Beacon, Probe Response, Association Response, and Reassociation Response MMPDUs transmitted within the BSS.

Stations should set the DSSS-OFDM subfield to 1 in transmitted Association Request and Reassociation Request MMPDUs when the MIB attribute *dot11DSSS-OFDMOptionImplemented* and *dot11DSSS-OFDMOptionEnabled* are true. Otherwise, stations should set the DSSS-OFDM subfield to 0 in transmitted Association Request and Reassociation Request MMPDUs. Unused bits of the Capability Information field are reserved.

Current AP Address Field

The 6-byte Current AP Address field, found only in a reassociation request frame, indicates to the new access point the access point that the station is currently associated with. Figure 6.26 illustrates the Current AP Address field format.

FIGURE 6.26 Current AP Address Field Format

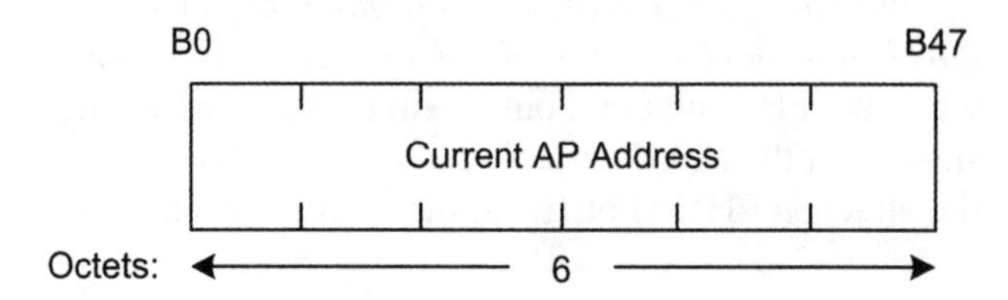

Listen Interval Field

The 16-bit Listen Interval field value identifies, in units of Beacon Interval, how often a station will wake to listen to beacon frames when operating in power-save mode. Large values may not be practical because of the amount of buffer space needed at the access point and the resulting lower throughput between the access point and the station. Figure 6.27 illustrates the Listen Interval field format.

FIGURE 6.27 Listen Interval Field Format

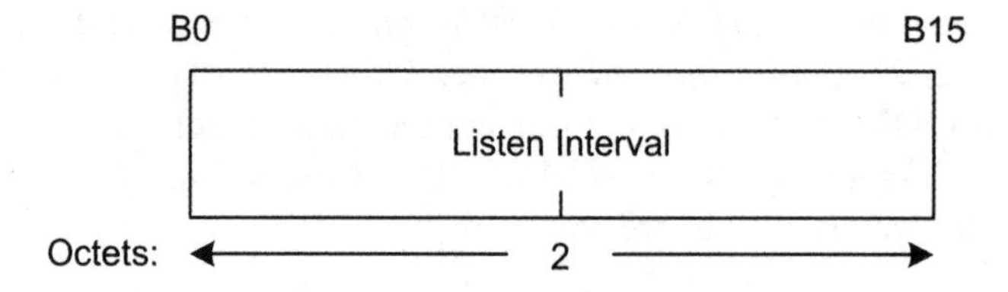

Reason Code Field

The 16-bit Reason Code field indicates (via a numbered code) why a station or access point is generating an unsolicited disassociation or deauthentication. Figure 6.28 shows the reason codes specified by section 7.3.1.7 of the 802.11 standard, and Figure 6.29 illustrates the Reason Code field format.

FIGURE 6.28 802.11 Reason Codes

Reason Code	Meaning
0	Reserved
1	Unspecified reason
2	Previous authentication no longer valid
3	Deauthenticated because sending station is leaving (or has left) IBSS or ESS
4	Disassociated due to inactivity
5	Disassociated because the access point is unable to handle all currently associated stations
6	Class 2 frame received from non-authenticated station
7	Class 3 frame received from non-associated station
8	Disassociated because sending station is leaving (or has left) BSS
9	Station requesting (re)association is not authenticated with responding station
10 – 65,535	Reserved

FIGURE 6.29 Reason Code Field Format

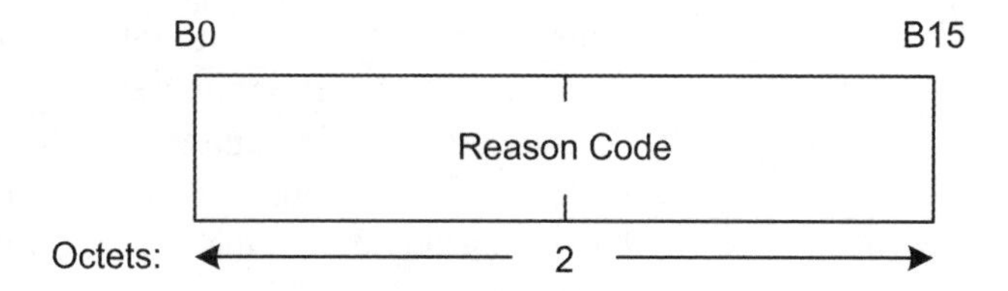

Status Code Field

The 16-bit Status Code field indicates the status of a particular operation. Figure 6.30 shows the status codes specified by section 7.3.1.9 of the 802.11g standard, and Figure 6.31 illustrates the Status Code field format.

FIGURE 6.30 802.11 Status Codes

Status Code	Meaning
0	Successful
1	Unspecified failure
2 – 9	Reserved
10	Cannot support all requested capabilities in the Capability Information field
11	Reassociation denied due to inability to confirm that association exists
12	Association denied due to reason outside the scope of this standard
13	Responding station does not support the specified authentication algorithm
14	Received an Authentication frame with authentication transaction sequence number out of expected sequence
15	Authentication rejected because of challenge failure
16	Authentication rejected due to timeout waiting for next frame in sequence
17	Association denied because the access point is unable to handle additional associated stations
18	Association denied due to requesting station not supporting all of the data rates in the BSSBasicRateSet parameter
19	Association denied due to requesting station not supporting the Short Preamble option
20	Association denied due to requesting station not supporting the PBCC Modulation option
21	Association denied due to requesting station not supporting the Channel Agility option
22 - 24	Reserved
25	Association denied due to requesting station not supporting the Short Slot Time option
26	Association denied due to requesting station not supporting the DSSS-OFDM option
27 – 65,535	Reserved

FIGURE 6.31 Status Code Field Format

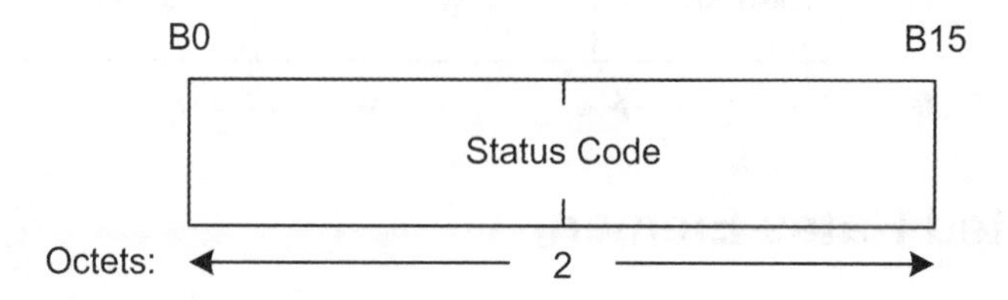

Timestamp Field

The 64-bit Timestamp field contains the timer value at the sending station or access point when it transmits the frame. In a BSS, receiving stations update their timing synchronization function (TSF) with the timestamp value automatically. In an IBSS, receiving stations update their TSF with the timestamp value if it is a value greater than their own timer. Figure 6.32 illustrates the Timestamp field format.

FIGURE 6.32 Timestamp Field Format

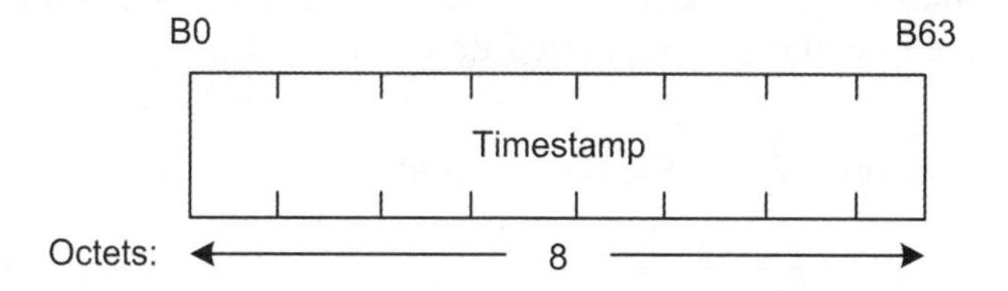

Service Set Identifier (SSID) Element

The variable length SSID field contains the identity of the extended service set (ESS). The maximum length is 32 bytes, and when the SSID has a length of zero, it is considered to be the broadcast SSID. A Probe Request frame having a broadcast SSID causes all access points to respond with a Probe Response frame. Most access points and wireless LAN switches have configuration settings to allow the administrator to override this functionality. Figure 6.33 illustrates the Service Set Identifier element format.

FIGURE 6.33 Service Set Identifier Element Format

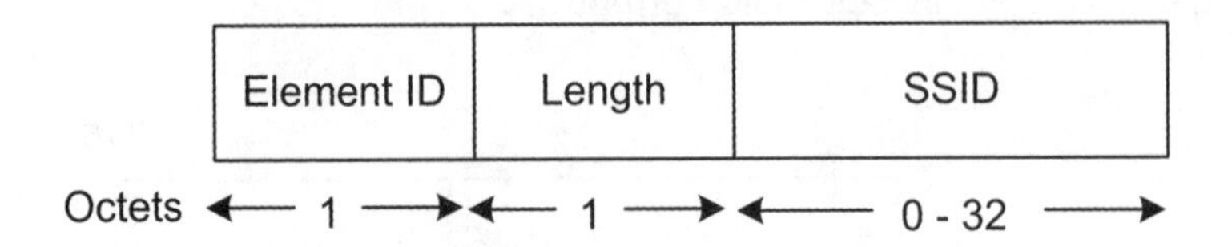

Supported Rates Element

The Supported Rates element specifies up to eight rates in the *OperationalRateSet* parameter. The information field is encoded as 1 to 8 octets, where each octet describes a single supported rate. If the number of rates in the *OperationalRateSet* exceeds eight, then an Extended Supported Rate element will be generated to specify the remaining supported rates. The use of the Extended Supported Rates element is optional otherwise.

The BSS basic rate set information in Beacon and Probe Response management frames is delivered to the management entity in a station via the *BSSBasicRateSet* parameter. It is used by the management entity in a station to avoid associating with a BSS if the station cannot receive and transmit all the data rates in the BSS basic rate set. Figure 6.34 illustrates the format of the Supported Rates Element.

FIGURE 6.34 Supported Rates Element Format

	Element ID	Length	Extended Supported Rates
Octets	1	1	1-8

If the DSSS-OFDM bit is set to 1 in the transmitted Capability Information field of an MMPDU, then any supported rates transmitted in that frame that include rates that are common to both DSSS-OFDM and ERP-OFDM will be interpreted by receiving and transmitting stations to indicate support for both DSSS-OFDM and ERP-OFDM at the indicated rate. However, if any of those rates are indicated as basic (a required rate which is indicated in the BSSBasicRateSet), then the basic rate designation shall be interpreted by the receiving and transmitting stations to apply only for the ERP-OFDM modulation and rate. If the PBCC bit is set to 1 in the transmitted capability field of an MMPDU, then any supported rates transmitted in that frame that include rates that are common to both PBCC and CCK shall be interpreted by receiving and transmitting stations to indicate support for both PBCC and CCK at the indicated rate. However, if any of those rates are indicated as basic, then the basic rate designation shall be interpreted by receiving and transmitting stations to apply only for the CCK modulation and rate. That is, if the rate is indicated as basic, the basic designation does not apply to DSSS-OFDM, PBCC, or ERP-PBCC.

Extended Supported Rates Element

The Extended Supported Rates element specifies the rates that are not carried in the Supported Rates element. Figure 6.35 illustrates the Extended Supported Rates Element format.

FIGURE 6.35 Extended Supported Rates Element Format

	Element ID	Length	Extended Supported Rates
Octets	1	1	1-255

The information field is encoded as 1 to 255 octets where each octet describes a single supported rate. The Extended Supported Rates element is found in transmitted Beacon, Probe Request, Probe Response, Association Request, Association Response, Reassociation Request, and Reassociation Response management frames. The Extended Supported Rates element in other management frame types is ignored by receiving stations. For stations supporting eight or fewer data rates, this element is optional for inclusion in all of the frame types that include the Supported Rates element. For stations supporting more than eight data rates, this element should be included in all of the frame types that include the supported rates element.

FH Parameter Set Element

The FH Parameter Set element contains the set of parameters necessary to allow synchronization for stations using a frequency-hopping (FH) PHY. The information field contains dwell time, hop set, hop pattern, and hop index parameters. The total length of the information field is 5 octets. Figure 6.36 illustrates the FH Parameter Set Element format.

FIGURE 6.36 FH Parameter Set Element Format

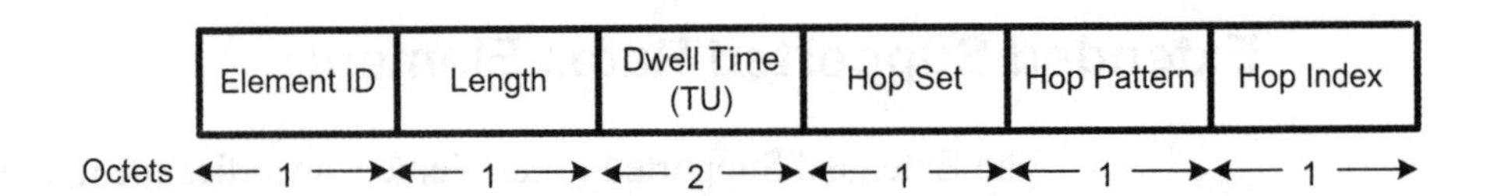

The dwell time field is 2 octets in length and contains the dwell time in TU. The hop set field identifies the current set of hop patterns and is a single octet. The hop pattern field identifies the current pattern within a set of hop patterns and is a single octet. The hop index field selects the current index within a pattern and is a single octet.

DS Parameter Set Element

The DS Parameter Set element contains information to allow channel number identification for stations using a direct sequence spread spectrum (DSSS) PHY. The information field contains a single parameter containing the Channel Number (see Figure 6.37 for values). The length of the Channel Number parameter is 1 octet as illustrated in Figure 6.38.

FIGURE 6.37 Channel Number Values

Channel ID	Frequency
1	2412 MHz
2	2417 MHz
3	2422 MHz
4	2427 MHz
5	2432 MHz
6	2437 MHz
7	2442 MHz
8	2447 MHz
9	2452 MHz
10	2457 MHz
11	2462 MHz
12	2467 MHz
13	2472 MHz
14	2484 MHz

FIGURE 6.38 DS Parameter Set Element Format

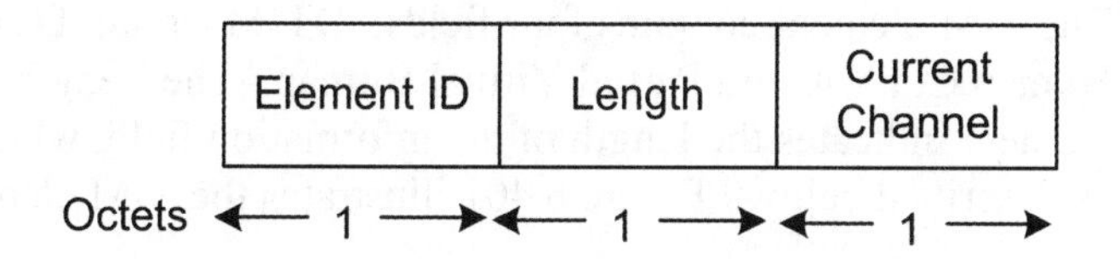

CF Parameter Set Element

The CF Parameter Set element contains the set of parameters necessary to support the PCF. The information field contains the *CFPCount*, *CFPPeriod*, *CFPMaxDuration*, and *CFPDurRemaining* fields. The total length of the information field is 6 octets. Figure 6.39 shows the format of the CF Parameter Set element.

FIGURE 6.39 CF Parameter Set Element Format

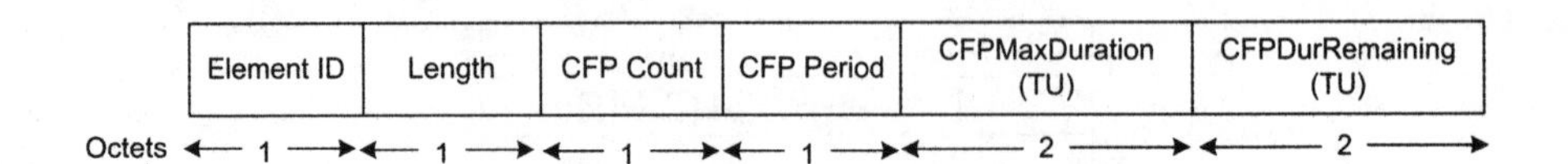

CFPCount indicates how many DTIMs (including the current frame) appear before the next CFP starts. A *CFPCount* of 0 indicates that the current DTIM marks the start of the CFP. *CFPPeriod* indicates the number of DTIM intervals between the start of CFPs. The value is an integral number of DTIM intervals. *CFPMaxDuration* indicates the maximum duration, in TU, of the CFP that may be generated by this PCF. This value is used by STAs to set their NAV at the TBTT of beacons that begin CFPs. *CFPDurRemaining* indicates the maximum time, in TU, remaining in the present CFP, and is set to zero in CFP Parameter elements of beacons transmitted during the contention period. The value of *CFPDurRemaining* is referenced to the immediately previous TBTT. This value is used by all stations to update their NAVs during CFPs.

Traffic Indication Map (TIM) Element

The TIM element contains four fields: DTIM Count, DTIM Period, Bitmap Control, and Partial Virtual Bitmap. The Length field for this element indicates the length of the information field, which is constrained as described below. Figure 6.401 illustrates the TIM element format.

FIGURE 6.40 Traffic Indication Map Element Format

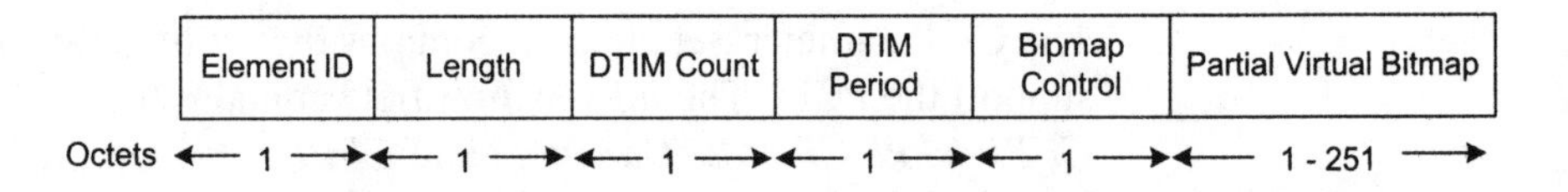

The DTIM Count field indicates how many beacons (including the current frame) appear before the next DTIM. A DTIM Count of 0 indicates that the current TIM is a DTIM. The DTIM count field is a single octet.

The DTIM Period field indicates the number of beacon intervals between successive DTIMs. If all TIMs are DTIMs, the DTIM Period field has the value 1. The DTIM Period value 0 is reserved. The DTIM period field is a single octet.

The Bitmap Control field is a single octet. Bit 0 of the field contains the Traffic Indicator bit associated with Association ID 0. This bit is set to 1 in TIM elements with a value of 0 in the DTIM Count field when one or more broadcast or multicast frames are buffered at the access point. The remaining 7 bits of the field form the Bitmap Offset.

The traffic-indication virtual bitmap, maintained by the access point that generates a TIM, consists of 2008 bits (0-2007). Each bit in the traffic-indication virtual bitmap corresponds to traffic buffered for a specific station within the BSS that the access point is prepared to deliver at the time the beacon frame is transmitted. Bit number N is 0 if there are no directed frames buffered for the station whose Association ID is N. If any directed frames for that station are buffered and the access point is prepared to deliver them, bit number N in the traffic indication virtual bitmap is 1. A point coordinator may decline to set bits in the TIM for CF-Pollable stations it does not intend to poll. In the event that all bits other than bit 0 in the virtual bitmap are 0, the Partial Virtual Bitmap field is encoded as a single octet equal to 0, and the Bitmap Offset field is 0.

IBSS Parameter Set Element

The IBSS Parameter Set element contains the set of parameters necessary to support an IBSS. The information field contains the ATIM Window parameter. The ATIM Window field is 2 octets in length and contains the ATIM Window length in TU. Figure 6.41 illustrates the format of the IBSS Parameter Set element.

FIGURE 6.41 IBSS Parameter Set Element Format

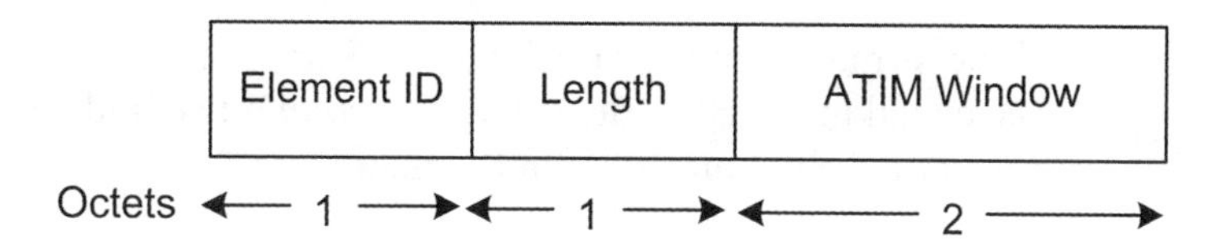

Challenge Text Element

The Challenge Text element contains the challenge text within Authentication exchanges. The element information field length is dependent upon the authentication algorithm and the transaction sequence number. Figure 6.42 illustrates the Challenge Text element format.

FIGURE 6.42 Challenge Text Element Format

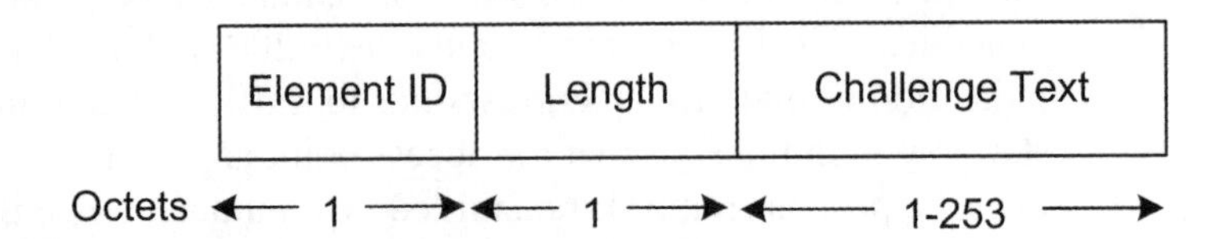

ERP Information Element

The ERP Information element contains information on the presence of Clause 15 (802.11 DSSS) or Clause 18 (802.11b DSSS) stations in the BSS that are not capable of Clause 19 ERP-OFDM (802.11g) data rates. It also contains the requirement of the ERP Information element sender (access point in a BSS or station in an IBSS) as to the use of protection mechanisms to optimize BSS performance and as to the use of long or short Barker preambles. Figure 6.43 shows the format of the ERP Information Element.

FIGURE 6.43 ERP Information Element Format

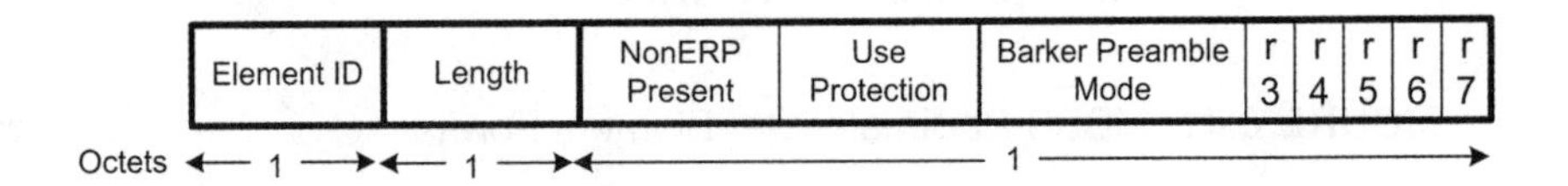

If one or more NonERP (802.11 DSSS or 802.11b DSSS) stations are associated in the BSS, the *Use_Protection* bit should be set to 1 in transmitted ERP Information Elements.

In an IBSS, the setting of the *Use_Protection* bit is left to the station. In an IBSS, there is no uniform concept of association; therefore, a typical algorithm for setting the *Use_Protection* bit will take into account the traffic pattern and history on the network. If a member of an IBSS detects one or more NonERP stations that are members of the same IBSS, then the *Use_Protection* bit should be set to 1 in the ERP Information Element of transmitted Beacon and Probe Response frames.

The *NonERP_Present* bit should be set to 1 when a NonERP station is associated with the BSS. Examples of when the *NonERP_present* bit may additionally be set to 1 include, but are not limited to, when:

1. A NonERP infrastructure or independent BSS is overlapping (a NonERP BSS may be detected by the reception of a beacon where the supported rates contain only IEEE 802.11 Clause 15 or Clause 18 rates).
2. In an IBSS, if a beacon frame is received from one of the IBSS participants where the supported rate set contains only Clause 15 or Clause 18 rates.
3. A management frame (excluding a Probe Request) is received where the supported rate set includes only Clause 15 or Clause 18 rates.

ERP access points and ERP stations should invoke the use of a protection mechanism after transmission or reception of the *Use_Protection* bit with a value of 1 in an MMPDU to or from the BSS that the ERP access point or ERP station has joined or started. ERP access points and ERP stations may additionally invoke protection mechanism use at other times. ERP access points and ERP stations may disable protection mechanism use after transmission or reception of the *Use_Protection* bit with a value of 0 in an MMPDU to or from the BSS that the ERP access point or ERP station has joined or started.

When there are no NonERP stations associated with the BSS and the ERP Information Element sender's *dot11ShortPreambleOptionImplemented* MIB variable is set to true, then the *Barker_Preamble_Mode* bit may be set to 0. The *Barker_Preamble_Mode* bit shall be set to 1 by the ERP Information Element sender if one or more associated NonERP stations are not short-preamble-capable, as indicated in their Capability Information field, or if the ERP Information Element sender's *dot11ShortPreambleOptionImplemented* MIB variable is set to false.

If a member of an IBSS detects one or more non-short preamble-capable stations that are members of the same IBSS, then the *Barker_Preamble_Mode* bit should be set to 1 in the transmitted ERP Information Element.

ERP access points and ERP stations should use long preambles when transmitting Clause 15, Clause 18, and Clause 19 frames after transmission or reception of an ERP Information Element with a *Barker_Preamble_Mode* value of 1 in an MMPDU to or from the BSS that the ERP access point or ERP station has joined or started. ERP access points and ERP stations should use long preambles regardless of the value of the short preamble capability bit from the same received or transmitted MMPDU that contained the ERP Information Element. ERP access points and ERP stations may additionally use long preambles when transmitting Clause 15, Clause 18, and Clause 19 frames at other times. ERP access points and ERP stations may use short preambles when transmitting Clause 15, Clause 18, and Clause 19 frames after transmission or reception of an ERP Information Element with a *Barker_Preamble_Mode* value of 0 in an MMPDU to or from the BSS that the ERP access point or ERP station has joined or started, regardless of the value of the short preamble capability bit from the same received or transmitted MMPDU. NonERP stations and NonERP access points may also follow the rules given in this paragraph. Recommended behavior for setting the *Use_Protection* bit is contained in section 9.10 of the 802.11g standard.

Bits r3 through r7 are reserved, set to 0, and are ignored on reception. Note that the length of this element is flexible and may be expanded in the future.

Summary

802.11 Management frames hold a wealth of information about the goings-on in an ESS. Understanding the purpose and contents of each frame type is vital in troubleshooting, site surveying, performance and security analysis, and other daily analysis tasks. The rules of management frame use are very complex. There have been many changes and additions between 802.11 and 802.11g. Trying to reconcile each of the additions and changes against previous standards, and then understand the rules for mixed 802.11/b/g environments will test the patience of the most seasoned analyst.

A good place to start learning management frame behavior is with a protocol analyzer, an access point, and a client device. Beacons and probes alone hold a tremendous amount of information that takes time to digest. Watching the client device authenticate and associate step-by-step, seeing corrupt frames in the display, realizing that frames that should be in the display are not always captured, and taking a detailed look at the contents of each frame type will take you from beginner to intermediate analyst in a short period of time.

Key Terms

Before taking the exam, you should be familiar with the following terms:

Announcement Traffic Indication Message (ATIM) frame

Association Request frame

Association Response frame

Authentication frame

Beacon frame

Deauthentication frame

Disassociation frame

Probe Request frame

Probe Response frame

Reassociation Request frame

Reassociation Response frame

Review Questions

1. Which service within an access point allows the access point to function as a portal device?

2. The Extended Supported Rates Information Element is present in a Beacon frame when the number of supported data rates exceeds _____ rates?

3. To which management frames types does the 802.11g standard add the ERP Information Element?

4. Which TIM field indicates how many beacons appear before the next DTIM?

5. The maximum number of stations that can associate to an access point according to the 802.11 standard is ______.

6. When the Privacy bit in the Capability Information field of a beacon is set to 1, what is being indicated about the BSS?

7. A Disassociation frame contains which frame body field(s)?

8. The _______ field value identifies, in units of Beacon Interval, how often a station will wake to listen to beacon frames when operating in power-save mode.

9. When using Shared Key authentication, are any of the frames in the authentication frame exchange encrypted?

10. How does a station indicate to an access point that it is pollable?

11. An ATIM frame contains which information elements and fixed fields in its frame body?

12. What is the difference in the frame body contents between an association frame and a reassociation frame?

CHAPTER

7

802.11 Control and Data Frames

CWAP Exam Objectives Covered:

- Distinguish the intended purpose of each 802.11 MAC layer frame type
    - Control frames
    - Data frames
- Explain the structure of each 802.11 MAC layer frame type
    - Control, Management, and Data frame payload contents and sizes
- Explain the structure of each 802.11 MAC layer frame type
    - Modifications for 802.11e

In This Chapter

Control Frames

Control frames provide the functionality to assist in the delivery of unicast data and unicast management frames. Control frames do not have a frame body[1], and must be transmitted at a rate compliant with the Multirate Support sections of the 802.11 series of standards[2]. This topic will be discussed in detail in a later section. The following sections define the structure of each control frame subtype:

Virtual Carrier-Sense Mechanism

Both virtual and physical carrier sense is required by the 802.11 standard. The virtual carrier-sense mechanism is achieved by distributing reservation information announcing the impending use of the medium for the rest of the frames in a frame exchange sequence. The exchange of RTS and CTS frames prior to the actual data frame is one means of distribution of this medium reservation information. The RTS and CTS frames contain a Duration/ID field that defines the period of time that the medium is to be reserved to transmit the actual data frame and the returning ACK frame. All stations within the reception range of either the station which transmits the RTS or the station which transmits the CTS will learn of the medium reservation. Thus, a station can be unable to receive from one of the two stations, yet still know about the impending use of the medium to transmit a data frame and its acknowledgment frame.

The RTS/CTS exchange also performs both a type of fast collision inference and a transmission path check. If the return CTS is not detected by the station originating the RTS, the originating station may repeat the process (after observing the other medium access rules) more quickly than if the long data frame had been transmitted and a return ACK frame had not been detected. Another advantage of the RTS/CTS mechanism occurs where multiple BSSs utilizing the same channel overlap. The medium reservation mechanism works across the BSA boundaries. The RTS/CTS mechanism may also improve operation in a typical situation where all

---

[1] IEEE 802.11 – 1999 (R2003) – Section 7.2.1
[2] IEEE 802.11g – 2003 – Section 9.6

stations can receive from the access point, but cannot receive from all other stations in the BSA. The RTS/CTS mechanism is not used for MPDUs with broadcast and multicast immediate addresses because multicast frames are never acknowledged nor retransmitted. The RTS/CTS mechanism need not be used for every data frame transmission.

Because the additional RTS and CTS frames add overhead inefficiency, the mechanism is not always justified, especially for short data frames. The use of the RTS/CTS mechanism is under control of the dot11RTSThreshold attribute. This attribute may be set on a per-station basis. This mechanism allows stations to be configured to use RTS/CTS either always, never, or only on frames longer than a specified length. A station configured not to initiate the RTS/CTS mechanism will still update its virtual carrier-sense mechanism with the duration information contained in a received RTS or CTS frame, and will always respond to an RTS addressed to it with a CTS. The medium access protocol allows for stations to support different sets of data rates. All stations will receive all the data rates in aBasicRateSet and transmit at one or more of the aBasicRateSet data rates. To support the proper operation of the RTS/CTS and the virtual carrier-sense mechanism, all stations should be able to detect the RTS and CTS frames.

Request-to-Send (RTS) Frame

The RTS frame is a 20-byte frame used to reserve the medium in cases where there are hidden nodes, either due to distance, or signal blockage or in cases where ERP-OFDM frames need protection from NonERP stations. Users can configure their wireless client utilities to initiate an RTS/CTS frame sequence always, never, or only on frames longer than a specified length. The Duration field, measured in microseconds, in a RTS frame is the amount of time the medium should be reserved to encompass a Clear-to-Send (CTS) frame, a data frame, an ACK frame, and three short interframe space (SIFS) intervals. In the correct order, this calculation would be SIFS-CTS-SIFS-DATA-SIFS-ACK. The duration value sets the NAV on other stations preventing them from transmitting long enough for the entire frame exchange sequence to complete. The RTS frame is always sent directly to an immediate receiver and is never relayed through an access point, and is preceded by DIFS.

FIGURE 7.1 RTS Frame

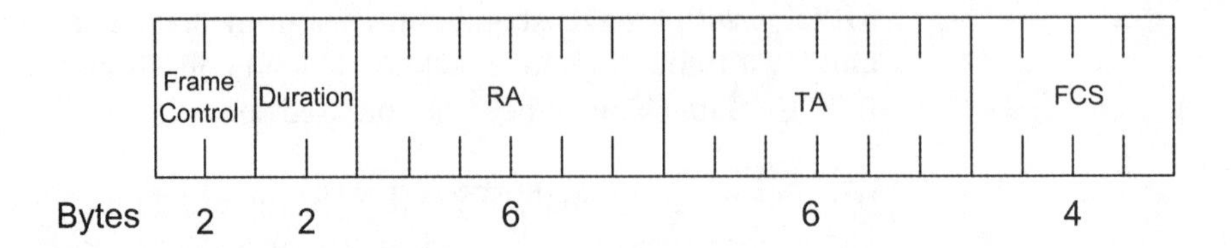

Clear-to-Send (CTS) Frame

After receiving a RTS addressed to it, a station sends a 14-byte CTS frame back to the originating station. The CTS frame's duration field informs those stations in its immediate area to set their NAV for a value equal to SIFS+DATA+SIFS+ACK. These nearby stations need to make this setting to their NAV in case one of these stations did not hear the RTS frame and appropriately set its NAV previously. CTS frames are preceded by SIFS, are always sent to an immediate receiver, and have the same frame structure as an ACK frame.

FIGURE 7.2 CTS Frame

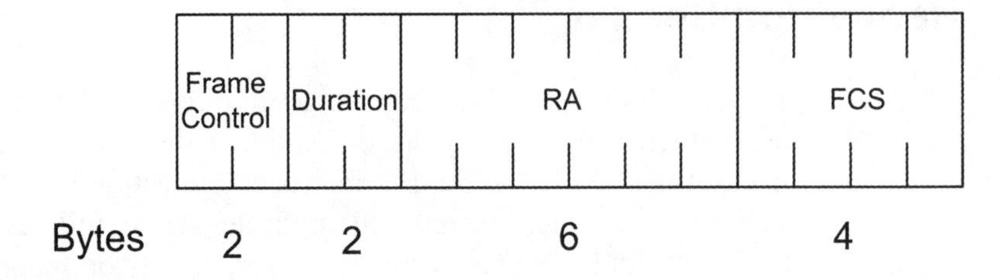

Acknowledgement (ACK) Frame

A station receiving an error-free directed data or directed management frame must send a 14-byte ACK frame to the transmitting station to acknowledge successful reception. Multicast and broadcast frames are never acknowledged. PS-Poll control frames may be acknowledged with an ACK frame, but also with a data frame. The value of the Duration field in the ACK frame, in microseconds, is equal to 0 if the More Fragment bit in the Frame Control field of the previous data or management frame is set to 0. If the More Fragment bit of the previous data or management frame is set to 1, then the Duration field specifies the amount of time from the Duration field of the previous data or management frame minus the time required to transmit the ACK frame and its associated SIFS interval.

FIGURE 7.3 ACK Frame

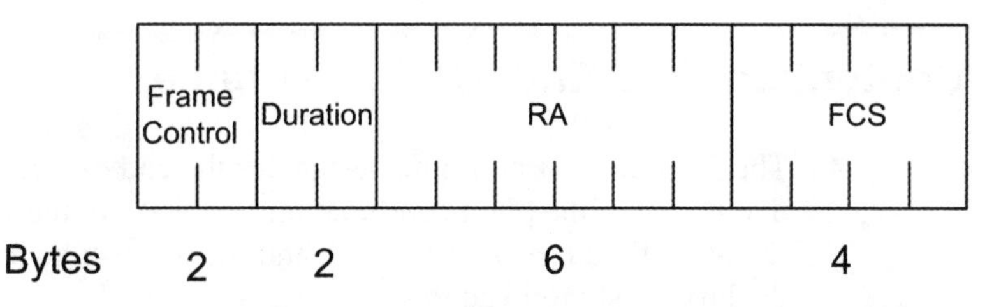

Do not confuse 802.11 ACKs with TCP or higher layer ACKs. Some wireless protocol analyzers are capable of decoding layers 2-7 and may show both 802.11 ACKs and TCP ACKs if encryption is not used.

Power Save Poll (PS-Poll) Frame

When a client in power-save mode discovers (through the TIM in a beacon) that the access point has data frames buffered for it, the client station should send the 20-byte PS-Poll frame to the access point to request that the data frames be sent. The access point must respond in a SIFS interval with either an ACK frame or a data frame. The power-save station must then acknowledge the data frame. The Duration/ID field of the PS-Poll frame contains the AID of the station, with the two most significant bits set to 1. All other stations receiving the PS-Poll frame update their NAV as though the duration field contained enough time for a SIFS interval and an ACK frame. This is the case even if the access point responds with a data frame rather than an ACK frame; the data frame will not have full NAV protection.

FIGURE 7.4 PS-Poll Frame

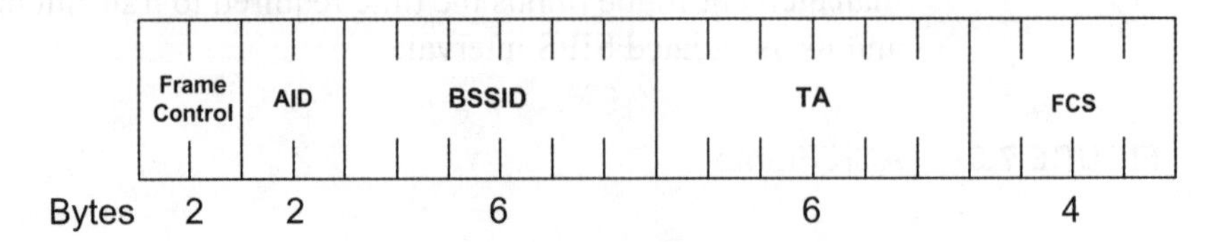

Contention-Free End (CF-End) Frame

The 20-byte CF-End frame designates the end of a contention-free period that is part of the point coordination function. In these frames, the Duration field is always set to 0, and the receiver address (RA) contains the broadcast group address.

FIGURE 7.5 CF-End Frame

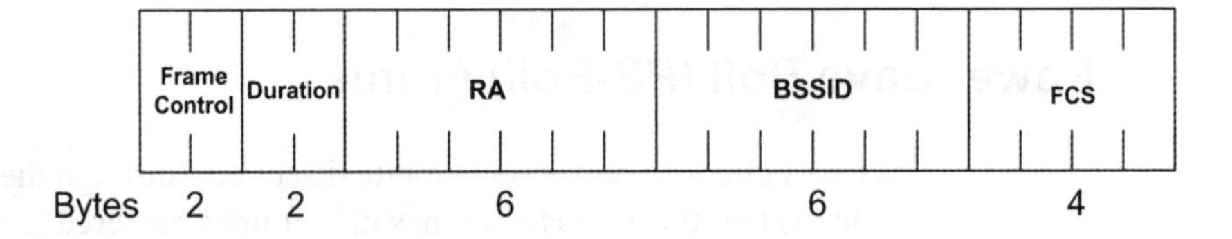

CF-End+CF-ACK Frame

The 20-Byte CF-End+CF-ACK frame designates the end of a contention-free period that is part of the point coordination function and acknowledges a directed data or management frame previously received by the point coordinator. In these frames, the Duration field is always set to 0, and the receiver address (RA) contains the broadcast group address.

FIGURE 7.6 CF-End+CF-ACK Frame

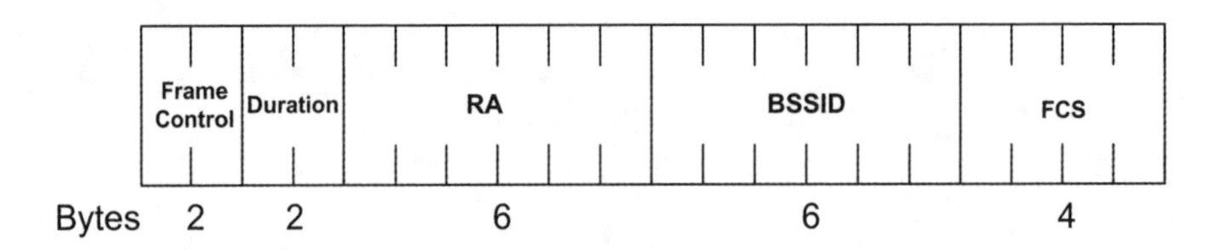

Control Frame Summary

Figure 7.7 summarizes each Control frame type, size, and contents.

FIGURE 7.7 Control Frame Quick Reference Chart

RTS

2	2	6	6	4
FC	D	RA	TA	FCS

CTS or ACK

2	2	6	4
FC	D	RA	FCS

PS-Poll

2	2	6	6	4
FC	AID	BSSID	TA	FCS

CF-End

2	2	6	6	4
FC	D	RA	BSSID	FCS

CF-End + CF-ACK

2	2	6	6	4
FC	D	RA	BSSID	FCS

Data Frames

There are eight data frame subtypes in two groups. The first group may carry data and consists of:

- Simple Data
- Data with contention-free acknowledgement (Data+CF-ACK)
- Data with CF-Poll (Data+CF-Poll)
- Data with CF-ACK and CF-Poll (Data+CF-ACK+CF-Poll)

The second group will not carry data and consists of:

- CF-ACK
- CF-Poll
- CF-ACK+CF-Poll
- Null Function

The first group of data frames carries a nonzero number of data bytes. The second group of data frames carries no data bytes at all. The data frame carries the MSDU in its frame body which is requested to be delivered by the upper layer protocols, LLC or DSS. The MAC header of the MPDU is made up of a control field, a duration/ID field, three or four address fields, a sequence control field, a frame body field, and the frame check sequence field.

Simple Data Frame

The Simple Data frame encapsulates the upper layer protocols packets, delivering them from one IEEE 802.11 station to another. It may appear in both the contention period and contention-free period. The Address 4 field is only used in data frames traversing a Wireless Distribution System (WDS) between two access points in an Extended Service Set (ESS). Unfragmented data frames have a duration field value equal to the amount of time to which other stations should set their NAV to allow for no interference with the subsequent SIFS+ACK. Fragmented data frames and their subsequent ACKs act as a virtual RTS/CTS exchange. Data fragments that are the last fragment have a duration value of SIFS+ACK, and data fragments that are not the last fragment have a duration of SIFS+ACK+SIFS+DATA+SIFS+ACK. The duration field value of multicast frames is always zero.

Data + CF-ACK

The Data+CF-ACK frame is identical to the simple data frame, with the following exceptions. The Data+CF-ACK frame may be sent only during a CFP, and it is never used in an IBSS. The acknowledgement carried in this frame is acknowledging the previously received data frame, which may have been from a different client than the receiver of the current frame.

Data + CF-Poll

The Data+CF-Poll frame is identical to the simple data frame, with the following exceptions. The Data+CF-Poll frame may be sent only by the point coordinator during a CFP. This frame is never sent by a mobile station, and it is never used in an IBSS. This frame is used by the point coordinator to deliver data to a mobile station and simultaneously request that the mobile station send a data frame that it may have buffered, when the current reception is completed

Data + CF-ACK + CF-Poll

The Data+CF-ACK+CF-Poll frame is identical to the simple data frame, with the following exceptions. The Data+CF-ACK+CF-Poll frame may be sent only by the PC during a CFP. This frame is never sent by a mobile station, and it is never used in an IBSS. This frame combines the functions of both the Data+CF-ACK and Data+CF-Poll frames into a single frame.

Null Function

The Null Function Frame, a data frame that contains no frame body and thus no data, is used to allow a station that has nothing to transmit to be able to complete a frame exchange sequence necessary for changing its power management mode. The sole purpose for this frame is to carry the power management bit in the frame control field to the access point, when a station changes its power management mode.

Some vendors also have stations send this frame (with the Power Management field set to 1) prior to implementing passive or active scanning of other channels in search for better access point signals. This configuration causes the access point to buffer any directed frames while the station is tuned to a different channel. Once the station tunes back to the primary channel, then the station sends another Null Function frame (with the Power Management field set to 0) to inform the access point to begin sending frames again to the station.

CF-ACK

The CF-ACK frame may be used by a mobile station that has received a data frame from the point coordinator during the CFP to acknowledge the correct receipt of that frame. Because this frame is 29 bytes long, it is more efficient to use an acknowledgement control frame (ACK frame). Either frame will provide the required acknowledgement to the point coordinator.

CF-Poll

The CF-Poll frame is used by the point coordinator to request that a mobile station send a pending data frame during the CFP. The point coordinator will send this frame, rather than the Data+CF-Poll, when it has no data to be sent to the mobile station.

CF-ACK + CF-Poll

The CF-ACK+CF-Poll is used by the point coordinator to acknowledge a correctly received frame and to solicit a pending frame from a mobile station. The acknowledgement and the solicitation may be intended for disparate mobile stations.

Frame Transmission Rates

Some PHYs have multiple data transfer rate capabilities that allow implementations to perform dynamic rate switching with the objective of improving performance. The algorithm for performing rate switching is beyond the scope of the 802.11 standards, but in order to ensure coexistence and interoperability on multirate capable PHYs, the 802.11 standards define a set of rules to be followed by all 802.11 stations.

All control frames that initiate a frame exchange sequence will be transmitted at one of the rates in the BSSBasicRateSet, unless the transmitting station's protection mechanism is enabled, and the control frame is a protection mechanism frame; in which case, the control frame will be transmitted at a rate according to the separate rules for determining the rates of transmission of protection frames in section 9.10 of the 802.11g standard.

Section 9.10 says that protection mechanisms frames should be sent using one of the mandatory Clause 15 (802.11 DSSS) or Clause 18 (802.11b DSSS) rates and using one of the mandatory Clause 15 or Clause 18 waveforms (modulations), so all stations in the BSA will know the duration of the exchange even if they cannot detect the ERP-OFDM signals using their CCA function. Note that when using the Clause 19 (802.11g) options, ERP-PBCC or DSSS-OFDM, there is no need to use protection mechanisms, as these frames start with a DSSS header.

All frames with multicast or broadcast in the Address-1 field should be transmitted at one of the rates included in the BSS basic rate set, regardless of their type or subtype. Data and/or management MPDUs with a unicast receiver in Address-1 should be sent on any supported data rate selected by a rate switching mechanism. No station should transmit a unicast frame at a rate that is not supported by the destination station, as reported in any Supported Rates and Extended Supported Rates element in the management frames. For Data+CF-ACK, Data+CF-Poll+CF-ACK, and CF-Poll+CF-ACK frames, the rate chosen for transmission should be supported by both the addressed recipient station and the station to which the ACK is intended. Under no circumstances should a station initiate transmission of a data or management frame at a data rate higher than the greatest rate in the *OperationalRateSet*.

To allow the transmitting station to calculate the contents of the Duration/ID field, a station responding to a received frame should transmit its control response (either CTS or ACK frames) at the highest rate in the *BSSBasicRateSet* that is less than or equal to the rate of the immediately previous frame in the frame exchange sequence (as defined in section 9.7 of the 802.11 standards) and that is of the same modulation type as the received frame. If no rate in the basic rate set meets these conditions, then the control frame sent in response to a received frame should be transmitted at the highest mandatory rate of the PHY that is less than or equal to the rate of the received frame, and that is of the same modulation type as the received frame. In addition, the control response frame should be sent using the same PHY options as the received frame, unless they conflict with the requirement to use the *BSSBasicRateSet*.

An alternative rate for the control response frame may be used, provided that the duration of the control response frame at the alternative rate is the same as the duration of the control response frame at the originally chosen rate and the alternative rate is in either the *BSSBasicRateSet* or the mandatory rate set of the PHY and the modulation of the control response frame at the alternative rate is the same type as that of the received frame.

802.11e Frame Format Modification

The 802.11e standard adds management, control, and data frame types beyond those specified in the original 802.11 standard. Additionally, a 2 byte Quality of Service (QoS) Control field is added as a header field in the MAC frame as shown in Figure 7.8. The 802.11e standard defines an additional station type, the QSTA, meaning, "station capable of quality of service" and an additional access point type, the QAP, meaning, "access point capable of quality of service." These station and access point types will be able to take advantage of the QoS header field and additional frame types.

FIGURE 7.8 802.11e Frame Format Modification

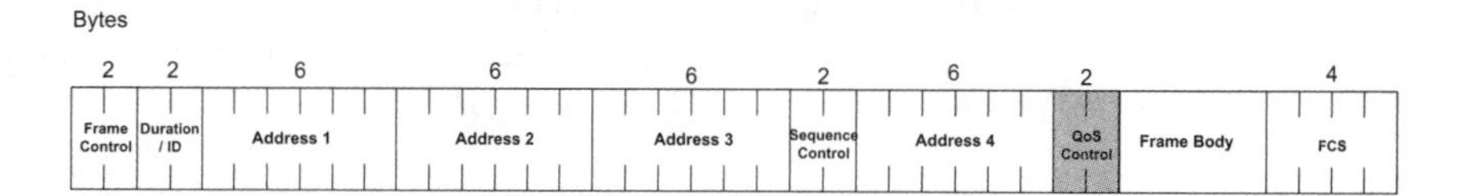

Summary

Control frames are used to manage use of the wireless medium. There are many types of control frames: some for reserving the medium, some for signaling completion of medium use, and others for requesting data traffic. Physical carrier sense is the mechanism used by the physical layer to determine the power level on the wireless medium. This allows the radio to know whether or not another station is presently transmitting. Virtual carrier sense is a MAC layer mechanism for reserving the medium through use of special frame types. In this way, stations are told the medium is busy when in reality it may or may not be.

Data frames contain the upper layer information in their frame body. Data frames, often called MPDUs, are transmitted between immediate wireless receivers. Their payloads, called MSDUs, may be "relayed" through an access point in order to go from one wireless station to another wireless station or they may be placed into Ethernet frames for deliver to wired stations. In DCF mode, only Simple Data frames and Null Function Data frames are used, but there are various types for use in PCF mode. Null Function Data frames are used differently in PCF and DCF modes.

Key Terms

Before taking the exam, you should be familiar with the following terms:

ACK frames

Basic Service Area (BSA)

Clauses

IEEE 802.11e

Medium reservation

Null Function frames

Physical Carrier Sense

Protection frames

PS-Poll frame

Virtual Carrier Sense

Review Questions

1. The 802.11 standard requires what two types of carrier sense mechanisms?

2. Name two different protection mechanisms specified in the 802.11g standard.

3. What additional field was added to the 802.11 MAC frame in the 802.11e standard and how many bytes does it include?

4. All frames with multicast or broadcast in the Address-1 field should be transmitted at one of the rates included in the _______________, regardless of their type or subtype.

5. The CF-End and CF-End+Ack frames always contain what type of address in the Receiver Address (RA) field?

6. What is the 802.11g rule for the modulation and data rate of transmitting protection mechanism frames?

CHAPTER

8

802.11 PHY Layers

CWAP Exam Objectives Covered:

- Explain PHY Layer terminology used in the 802.11 series of standards
- Describe the PLCP Layer (802.11a/b/g)
  - Purpose
  - Preambles and Headers
  - Payloads
- Describe the PMD Layer

In This Chapter

PHY Architecture

PHY Operations

DSSS PHY

ERP-OFDM PHY

DSSS-OFDM PHY

Transmit Procedure

Receive Procedure

Physical Layer Architecture

This section focuses on operation of items specified by the 802.11 series of standards for the physical layer. These items will include the PLCP and PMD sublayers, management layer entities, and generic management primitives. An in-depth understanding of how the physical layer operates and how it interfaces with the MAC layer is vitally important to the analyst's understanding of information gathered by a wireless protocol analyzer.

PLCP Sublayer

The MAC layer communicates with the Physical Layer Convergence Protocol (PLCP) sublayer via primitives (a set of "instructive commands" or "fundamental instructions") through a service access point (SAP). When the MAC layer instructs it to do so, the PLCP prepares MAC protocol data units (MPDUs) for transmission. The PLCP minimizes the dependence of the MAC layer on the PMD sublayer by mapping MPDUs into a frame format suitable for transmission by the PMD. The PLCP also delivers incoming frames from the wireless medium to the MAC layer. The PLCP sublayer is illustrated in Figure 8.1.

The PLCP appends a PHY-specific preamble and header fields to the MPDU that contain information needed by the Physical layer transmitters and receivers. The 802.11 standard refers to this composite frame (the MPDU with an additional PLCP preamble and header) as a PLCP protocol data unit (PPDU). The MPDU is also called the PLCP Service Data Unit (PSDU), and is typically referred to as such when referencing physical layer operations. The frame structure of a PPDU provides for asynchronous transfer of PSDUs between stations. As a result, the receiving station's Physical layer must synchronize its circuitry to each individual incoming frame.

PMD Sublayer

Under the direction of the PLCP, the Physical Medium Dependent (PMD) sublayer provides transmission and reception of Physical layer data units between two stations via the wireless medium. To provide this service, the PMD interfaces directly with the wireless medium (that is, RF in the air) and provides modulation and demodulation of the frame transmissions. The PLCP and PMD sublayers communicate via primitives, through a SAP, to govern the transmission and reception functions. The PMD sublayer is illustrated in Figure 8.1.

Management Layer Entities

Both MAC and PHY layers conceptually include management entities, called the MAC sublayer management entity and the PHY sublayer management entity. These entities are referred to as the MAC Layer Management Entity (MLME), and the Physical Layer Management Entity (PLME). These entities provide the layer management service interfaces through which layer management functions may be invoked. In order to provide correct MAC operation, a station management entity (SME) shall be present within each station. The SME is a layer-independent entity that may be viewed as residing in a separate management plane or as residing "off to the side." The exact functions of the SME are not specified in the 802.11 standard, but in general this entity may be viewed as being responsible for such functions as the gathering of layer-dependent status from the various layer management entities, and similarly setting the value of layer-specific parameters. The SME would typically perform such functions on behalf of general system management entities and would implement standard management protocols. Figure 8.1 depicts the relationship among management entities.

FIGURE 8.1 802.11 Physical and MAC Layer Architecture

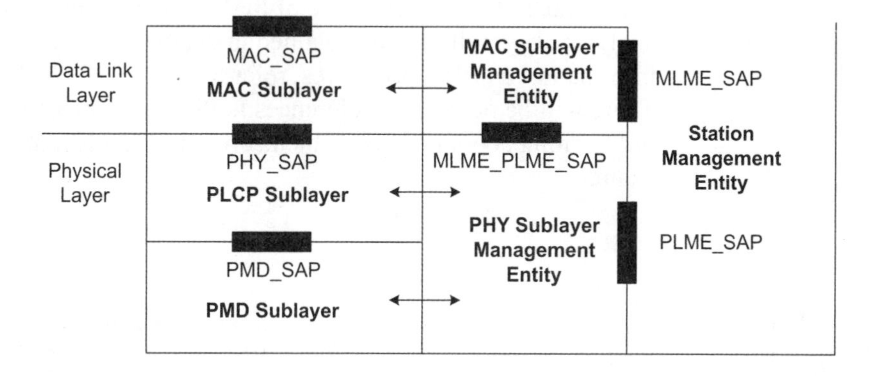

The various entities within this model interact in various ways. Particular interactions are defined explicitly within the 802.11 standard, via a service access point (SAP) across which defined primitives are exchanged. Other interactions are not defined explicitly within the 802.11 standard, such as the interfaces between MAC and MLME and between PLCP and PLME. The specific manner in which these MAC and PHY management entities are integrated into the overall MAC and PHY layers is not specified within the 802.11 standard.

Generic Management Primitives

The management information specific to each layer is represented as a management information base (MIB) for that layer. The MAC and PHY layer management entities are viewed as "containing" the MIB for that layer. The generic model of MIB-related management primitives exchanged across the management SAPs is to allow the SAP user-entity to either GET the value of a MIB attribute, or to SET the value of a MIB attribute.

The practical usage example of management primitives is when the user configures an access point or a mobile station's wireless utilities. This is done through a configuration interface such as CLI, GUI, SNMP, or custom software. Configuration of the access point's features through its web interface, for example, will SET a MIB attribute value to perhaps true/false or to some logical value.

Physical Layer Service Primitives

Due to lack of direct relevance of PHY service primitives to protocol analysis, they will not be explained in detail in this text. For more information on PHY primitives, refer to 802.11-1999 (R2003), Clause 12. There will be occasional references to these primitives within this text, but learning about primitives themselves is not relevant for the CWAP exam.

Physical Layer Operations

The general operation of the various Physical layers is very similar. To perform PLCP functions, the 802.11 standard specifies the use of state machines. Each state machine performs one of the following functions:

- Carrier Sense/Clear Channel Assessment (CS/CCA)
- Transmit (Tx)
- Receive (Rx)

Carrier Sense/Clear Channel Assessment (CS/CCA)

Carrier Sense/Clear Channel Assessment is used to determine the state of the medium. The CS/CCA procedure is executed while the receiver is turned on and the station is not currently receiving or transmitting a packet. The CS/CCA procedure is used for two specific purposes: to detect the start of a network signal that can be received (CS) and to determine whether the channel is clear prior to transmitting a packet (CCA).

Transmit (Tx)

Transmit (Tx) is used to send individual octets of the data frame. The transmit procedure is invoked by the CS/CCA procedure immediately upon receiving a PHY-TXSTART.request (TXVECTOR) from the MAC sublayer. The CSMA/CA protocol is performed by the MAC with the PHY PLCP in the CS/CCA procedure prior to executing the transmit procedure.

Receive (Rx)

Receive (Rx) is used to receive individual octets of the data frame. The receive procedure is invoked by the PLCP CS/CCA procedure upon detecting a portion of the preamble sync pattern followed by a valid SFD and PLCP Header. Although counter-intuitive, the preamble and PLCP header are not "received". Only the MAC frame is "received".

The following sections describe how each of the PLCP functions is used for transferring data between the MAC and Physical layers.

Carrier Sense Function

The Physical layer implements the carrier sense operation by directing the PMD to check to see whether the medium is busy or idle. The PLCP performs the following sensing operations if the station is not transmitting or receiving a frame:

- *Detection of incoming signals* - The PLCP within the station will sense the medium continually. When the medium becomes busy, the PLCP will read in the PLCP preamble and header of the frame to attempt synchronization of the receiver to the data rate of the signal.
- *Clear channel assessment* - The clear channel assessment operation determines whether the wireless medium is busy or idle. If the medium is idle, the PLCP will send a *PHY-CCA.indicate* primitive (with its status field indicating idle) to the MAC layer. If the medium is busy, the PLCP will send a *PHY-CCA.indicate* primitive (with its status field indicating busy) to the MAC layer. The MAC layer can then make a decision on whether to send a frame.

Stations and access points that are 802.11-compliant store the clear channel assessment operating mode in the Physical layer MIB attribute *aCCAModeSuprt*. A developer can set this mode through station initialization procedures. Figure 8.2 shows an example of configuring the different CCA operating modes.

FIGURE 8.2 Configuring CCA Operating Modes on a Mobile Station

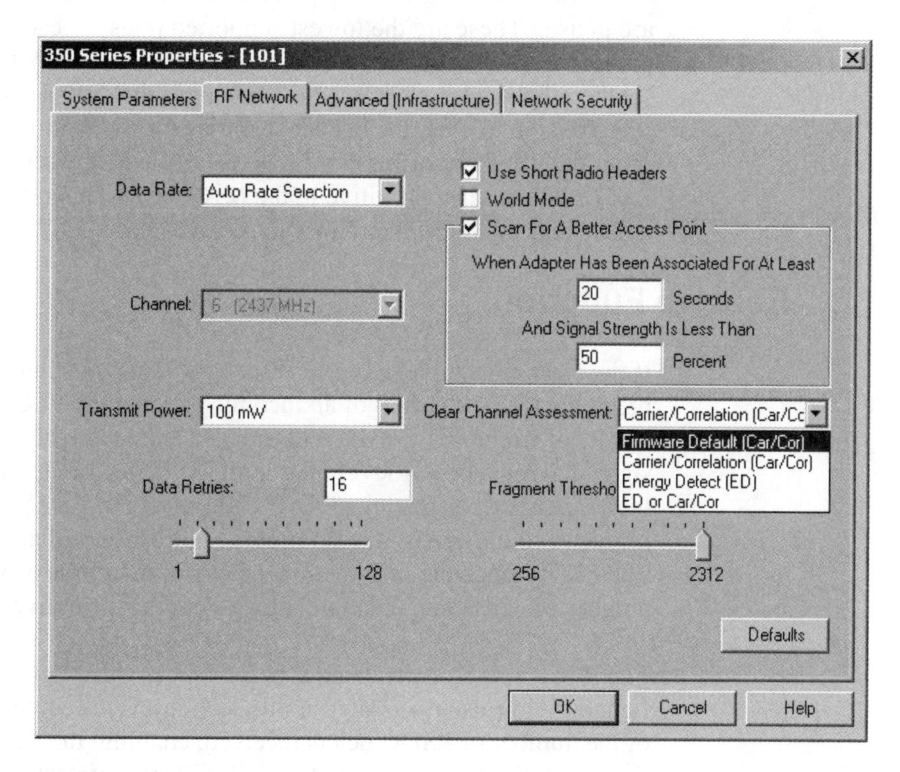

Transmit Function

The PLCP will switch the PMD to transmit mode after receiving the *PHY-TXSTART.request* primitive from the MAC layer. The MAC layer sends the number of octets (0-4095) and the data rate instruction along with this request. The PMD responds by sending the preamble of the frame at the antenna within 20 microseconds.

Multicast and broadcast frames are generally sent at the lowest basic data rate.

The transmitter sends the preamble at 1 Mbps (802.11 or 802.11b DSSS) or 6 Mbps (802.11a or 802.11g ERP-OFDM). The PHY header is then sent at 1 Mbps (802.11 or 802.11b DSSS) when long preambles are in use, 2 Mbps (802.11b DSSS) when short preambles are in use, or 6 Mbps (802.11a or 802.11g OFDM) when fixed 12-symbol OFDM preambles are in use. These are the lowest supported rates for each PHY and provide a specific common data rate at which receivers listen. After sending the header, the transmitter changes the data rate of the transmission to what the header specifies for transmitting the PSDU. After the PSDU transmission takes place, the PLCP sends a *PHY-TXSTEND.confirm* primitive to the MAC layer, shuts off the transmitter, and switches the PMD circuitry to receive mode.

Receive Function

If the clear channel assessment discovers a busy medium and valid preamble (Sync & SFD) of an incoming frame, the PLCP will monitor the header of frame. The PMD will indicate a busy medium when it senses a signal having a power level of at least -85 dBm. If the PLCP determines the header is error free, the PLCP will send a *PHY-RXSTART.indicate* primitive to the MAC layer to provide notification of an incoming frame. The PLCP sends the information it finds in the frame header (such as the number of octets and data rate) along with this primitive.

The PLCP sets an octet counter based on the value in the PPDU header's Length field (discussed later in this section). This counter will keep track of the number of PSDU octets received, enabling the PLCP to know when the end of the frame occurs. As the PLCP receives data, it sends octets of the PSDU to the MAC layer via *PHY-DATA.indicate* messages. After receiving the final octet, the PLCP sends a *PHY-RXEND.indicate* primitive to the MAC layer to indicate the final octet of the frame.

The receive function will operate with single or multiple antenna diversities. You can select the level of diversity (that is, the number of antennas) via access point and radio card parameters. The strength of the transmitted signal decreases as it propagates to the destination. Many factors, such as the distance, heat, rain, fog, and obstacles may cause this signal degradation. Multipath propagation can also lessen the signal strength at the receiver. Diversity is a method of improving reception by receiving the signal on multiple antennas and processing the superior signal.

DSSS PHY

The IEEE 802.11b Direct Sequence Spread Spectrum (DSSS) Physical layer delivers frames at 1, 2, 5.5, and 11 Mbps rates in the 2.4 GHz ISM band. The original 802.11 Clause 15 DSSS standard specified only 1 and 2 Mbps data rates using only long preambles. The only coding/modulation used in 802.11 Clause 15 is Barker code with DBPSK (1 Mbps) and DQPSK (2 Mbps). Figure 8.3 illustrates the construction of the DSSS PLCP Protocol Data Unit (PPDU), which includes a long preamble, the header, and the MPDU (PSDU) as specified in the 802.11 standard. The preamble and the header are both transmitted at 1 Mbps when using the long preamble format. The MPDU is transmitted at the data rate specified by the transmitting station (or access point). The preamble enables the receiver to synchronize to the incoming signal properly before the actual content of the frame arrives. The header provides information about the frame, and the PSDU is the MPDU the transmitting station is sending.

FIGURE 8.3 DSSS PPDU, 802.11-1999 (R2003)

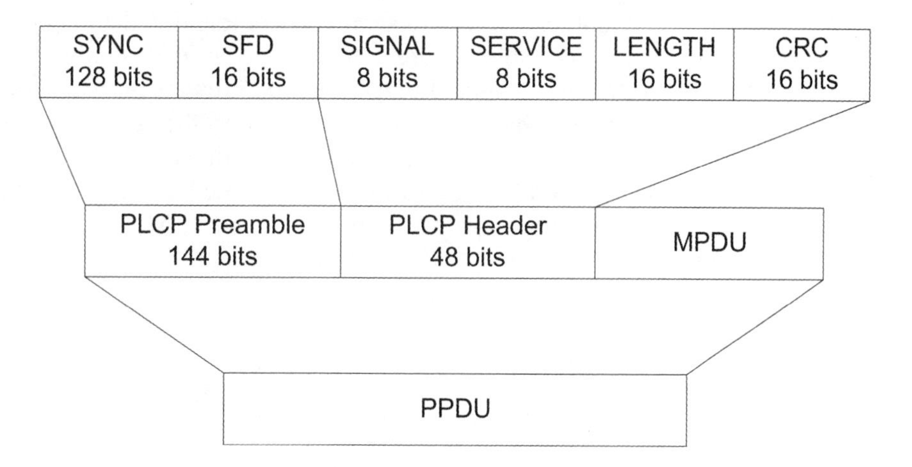

The 802.11b standard further specifies rates of 5.5 and 11 Mbps, each using CCK modulation. The option of a short preamble was introduced in the 802.11b standard, giving the administrator two configuration options.

Figure 8.4 illustrates the same DSSS PPDU specified in the 802.11 standard, but with the newly supported MPDU data rates.[1]

FIGURE 8.4 802.11b, DSSS PPDU, Long Preamble

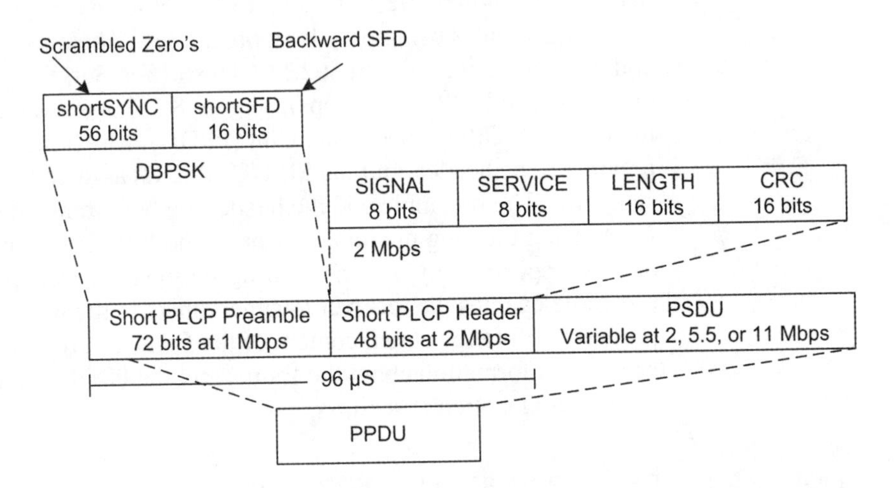

Figure 8.5 illustrates the optional 802.11b DSSS PPDU with the short preamble.[2] Instead of scrambled 1s found in the long preamble Sync field, scrambled 0s are used. The Sync field is only 56 bits instead of the 128 in the original PPDU. The SFD field is presented in reverse bit order. The preamble is transmitted at 1 Mbps, the header at 2 Mbps, and the PSDU (MPDU) at the data rate specified by the transmitting station (or access point). DSSS PPDUs transmitted using short preambles only support PSDU data rates of 2, 5.5, and 11 Mbps.

[1] 802.11b – 1999 (Cor 2001), Section 18.2.2.1
[2] 802.11b – 1999 (Cor 2001), Section 18.2.2.2

FIGURE 8.5 802.11b, DSSS PPDU, Short Preamble

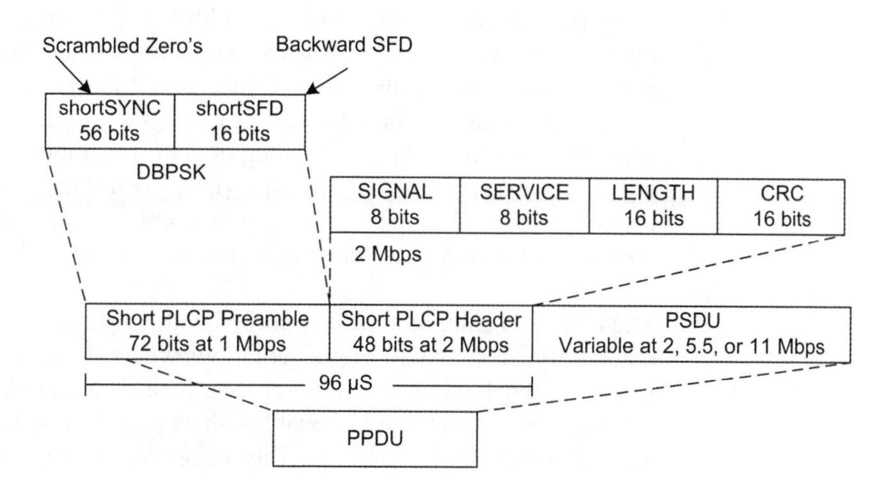

DSSS Preamble

The preamble is the first of three parts of a PPDU. The preamble consists of two parts: The Synchronization (Sync) field and Start Frame Delimiter (SFD) field.

The Sync field consists of a string of 0s or 1s, alerting the receiver that a potentially receivable signal is present. A receiver will begin to synchronize with the incoming signal after detecting the Sync. Consider that receivers may not receive the entire Sync field, but rather only catch part of it. Since the Sync field is a continuous stream of 0s or 1s, it really does not matter where in the stream the receiver realizes that there is a Sync signal being transmitted so long as it synchronizes before the SFD arrives.

The Start Frame Delimiter field defines the beginning of a frame. The bit pattern for this field is always 1111001110100000 when using long preambles and reversed when using short preambles. These patterns are unique to the DSSS PLCP.

Starting with 802.11b, short preambles were optional, and there were various implementations of short preambles in the market. For example, some access points implemented short preambles as, "short preambles only." Other access points implemented short preambles as "short or long preambles are ok." In a, "short preambles only" implementation where the access point is configured for short preambles, a station using long preambles will not be able to associate. In a, "short or long preambles are ok" implementation where the access point is configured for short preambles, stations using either long or short preambles may associate, but the lowest common denominator (long preambles) is always used in the BSS. This is to say that if a long preamble station enters the BSS, the access point will declare that all stations must now use long preambles.

The 802.11g standard made support of both long and short preambles mandatory, such that all implementations where the access point has short preambles enabled mean, "short or long preambles are ok." To see whether the access point has enabled short preamble support, see the *Short Preamble* bit of the Capability Information fixed field.

Beacons & Probe Responses

When only ERP stations are present in the BSS, the access point uses an OFDM PHY (and thus OFDM preambles) for the beacon frames. When a NonERP station associates to the BSS, the access point uses the DSSS PHY (and thus DSSS preambles) for the beacon frames. When the NonERP stations are all short-preamble capable, the access point sends the beacon with a short preamble. When any of the NonERP stations are long-preamble-only capable, the access point sends the beacon using a long preamble. When a NonERP station sends a probe request frame to the access point using a long preamble, the access point must reply with a probe response frame using a long preamble. When a NonERP station sends a probe request frame to the access point using a short preamble, the access point must reply with a probe response frame using a short preamble. This is sometimes considered the "preamble echo" rule, though it is not called by this name in the 802.11 series of standards.

DSSS Header

Signal Field

The Signal field identifies the type of modulation that the receiver must use to demodulate the signal. The value of this field is equal to the data rate divided by 100 Kbps. The only two possible values allowed in the original 802.11 standard were:

Data Rate	Signal Field Value
1Mbps	00001010
2Mbps	00010100

For 802.11 b, the four possible values were:

Data Rate	Signal Field Value
1Mbps	00001010
2Mbps	00010100
5.5Mbps	00110111
11Mbps	01101110

Regardless of the rate or preamble used with DSSS-OFDM the Signal field is set to a 3 Mbps value. That is, the eight-bit value is set to 00011110. With DSSS-OFDM, an optional 802.11g PHY, this value is simply a default setting used for BSS compatibility and to ensure that NonERP stations read the length field and defer the medium for that amount of time even though they cannot demodulate the MPDU due to unsupported rates.

Service Field (802.11 & 802.11b)

The 802.11 standard reserves the Service field for future use; however, a value of 00000000 means 802.11 compliance.[1] The 802.11b standard made use of the Service field as shown in Figure 8.6.[2]

---

[1] 802.11–1999 (R2003), Section 15.2.3.4

[2] 802.11b – 1999 (Cor 2001), Section 18.2.3.4

FIGURE 8.6 802.11b Service Field

b0	b1	b2	b3	b4	b5	b6	b7
Reserved	Reserved	Locked clocks bit 0 = not locked 1 = locked	Modulation Selection bit 0 = CCK 1 = PBCC	Reserved	Reserved	Reserved	Length extension bit

Bit 7 (is used to extend the Length header field). Since both PBCC, an optional modulation type specified in 802.11b and 802.11g, and CCK modulations are supported in 802.11b, bit 3 is used to indicate whether PBCC or CCK is in use. Bit 2 is used to indicate that the transmit frequency and symbol clocks are derived from the same oscillator. This *Locked Clock* bit is set by the PHY layer based on its implementation configuration.

Service Field (802.11g)

Three bits of the Service field have been defined to support the optional modes of the 802.11g standard.[1] Figure 8.7 illustrates the bits within the Service field. Bits b0, b1, and b4 are reserved and are set to 0. Bit b2 is used to indicate that the transmit frequency and symbol clocks are derived from the same oscillator, the same as with 802.11b. For all ERP systems, the Locked Clock Bit is set to 1. Bit b3 is used to indicate if the data is modulated using the optional ERP-PBCC modulation. Bit b3 is defined in section 18.2.3.4 of the 802.11b standard with the caveat that the ERP-PBCC mode now has the additional optional rates of 22 and 33 Mbps in the 802.11g standard.[2] Bits b5, b6, and b7 are used to resolve data field length ambiguities for the optional ERP-PBCC-11, ERP-PBCC-22, and ERP-PBCC-33 modes. These bits are fully defined in 802.11g, Section 19.6. Bit b7, the Length Extension Bit, is also used to resolve data field length ambiguities for the CCK 11 Mbps per 802.11b, Section 18.2.3.5. Bits b3, b5, and b6 are set to 0 for CCK.

---

[1] 802.11g – 2003, Section 19.3.2.1
[2] 802.11g – 2003, Section 19.3.3.2

FIGURE 8.7 802.11g DSSS-OFDM Service Field

b0	b1	b2	b3	b4	b5	b6	b7
Reserved	Reserved	Locked clocks bit 0 = not locked 1 = locked	Modulation Selection bit 0 = Not ERP-PBCC 1 = PBCC	Reserved	Length extension bit (ERP-PBCC)	Length extension bit (ERP-PBCC)	Length extension bit

Length Field

OFDM PHYs treat the Length field as number of octets to transfer between MAC and PLCP as stated above. The DSSS and DSSS-OFDM PHYs are different and treat the Length field as number of microseconds required to transmit the PSDU. The DSSS and DSSS-OFDM PHYs calculate the Length field based on the number of octets presented by the MAC to the PLCP. Note that the length extension bits in the Signal field are not needed or used for DSSS-OFDM. Radio receivers spend most of their time doing carrier sense and clear channel assessment (CS/CCA). If the DSSS PHY header is successfully decoded, the receiver knows how long to spend receiving the rest of this frame. If the signal drops or is otherwise corrupted during that time, CS/CCA alone might conclude that the channel has returned to idle when in fact another station in a better position to receive is still successfully receiving. This means that the DSSS PHY header's Length field is effectively telling the MAC layer how long to consider the medium busy.

The DSSS PHY header's Length field will never show up on a protocol analyzer. It is unrelated to the duration field and NAV timers at the MAC layer.

In packetized RF data transmissions systems, transmitted messages are susceptible to various types of bit errors due to noise, interference, data collisions, and multipath in a given RF channel. The main purpose of error detection algorithms is to enable an RF receiver of a transmitted message to determine if the message is corrupted. There are various types of error detection algorithms to choose from. The most common method for detecting bit errors in messages is through the use of CRCs (Cyclic Redundancy Codes). CRCs are very useful in detecting single bit errors, multiple bit errors, and burst errors in packetized messages. In theory

CRCs could be thought of as simply taking a binary message and dividing it by a fixed binary number, with the remainder being the checksum, or more commonly the CRC. The CCITT CRC-16 is a standardized algorithm with origins to the CCITT standards body. The Signal, Service, and Length fields are all protected with a CCITT CRC-16 frame check sequence (FCS). The CRC operation is done at the transmitting station before scrambling. The Physical layer does not determine whether errors are present within the PSDU. CRC-16 detects all single and double-bit errors and ensures detection of 99.998% of all possible errors. Most experts feel CRC-16 is sufficient for data transmission blocks of 4 kilobytes or less.

DSSS PMD Sublayer

The DSSS PMD performs the actual transmission and reception of PPDUs under the direction of the PLCP. To provide this service, the PMD interfaces directly with the wireless medium (that is, RF in the air) and provides DSSS modulation and demodulation of the frame transmissions.

With direct sequence, the PLCP and PMD communicate via primitives, enabling the DSSS PLCP to direct the PMD when to transmit data, change channels, receive data from the PMD, and so on. The operation of the DSSS PMD translates the binary representation of the PPDUs into a radio signal suitable for transmission. The DSSS Physical layer performs this process by multiplying a radio frequency carrier by a pseudo-noise (PN) digital signal. The resulting signal appears as noise if plotted in the frequency domain. The wider bandwidth of the direct sequence signal enables the signal power to drop below the noise threshold without loss of information.

ERP-OFDM PHY

The two IEEE 802.11 Orthogonal Frequency Division Multiplexing (OFDM) Physical layers each deliver up to 54 Mbps data rates in the 2.4 GHz (802.11g) and 5GHz (802.11a) bands respectively. This section describes the architecture and operation of 802.11 OFDM.

The benefits of OFDM are high spectral efficiency, resiliency to RF interference, and lower multipath distortion. The orthogonal nature of OFDM allows subchannels to overlap, having a positive effect on spectral efficiency. The subcarriers transporting information are just far enough apart to avoid interfering with each other, theoretically.

ERP-OFDM PPDU

Figure 8.8 illustrates the format of an ERP-OFDM PPDU, used both in 802.11a[1] and 802.11g[2]. This is the only PLCP that 802.11a specifies, but one of several specified in the 802.11g standard. ERP-OFDM is, by far, the most often implemented PPDU in the 802.11g standard, and supports data rates of 6, 9, 12, 18, 24, 36, 48, and 54 Mbps. The ERP-OFDM PPDU has three parts: Preamble, Header, and Data Field.

FIGURE 8.8 ERP-OFDM PPDU (802.11a/g)

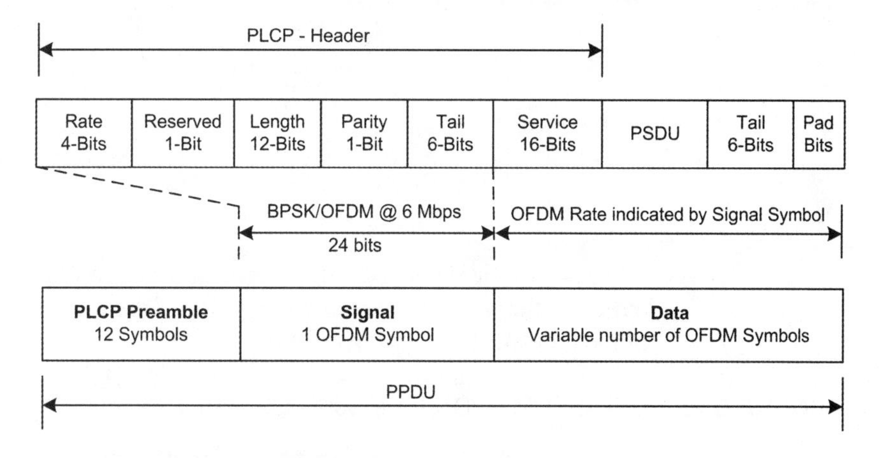

ERP-OFDM PPDU Preamble

The ERP-OFDM PPDU Preamble (Sync) enables the receiver to acquire an incoming OFDM signal (signal detect) and synchronize its demodulator. The preamble consists of 12 training symbols[3], ten of which are short and are used for establishing AGC (automatic gain

[1] 802.11a – 1999, Section 17.3.2
[2] 802.11g – 2003, Section 19.3.2.3
[3] 802.11a – 1999, Section 17.3.3

control), diversity selection, and the coarse frequency offset estimate of the carrier signal. The receiver uses the two long training symbols for channel and fine frequency offset estimation. With the ERP-OFDM preamble, it takes up to 16 microseconds to train the receiver after first detecting a signal on the RF medium.

ERP-OFDM PPDU Header

The ERP-OFDM Header consists of 4 rate bits, 1 reserved bit, 12 Length bits, 1 Parity bit, 6 Tail bits, and 16 Service bits. The Signal field is one symbol (24 bits) long, is not scrambled, and has the same contents as the entire PPDU header minus the Service subfield (16 bits). The Signal field is always transmitted at 6 Mbps using BPSK modulation. This section outlines the significance of each subfield within the Signal field and PPDU Header.

FIGURE 8.9 ERP-OFDM PPDU Header

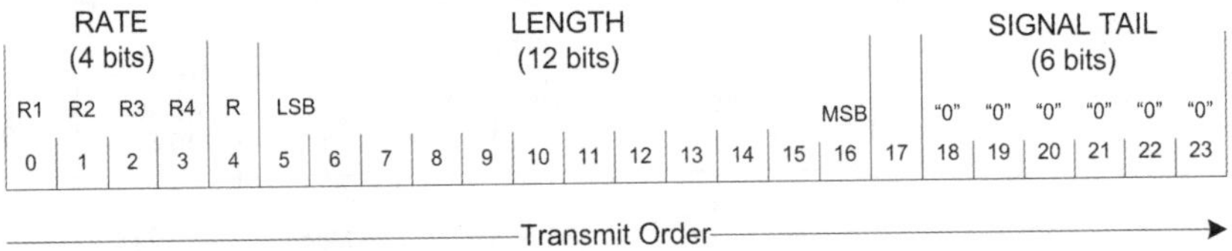

RATE (4 bits)					LENGTH (12 bits)													SIGNAL TAIL (6 bits)					
R1	R2	R3	R4	R	LSB										MSB		"0"	"0"	"0"	"0"	"0"	"0"
0	1	2	3	4	5	6	7	8	9	10	11	12	13	14	15	16	17	18	19	20	21	22	23

Transmit Order →

The Rate subfield consists of 4 bits as outlined in Figure 8.10, and indicates the modulation and coding rate of the rest of the PPDU, starting immediately after the Signal field.

FIGURE 8.10 ERP-OFDM PPDU Rate Subfield

Bits 1-4	Data Rate	Modulation
1101	6 Mbps	BPSK
1111	9 Mbps	BPSK
0101	12 Mbps	QPSK
0111	18 Mbps	QPSK
1001	24 Mbps	16QAM
1011	36 Mbps	16QAM
0001	48 Mbps	64QAM
0011	54 Mbps	64QAM

Some 802.11a chipset manufacturers are using proprietary techniques to combine OFDM channels for applications requiring data rates that exceed 54 Mbps.

The Reserved subfield (1 bit) is set to 0 since it is currently unused. The Length subfield (12 bits) indicates the number of octets in the PSDU that the MAC is currently requesting the PHY to transmit. The Parity subfield is a one bit positive (even) parity bit, based on the first 17 bits (0-16) of the frame (Rate, Reserved, and Length subfields). The Signal Tail subfield is 6 bits, each of which is always set to 0.

ERP-OFDM PPDU Data Field

The Data field consists of the Service subfield, PSDU, Tail subfield, and Pad Bits subfield. The Service subfield consists of 16 bits, with the first 7 bits as zeros to synchronize the descrambler in the receiver. The remaining 9 bits are reserved for future use and set to all 0s. As part of the Data field[1], the Service subfield is transmitted at the rate specified in the Signal field's Rate subfield.

FIGURE 8.11 ERP-OFDM PPDU Service Field

Scrambler Initialization							Reserved SERVICE Bits								
"0"	"0"	"0"	"0"	"0"	"0"	"0"	R	R	R	R	R	R	R	R	R
0	1	2	3	4	5	6	7	8	9	10	11	12	13	14	15

Transmit Order →

The PSDU is the data unit being sent down from the MAC layer for transmission on the wireless medium. The PSDU is transmitted at the data rate specified in the Signal field's Rate subfield and has a maximum length of 4095 octets[2].

The PPDU Tail subfield is 6 bits of 0, which are required to return the convolutional encoder to the "zero state." This procedure improves the

[1] 802.11a – 1999, Section 17.3.5
[2] 802.11a – 1999, Section 17.5.2, Table 93

error probability of the convolutional decoder, which relies on future bits when decoding and which may be not be available past the end of the message. The Tail field is produced by replacing six scrambled "zero" bits following the message end with six non-scrambled "zero" bits.

The Pad Bits subfield contains at least six bits, but it is actually the number of bits that make the Data field a multiple of the number of coded bits in an OFDM symbol (48, 96, 192, or 288).

A data scrambler using a 127-bit sequence generator scrambles all bits in the data field to randomize the bit patterns in order to avoid long streams of ones and zeros. Long streams of ones or zeros may create a DC bias voltage in the receiver circuitry, which may result in receiver errors. The data scrambler "balances" the number of ones and zeros being transmitted between stations.

ERP-OFDM PMD Sublayer

The ERP-OFDM PMD performs the actual transmission and reception of PPDUs under the direction of the PLCP. To provide this service, the PMD interfaces directly with the wireless medium and provides OFDM modulation and demodulation of the frame transmissions.

With ERP-OFDM, the PLCP and PMD communicate via primitives, enabling the DSSS PLCP to direct the PMD when to transmit data, change channels, receive data from the PMD, and so on. The operation of the ERP-OFDM PMD translates the binary representation of the PPDUs into a radio signal suitable for transmission. The ERP-OFDM Physical layer performs this process by dividing a high-speed serial information signal into multiple lower-speed sub-signals that the system transmits simultaneously at different frequencies in parallel.

DSSS-OFDM PHY

The 802.11g standard extended use of the DSSS PHY by specifying an optional PPDU type consisting of the same DSSS preamble and header, but accepting an ERP-OFDM PPDU as its PSDU. The IEEE calls this new PPDU type DSSS-OFDM. Both long and short preambles are supported with DSSS-OFDM, and no protection mechanisms are required by DSSS-OFDM stations when operating with DSSS stations present in the BSA. Figures 8.12 and 8.13 illustrate the construction of both long and short preamble format DSSS-OFDM PPDUs. The preamble and header transmission rates apply to DSSS-OFDM as with DSSS.

FIGURE 8.12 802.11g, DSSS-OFDM PPDU, Long Preamble

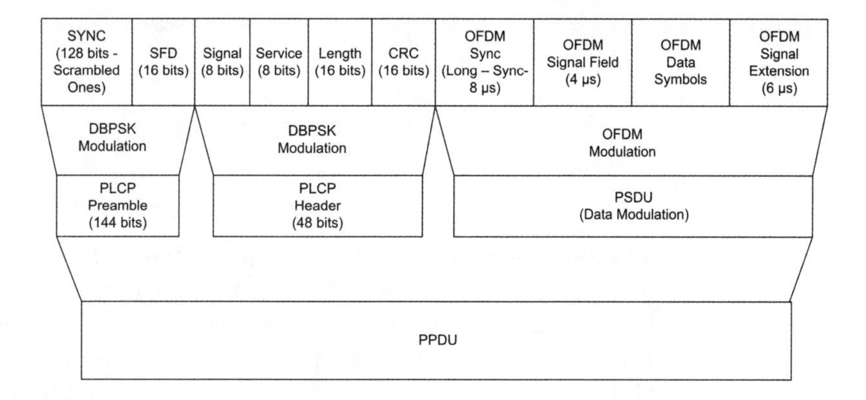

FIGURE 8.13 802.11g, DSSS-OFDM PPDU, Short Preamble

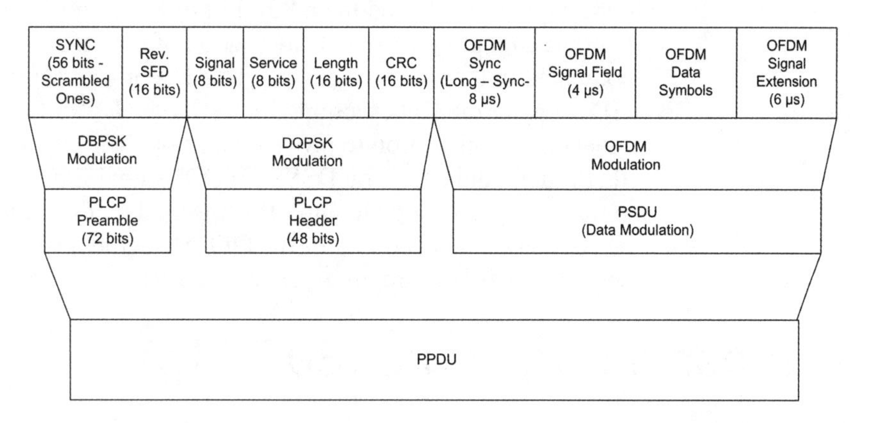

This section illustrates the format of the PSDU portion of the DSSS-OFDM PPDU. Figure 8.14 shows an expanded view of the DSSS-OFDM PSDU.

FIGURE 8.14 DSSS-OFDM PSDU Format

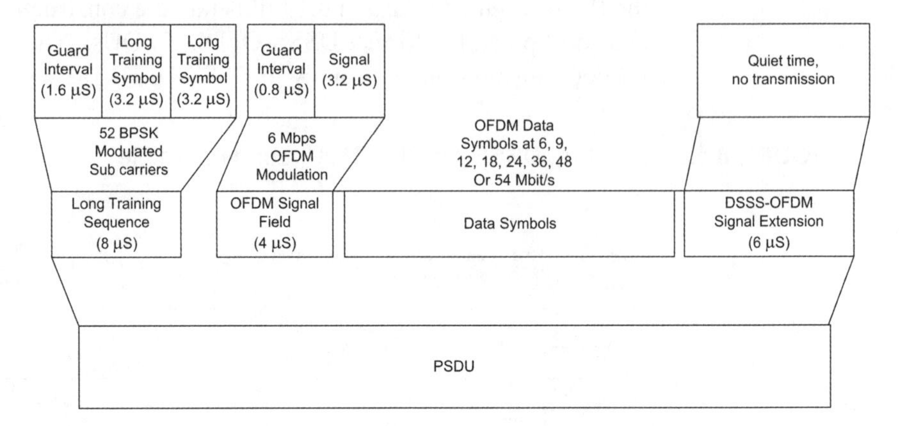

The PSDU is composed of four major sections. The first is the long sync training sequence that is used for acquisition of receiver parameters by the OFDM demodulator. The long sync training sequence for DSSS-OFDM is identical to the long training symbols of the 802.11a and 802.11g ERP-OFDM preamble. The second section is the OFDM Signal field that provides the demodulator information on the OFDM data rate and length of the OFDM data section. The Signal field for DSSS-OFDM is identical to the Signal field found in an 802.11a or 802.11g ERP-OFDM header. After the Signal field is the Data section of the PSDU. This section is modulated in the same way as any 802.11a or 802.11g ERP-OFDM PSDU. After the Data section, the PSDU for DSSS-OFDM appends a signal extension section to provide additional processing time for the OFDM demodulator. The DSSS-OFDM Signal Extension is a period of no transmission of 6 µs length. It is inserted to allow more time to finish the convolutional decoding of the OFDM segment waveform and still meet the 10 µs SIFS requirement of the ERP.

Transmit Procedure (802.11g)

The transmit procedure depends on the data rate and modulation format requested. For data rates of 1, 2, 5.5, 11, 22, and 33 Mbps, the PLCP transmit procedure is the same as for 802.11b. For the ERP-OFDM mandatory rates of 6, 12, and 24 and the optional rates of 9, 18, 36, 48, and 54 Mbps the PLCP transmit procedure is the same as for 802.11a. The transmit procedures for the optional DSSS-OFDM mode using the long or short PLCP preamble and header are the same as those described in 802.11b for the preamble and header and 802.11a for the PSDU.

Receive Procedure (802.11g)

An ERP receiver should be capable of receiving 1, 2, 5.5, and 11 Mbps PLCPs using either the long or short preamble formats described in 802.11b, and should be capable of receiving 6, 12, and 24 Mbps using the modulation and preamble described in 802.11a. The PHY may also implement the ERP-PBCC modulation at rates of 5.5, 11, 22, and 33 Mbps; the ERP-OFDM modulations at rates of 9, 18, 36, 48, and 54 Mbps; and/or the DSSS-OFDM modulation rates of 6, 9, 12, 18, 24, 36, 48, and 54 Mbps. A receiver should be capable of detecting the preamble type (ERP-OFDM, Short Preamble, or Long Preamble) and the modulation type. Upon the receipt of a PPDU, the receiver should first distinguish between the ERP-OFDM preamble and the single carrier modulations (long or short preamble). In the case where the preamble is an ERP-OFDM preamble, the PLCP receive procedure should follow that of the 802.11a standard. Otherwise, the receiver should then distinguish between the long preamble and short preamble as specified in the 802.11b standard. The receiver should then demodulate the Service field to determine the modulation type. For short preamble and long preamble using DSSS, CCK, or PBCC modulations, the receiver should then follow the receive procedure described in 802.11b.

A receiver that supports DSSS-OFDM is capable of receiving all rates specified by 802.11 DSSS (1 & 2 Mbps) and all mandatory rates in 802.11a (6, 12, & 24 Mbps) and 802.11b (1, 2, 5.5, & 11 Mbps). If the Signal field indicates 3 Mbps, the receiver should attempt to receive a DSSS-OFDM frame. The remaining receive procedures for a DSSS-OFDM-capable receiver are the same as those described in 802.11b, and they do not change apart from the ability to receive DSSS-OFDM in the PSDU.

Summary

The 802.11 series of physical layer specifications includes a variety of options that govern the transmission and reception of frames. There are several 802.11 series PHYs, such as FHSS, DSSS, HR-DSSS, ERP-OFDM, DSSS-OFDM, and ERP-PBCC. Each PHY layer has a particular PLCP, which defines framing, and PMD that defines signal modulation. Understanding the differences and interactions between each PHY will allow the analyst to better design, baseline, and troubleshoot WLANs of various types, even in mixed environments.

Key Terms

Before taking the exam, you should be familiar with the following terms:

Direct Sequence Spread Spectrum (DSSS)

DSSS-OFDM

ERP-OFDM

Frequency Shift Keying (FSK)

long preamble

Orthogonal Frequency Division Multiplexing (OFDM)

PLCP Header

Physical Layer Convergence Procedure (PLCP)

Physical Layer Service Primitives

Physical Medium Dependent (PMD)

Quadrature Amplitude Modulation (QAM)

service primitives

short preamble

Review Questions

1. The MAC sublayer and the PLCP sublayer coordinate transmission of frames to and from the wireless medium using what?

2. An ERP-OFDM PPDU is comprised of what three parts?

3. In the DSSS PLCP header, what purpose does the length field serve?

4. When a short-preamble-capable NonERP station associates with an 802.11g access point while all other stations are ERP-OFDM capable, what PHY does the access point use for transmitting beacons?

5. The scrambler and descrambler serve what purpose in wireless LAN transmitters and receivers?

6. An ERP-OFDM Preamble uses how many symbols for training the receiver?

7. When transmitting a DSSS PSDU across the wireless medium using short preambles, what is the lowest supported data rate for the PSDU?

8. A DSSS PLCP header is protected from in-transit bit-flipping attacks by which field?

9. DSSS-OFDM supports what two preamble lengths?

10. An ERP-OFDM header is transmitted at what data rate?

802.11 System Architecture

CWAP Exam Objectives Covered:

- This chapter describes the 802.11 system architecture to provide a foundation for exam objectives covered in later chapters.

CHAPTER

9

In This Chapter

There are many types of equipment used in the wireless LAN market, not just generic access points and client devices. Knowing the architecture of the various infrastructure devices allows an analyst to understand how the equipment should operate versus how it may currently be operating. Wireless LAN systems can be designed in many ways using a variety of equipment. You need to know what processes are happening on and between devices. Security design and protocols play into troubleshooting with analyzers, and knowing what should be seen on the analyzer is important when trying to locate a problem. This chapter covers many different architectures to help you understand the different ways analyzers can be used.

Access Point Architecture

The 802.11 standard does not specify functionality of and interaction with non-802.11 LANs. In most cases, however, the non-802.11 LAN is Ethernet (IEEE 802.3). For example, the portal function of an access point allows 802.11 mobile stations to interface with stations connected to an Ethernet network. For this connection to occur properly, access points must translate 802.11 frames into 802.3 (Ethernet) frames and vice versa. The 802.11 standard defines the portal and access point as two different elements, but they often reside within the same hardware. A pure access point (without a portal), which does not perform translation, is rare. The more common implementation is to have both the portal and access point elements within the same hardware. In fact, we generally use the term "access point" to mean a single piece of hardware that includes a portal.

Message Forwarding

Access points forward messages between the wired and wireless networks, operating as a translational bridge. The access points will forward messages in unicast frames if the destination MAC address is on the other side of the access point. For example, the access point will forward the message in a data frame from the wireless to the wired side of the access point if a mobile 802.11 station is sending the message to an Ethernet client residing on the wired network.

Access points will always forward messages in broadcast and multicast frames, unless configuration settings on the access point disable this function. Many network devices and applications periodically send "heartbeats" in multicast frames, which can place a large amount of overhead on both the wireless and wired side of the access point. If forwarding these types of messages causes significant degradation in throughput, then consider disabling the broadcast and multicast forwarding. Be sure, however, that disabling this functionality still allows applications to function properly.

Keep in mind that when using wireless protocol analyzers, you will not be able to view the broadcast/multicast frames as they originate on the wired side of the network. If you suspect that a problem lies on the wired side of the access point, then you'll need to utilize an Ethernet analyzer that can connect to and decode Ethernet networks.

Frame Translation

In addition to message forwarding, the access point performs frame translation between the wireless 802.11 medium and the wired (likely 802.3 Ethernet) medium. This is necessary to map the contents of fields of one protocol to fields of another protocol. This includes both the MAC frame header and the 802.2 LLC header that is part of every 802.11 frame body but not part of most 802.3 frame bodies.

To understand frame translation, it is helpful to understand Ethernet frame formats. Four different frame formats are used on an Ethernet network: LLC, SNAP, Novell Proprietary, and Type-2. The complete structure of these formats is not relevant at this point.

Frame formats relate to frame translation because the Type-2 and Novell Proprietary frame formats are only supported on Ethernet networks and not other 802 networks such as 802.11. It was to address this issue that the IEEE created the SNAP frame format. The SNAP format encapsulates the relevant parts of a Type-2 header in an 802-compliant frame, allowing protocols that want to use Type-2 to send traffic on networks that do not support Type-2 (such as 802.11).

If an 802.11 station talked to an 802.3 station using a frame format that was common to both, such as LLC, then translation would be very simple, but on Ethernet, TCP/IP protocol stacks nearly always use the Type-2 frame format. This means that more complex translation is almost always necessary when an 802.11 station talks to an 802.3 station (or vice versa) using TCP/IP.

The IETF RFC 1042, published in 1988, describes how TCP/IP protocols should be translated to SNAP frames so that they can be transmitted on 802 networks that do not support the Type-2 frame format. The frames are translated according to RFC 1042 and then the translating device (the access point, in the case of 802.11) bridges the frame onto the destination network. Although the RFC says nothing about bridging between these dissimilar data-links, bridging IP packets between Ethernet and 802.11 is not a difficult task.

Some protocols use the SNAP frame format natively on Ethernet. Instead of using SNAP as a fall-back method of encapsulating a Type-2 header, in this case, SNAP is the preferred frame format of the protocol. Translating these frames is simple when they go from Ethernet to 802.11, since both Ethernet and 802.11 support the SNAP header. When frames go from 802.11 to Ethernet, the access point must decide whether to pass the frames through as SNAP or convert them to Type-2. Remember that, if a Type-2 Ethernet transmitter talked to an 802.11 station, the access point would convert the Type-2 frames to SNAP. In the other direction, the access point would convert SNAP back to Type-2. But if the Ethernet station had been using SNAP originally, then converting to Type-2 would be the wrong decision, and the Ethernet station potentially wouldn't understand the received frame.

Some mechanism is needed to decide whether to convert 802.11 SNAP to 802.3 SNAP or convert from 802.11 SNAP to 802.3 Type-2. The logic for making this decision is known as "selective LLC translation," and is described in the IEEE 802.1H-1997 standard. Alternatively, an 802.11 access point may simply translate all Type-2 frames to 802.11 SNAP and all 802.11 SNAP frames to Type-2. In most cases this will work, since very few protocols use SNAP natively. This is the approach taken by many access point manufacturers.

802.1H provides a way to tunnel Type-2 based protocols through an intermediary data link to another Ethernet segment. This uses a special LLC/SNAP header of 0xAAAA030000F8 rather than the usual 0xAAAA03000000. The second bridge takes this special header to mean "convert back to Type-2 without regard to the protocol in the rest of the frame body." A wireless bridge may transmit such an "Ethernet tunnel" frame on an 802.11 network to another wireless bridge leading to a remote Ethernet segment.

Refer to Figures 9.1, 9.2, and 9.3 for illustrations of the frame translation types in action.

FIGURE 9.1 Example of ARP Frame (Version II Ethernet)

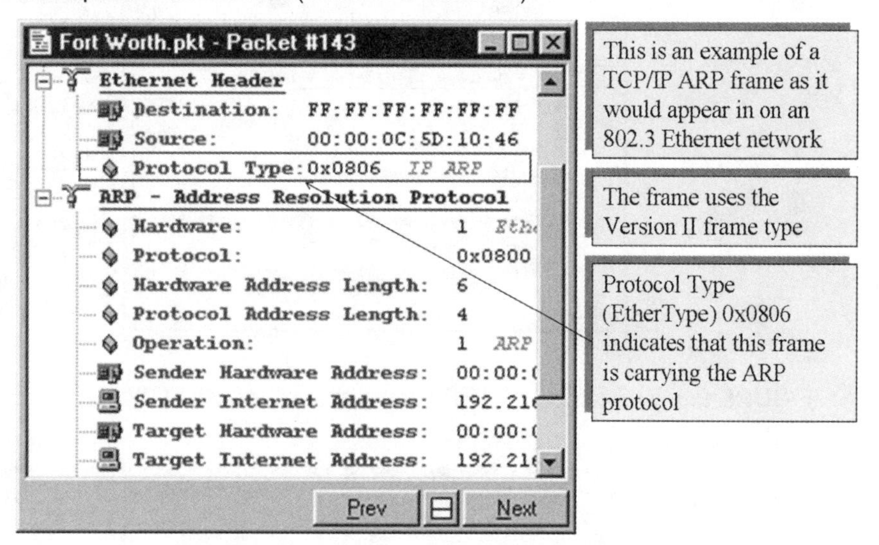

FIGURE 9.2 Example of ARP Frame (translated with RFC 1042)

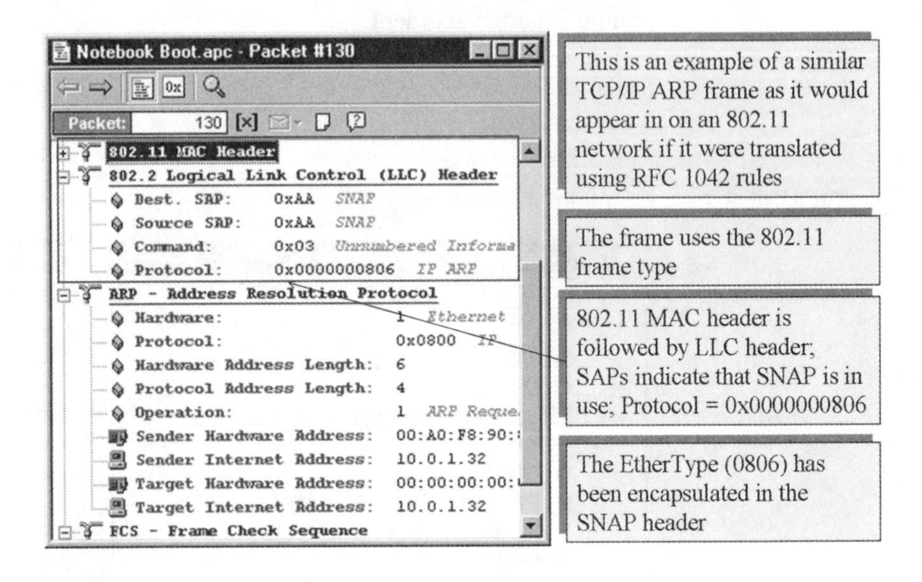

FIGURE 9.3 Example of ARP Frame (translated with 802.1H)

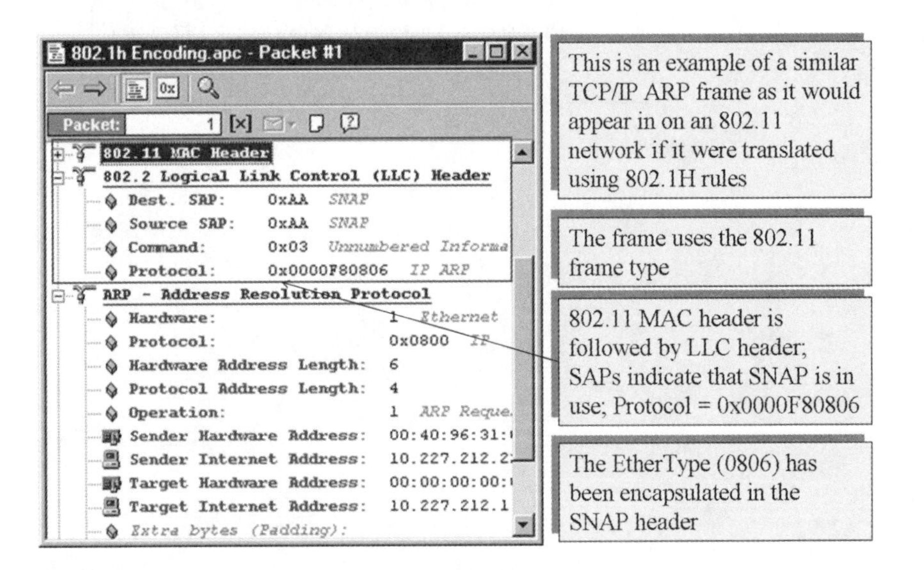

Proprietary Enhancements

Vendors incorporate specific proprietary enhancements into access points in order to compete in the market, and these enhancements often go far beyond what the 802.11 series of standards defines. As a result, these enhancements often cause interoperability problems that require detailed analysis to solve.

Auto-channel Selection

Some access points will automatically set the RF channel after monitoring the presence of RF activity on each of the available channels. The access point will then tune to the least congested channel. When using a wireless protocol analyzer, troubleshooting a conversation between a station and an access point that is changing channels by itself may become quite difficult. Access points that automatically change channels may additionally cause problems when performing a baseline analysis or when using wireless intrusion detection systems. Figure 9.4 illustrates an auto-channel selection feature in an access point.

FIGURE 9.4 Auto-channel Selection Feature

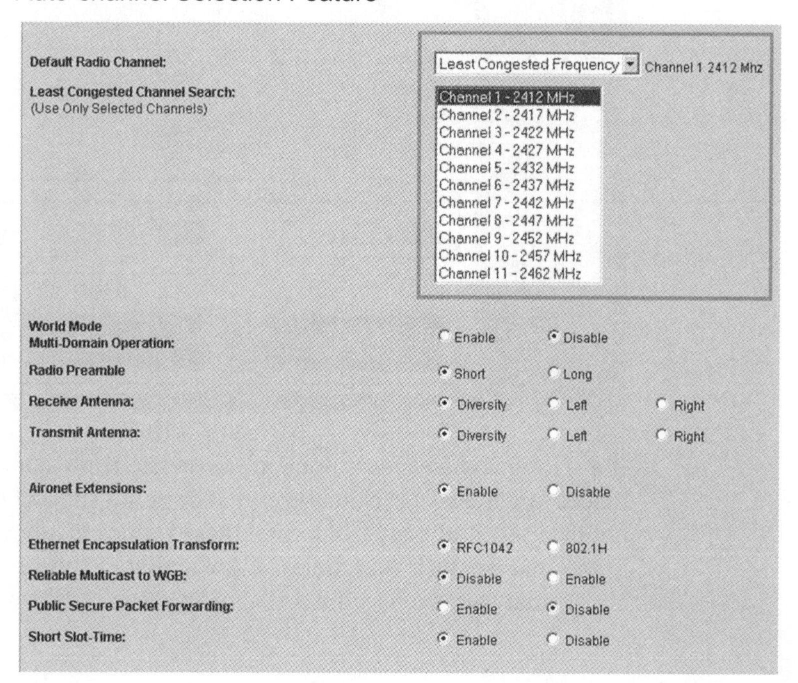

Performance Enhancements

Many of the SOHO class access points offer a form of "turbo" or "Super A/G" operation, which increases the data rate beyond what the particular 802.11 standard defines. This proprietary enhancement usually requires, however, that the radio cards and the access point be from the same vendor. In cases where "static turbo" mode is selected, associations with non-turbo stations will be blocked by the access point. Some access points have a "dynamic turbo" mode that allows a mix of both turbo and non-turbo mode stations, but turbo stations will be limited to 54 Mbps operation when non-turbo client stations are present and operating. Figure 9.5 illustrates an 802.11g turbo 108 Mbps connection centering on the center frequency of channel 6 (2347 MHz), achieving approximately 37.5 Mbps of throughput. Turbo mode can be used with 802.11g or 802.11a channels.

FIGURE 9.5 Turbo Mode Client Operation

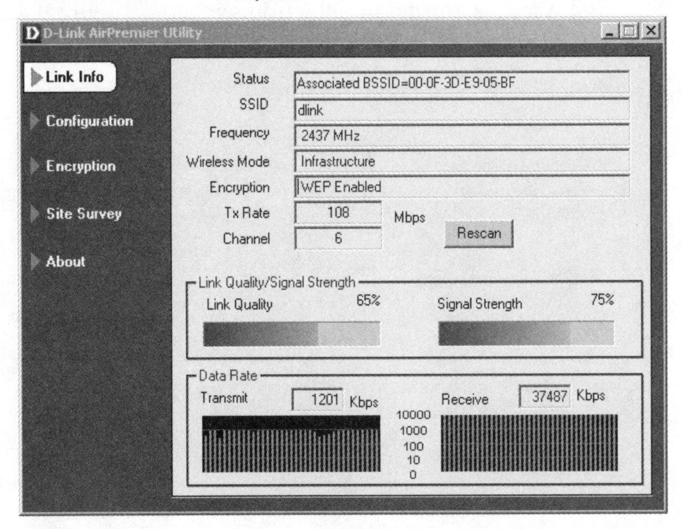

If a station and an access point are associated and using "Super A/G" modes, a wireless LAN analyzer that does not support these modes will not receive or attempt to interpret these frames in any way. It will appear to the analyzer as if no wireless LAN activity is present. However, use of an RF analyzer, such as the software spectrum analyzer shown below, will reveal the presence of a wireless LAN system using a large portion of the frequency band.

FIGURE 9.6 802.11g Spectrum Analysis

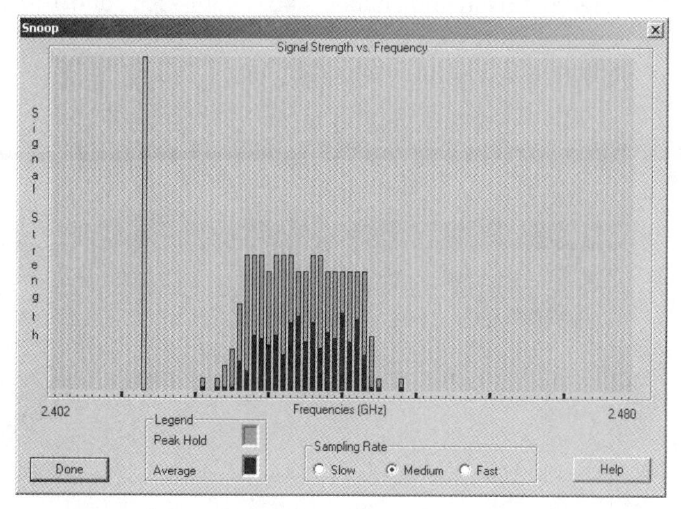

FIGURE 9.7 Turbo 802.11g Mode Spectrum Analysis

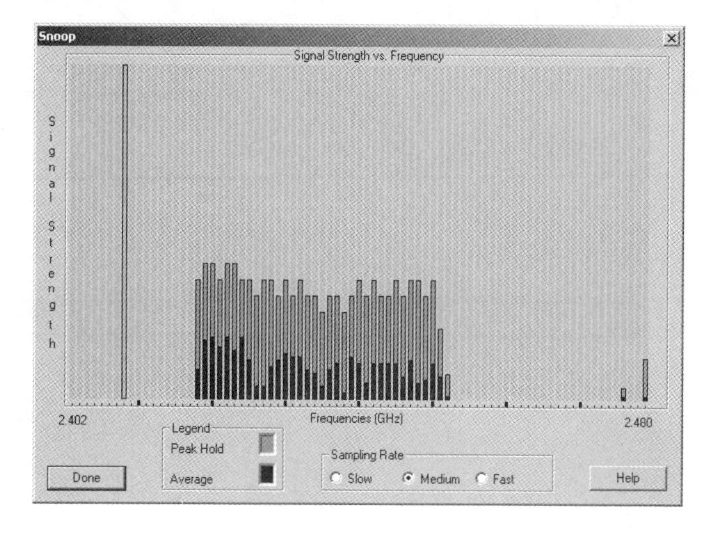

In Figure 9.6, only 20 MHz of the 2.4 GHz ISM band is in use. In Figure 9.7, twice as much spectrum is being used to produce double the throughput available with 802.11g standard compliant equipment. Most wireless analyzers support capturing on these designated turbo mode channels when using radio cards that support Super-A and Super-G modes. These designated channels are comprised of channel 6 in the 2.4 GHz ISM band and channels 42, 50, 58, 152, and 160 in the 5 GHz UNII bands. Figures 9.8 and 9.9 illustrate configuration of two analyzers to capture 802.11a turbo mode frames.

FIGURE 9.8 Turbo 802.11a Capture Configuration Example #1 – Commview for WiFi

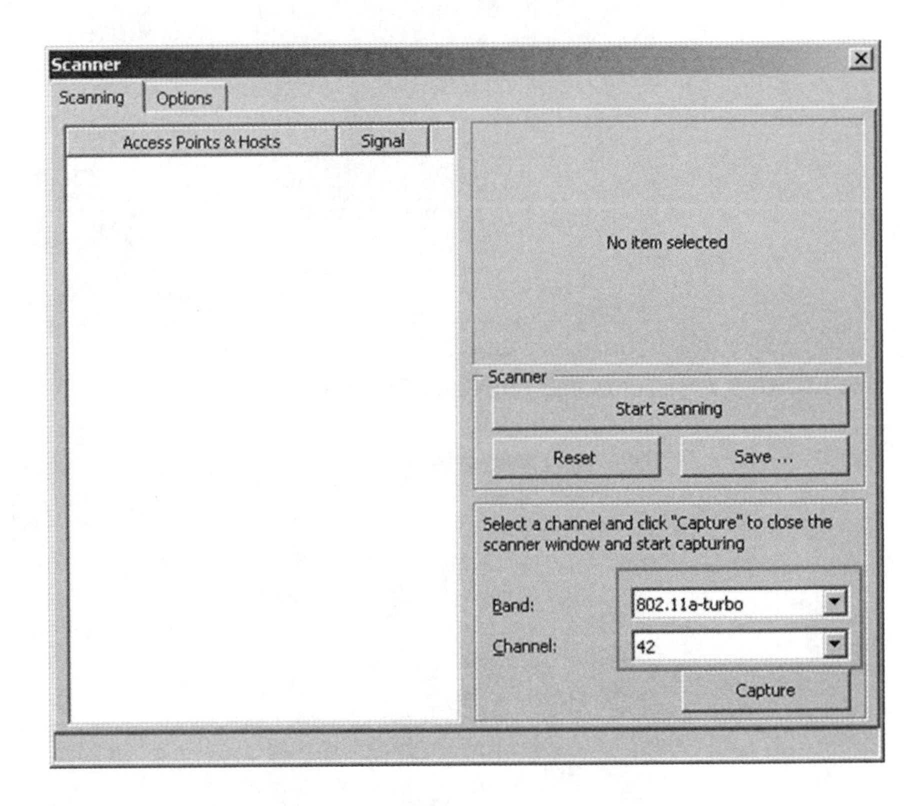

FIGURE 9.9 Turbo 802.11a Capture Configuration Example #2 – AiroPeek NX

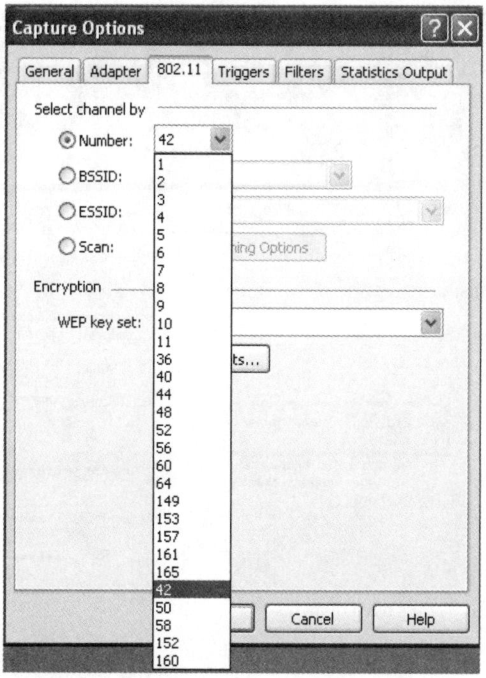

SSID Broadcasting

Some access points allow administrators to disable the broadcasting of the SSID in beacon frames. Additionally, these same access points often have functionality allowing the administrator to control whether the access point responds to probe request frames that have null (broadcast) SSID fields. Both of these features violate the 802.11 standard. They were created as a first response to the weaknesses of WEP and remain popular "security by obscurity" features despite the arrival of WPA in 2003 and the 802.11i amendment in 2004. Figure 9.10 illustrates an access point's configuration option of hiding the SSID.

FIGURE 9.10 SSID Broadcasting Option

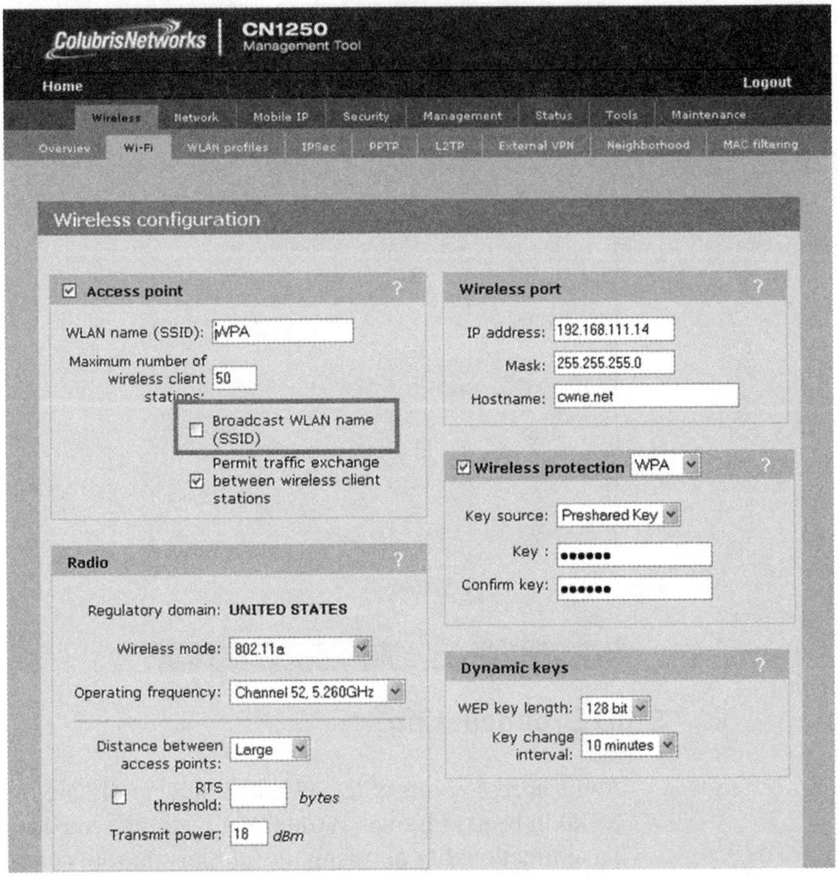

Figure 9.11 illustrates a beacon management frame with a null SSID field. When the SSID is removed from the beacon frames, however, the mobile stations may experience difficulties roaming from one access point to another. For instance, some radio cards will scan other channels and only identify candidate access points for reassociation if the SSID found in a beacon frame matches the one configured in the radio card. If the SSID is not within the beacon, then the radio card may ignore that particular access point as a possible access point to which to roam.

FIGURE 9.11 Beacon frame with null SSID field

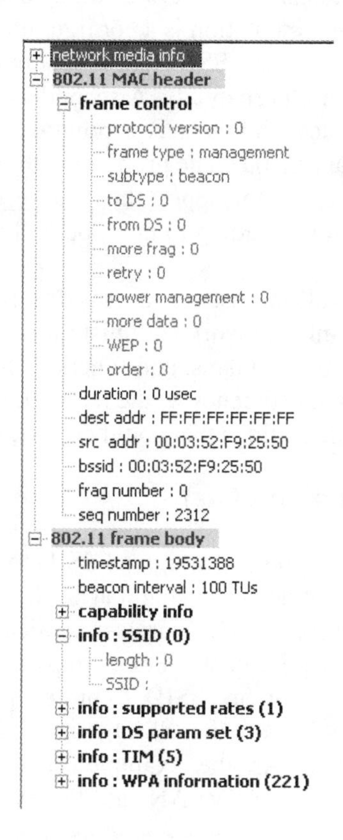

Multiple SSIDs

Some enterprise-class access points and wireless LAN switches support multiple simultaneous SSIDs. This feature logically divides the BSS with one BSSID and one series of beacons into subsets of stations that share a common cipher suite and multicast key. By binding each SSID to an Ethernet 802.1Q VLAN in the access point or wireless LAN switch, the members of one SSID form an extension of that VLAN's broadcast domain. Only one SSID is carried in the beacons; the other SSIDs are hidden. Many end users want to take advantage of this technology because using access points to support more than one security profile, such as public Internet access and inventory control, increases flexibility and keeps costs down by not requiring multiple wireless LAN systems.

Additionally, using multiple simultaneous SSIDs assures that existing systems are used to their fullest extent before adding additional hardware. An example of such an implementation is a configuration in which the administrator can choose a primary SSID that is sent in every beacon (assuming SSID broadcasting is enabled). A secondary SSID can be assigned to the radio interface, and the access point will respond to probe requests using the secondary SSID. The problem with the multiple SSID feature is that no standard exists to support this feature; therefore, vendors have varying implementations, leading to interoperability problems.

The use of multiple SSIDs allows users to access different networks through a single access point. Network administrators can assign different policies, security mechanisms, and functions for each SSID, increasing the flexibility and efficiency of the network infrastructure. The following are some possible settings that could be assigned to each SSID:

Virtual Local Area Networks (VLANs)

If the network uses VLANs, you can assign an SSID to VLAN-1, and the access point will group client devices using that SSID into VLAN1. This configuration enables the separation of wireless applications based on security and performance requirements. For example, you could enable encryption and authentication on one SSID, to protect private applications, and no security on another SSID to maximize open connectivity for public usage. Another example is using VoWLAN networks with QoS features on one VLAN, and Data with no QoS features on another VLAN.

SSID broadcasting

In some cases, such as public Internet access applications, access points are set to broadcast the SSID to enable user radio cards to automatically find available access points. For private applications, some people think it is sometimes better to not broadcast the SSID for security reasons: a visible SSID invites intruders by serving as an easy target. Multiple SSIDs means you can mix and match the broadcasting of SSIDs.

Maximum number of client associations

Some companies set the maximum number of users that can associate via a particular SSID, which makes it possible to control usage of particular applications. This control mechanism can help provide a somewhat limited form of bandwidth control for particular applications.

The use of multiple SSIDs means more flexibility when deploying a shared wireless LAN infrastructure. Instead of supporting only one type of application, possibly one that requires significant authentication and encryption, the wireless LAN can also maintain other applications that do not require such stringent controls. For example, the access point could support both public and operational users from a single access point.

The benefits of a shared infrastructure are certainly cost savings and enabling of mobile applications. Rather than having two separate wireless LANs (which probably is not feasible), a company can deploy one wireless LAN to satisfy all requirements. Sometimes a company needs to have several applications supported together to make the costs of deploying a wireless LAN feasible.

Repeater Access Point Operation

Access points, which require interconnecting cabling, generally play a dominant role for providing coverage in most wireless LAN deployments. Wireless repeater access points, though, are an alternative way to extend the range of an existing wireless LAN instead of adding more cable infrastructure for cable attached access points. Figure 9.12 illustrates the operation of a repeater. There are very few single purpose 802.11 wireless repeater access points on the market, but some access points have an optional repeater mode that can be manually enabled. One problem is that repeater access points can significantly decrease the performance capability of a wireless LAN, which is why wireless LAN analysts should be aware of them.

FIGURE 9.12 Repeater Access Point Architecture

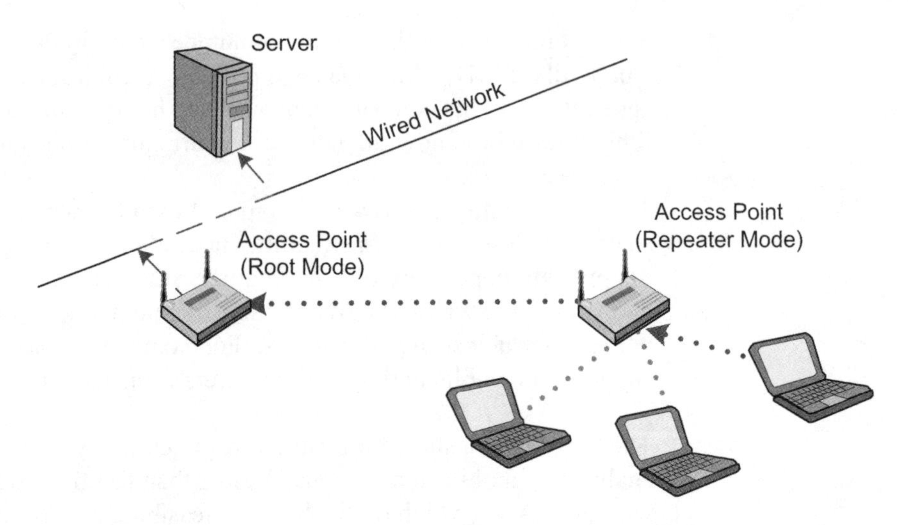

A wireless LAN repeater does not physically connect by wire to any part of the network. Instead, it receives radio signals (802.11 frames) from an access point, end user device, or another repeater. In general, a repeater has two important functions. First, it acts as an access point to downstream stations. Second, it performs wireless distribution system services, transmitting frames that originated at downstream stations to an associated upstream root access point.

This network configuration allows network designers to extend the range of the existing network infrastructure. Repeaters provide connectivity to remote areas that normally would not have wireless network access. You may have one access point in a home or small office that doesn't quite cover the entire area where users need connectivity, such as a basement or patio. The placement of a repeater between the areas that have RF coverage and those areas that do not have RF coverage, however, will provide connectivity throughout the entire space. The wireless repeater fills holes in coverage, enabling seamless roaming.

Most wireless LAN repeaters currently available today are actually built-in functions of access points, but stand alone repeaters do exist in the market. The advantage of the stand alone repeaters is that they are generally less expensive. One downside of wireless repeaters, though, is that they reduce throughput on the wireless LAN. A repeater must receive and subsequently retransmit each message on the same RF channel, which effectively doubles the number of frames that are sent on that channel. This functionality significantly lowers performance. This problem compounds when using multiple repeaters, so be sure to plan the use of repeaters sparingly.

Figures 9.13 and 9.14 illustrate repeater mode operation. All four address fields are used in MPDUs between the access points, and the ToDS and FromDS fields are both 1. A repeater network, where a non-root access point is repeating frames to a root access point is a wireless distribution system (WDS). Three addresses are used between each wireless client station and its associated access point. Acknowledgements (ACKs) have only one address. Frames transmitted between access points use four addresses.

FIGURE 9.13 Repeater Operation - Addressing

Packet	Address 1	Address 2	Address 3	Address 4	Data Rate	Size	Protocol
1	00:0D:65:C9:32:76	00:0A:8A:47:BF:4A	00:0A:8A:47:BC:1A		11.0	96	PING Req
2	00:0A:8A:47:BF:4A				11.0	14	802.11 Ack
3	00:0D:ED:A5:4F:70	00:0D:65:C9:32:76	00:0A:8A:47:BC:1A	00:0A:8A:47:BF:4A	11.0	102	PING Req
4	00:0D:65:C9:32:76				11.0	14	802.11 Ack
5	00:0A:8A:47:BC:1A	00:0D:ED:A5:4F:70	00:0A:8A:47:BF:4A		11.0	96	PING Req
6	00:0D:ED:A5:4F:70				11.0	14	802.11 Ack
7	00:0D:ED:A5:4F:70	00:0A:8A:47:BC:1A	00:0A:8A:47:BF:4A		11.0	96	PING Reply
8	00:0A:8A:47:BC:1A				11.0	14	802.11 Ack
9	00:0D:65:C9:32:76	00:0D:ED:A5:4F:70	00:0A:8A:47:BF:4A	00:0A:8A:47:BC:1A	11.0	102	PING Reply
10	00:0D:ED:A5:4F:70				11.0	14	802.11 Ack
11	00:0A:8A:47:BF:4A	00:0D:65:C9:32:76	00:0A:8A:47:BC:1A		11.0	96	PING Reply
12	00:0D:65:C9:32:76				11.0	14	802.11 Ack
13	00:0D:65:C9:32:76	00:0A:8A:47:BF:4A	00:0A:8A:47:BC:1A		11.0	96	PING Req
14	00:0A:8A:47:BF:4A				11.0	14	802.11 Ack
15	00:0D:ED:A5:4F:70	00:0D:65:C9:32:76	00:0A:8A:47:BC:1A	00:0A:8A:47:BF:4A	11.0	102	PING Req
16	00:0D:65:C9:32:76				11.0	14	802.11 Ack
17	00:0A:8A:47:BC:1A	00:0D:ED:A5:4F:70	00:0A:8A:47:BF:4A		11.0	96	PING Req
18	00:0D:ED:A5:4F:70				11.0	14	802.11 Ack
19	00:0D:ED:A5:4F:70	00:0A:8A:47:BC:1A	00:0A:8A:47:BF:4A		11.0	96	PING Reply
20	00:0A:8A:47:BC:1A				11.0	14	802.11 Ack
21	00:0D:65:C9:32:76	00:0D:ED:A5:4F:70	00:0A:8A:47:BF:4A	00:0A:8A:47:BC:1A	11.0	102	PING Reply
22	00:0D:ED:A5:4F:70				11.0	14	802.11 Ack
23	00:0A:8A:47:BF:4A	00:0D:65:C9:32:76	00:0A:8A:47:BC:1A		11.0	96	PING Reply
24	00:0D:65:C9:32:76				11.0	14	802.11 Ack
25	00:0D:65:C9:32:76	00:0A:8A:47:BF:4A	00:0A:8A:47:BC:1A		11.0	96	PING Req
26	00:0A:8A:47:BF:4A				11.0	14	802.11 Ack
27	00:0D:ED:A5:4F:70	00:0D:65:C9:32:76	00:0A:8A:47:BC:1A	00:0A:8A:47:BF:4A	11.0	102	PING Req
28	00:0D:65:C9:32:76				11.0	14	802.11 Ack
29	00:0A:8A:47:BC:1A	00:0D:ED:A5:4F:70	00:0A:8A:47:BF:4A		11.0	96	PING Req
30	00:0D:ED:A5:4F:70				11.0	14	802.11 Ack
31	00:0D:ED:A5:4F:70	00:0A:8A:47:BC:1A	00:0A:8A:47:BF:4A		11.0	96	PING Reply
32	00:0A:8A:47:BC:1A				11.0	14	802.11 Ack
33	00:0D:65:C9:32:76	00:0D:ED:A5:4F:70	00:0A:8A:47:BF:4A	00:0A:8A:47:BC:1A	11.0	102	PING Reply
34	00:0D:ED:A5:4F:70				11.0	14	802.11 Ack
35	00:0A:8A:47:BF:4A	00:0D:65:C9:32:76	00:0A:8A:47:BC:1A		11.0	96	PING Reply
36	00:0D:65:C9:32:76				11.0	14	802.11 Ack

FIGURE 9.14 Repeater Access Point Operation – Header Fields

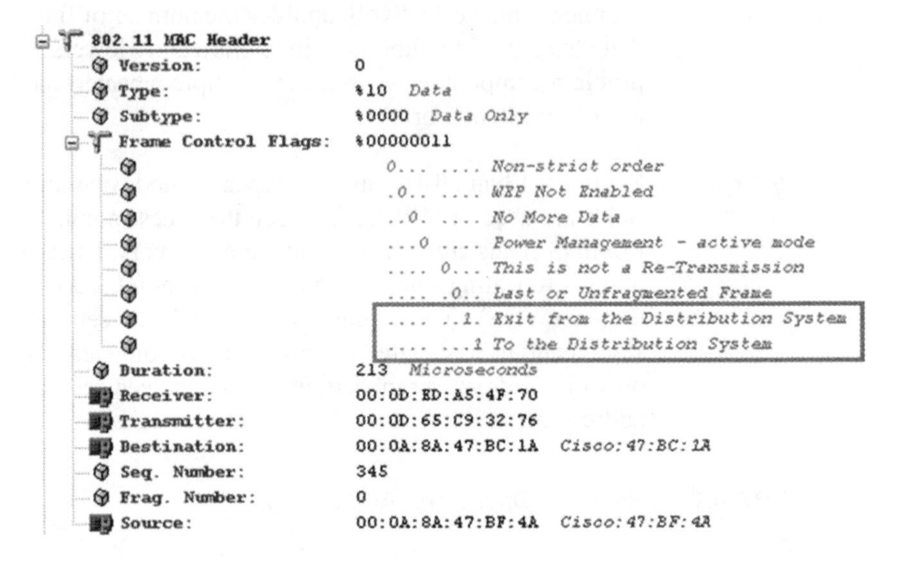

Generally speaking, wireless repeater access points are an excellent way to increase the radio range of an existing wireless LAN, especially if it is not practical to install an additional wired access point to fully cover the location. Do not install too many repeater access points, or you will see your overall wireless network throughput drop significantly.

Security Measures

Link layer (layer 2) security protocols are used directly between wireless stations and the access point. Cipher suites include CCMP, TKIP, and WEP. CCMP uses an AES block cipher, and TKIP and WEP use an RC4 stream cipher. CCMP is specified by the 802.11i amendment and the WPA2 interoperability standard. TKIP is specified by the WPA interoperability standard, and WEP is specified by the original 802.11 standard. WEP calls for out-of-band pre-sharing of keys, while WPA and WPA2 call for either pre-sharing of out-of-band keys or key management automation based on AAA service. Authentication and key management generally takes the form of 802.1X/EAP and RADIUS elements. EAP types may include EAP-TLS, EAP-TTLS, EAP-PEAP, EAP-LEAP, and many others.

Pre-Shared Keys

WEP and WPA-PSK are examples of data link layer wireless security protocols where a persistent shared key is used between the access point and the client device. The pre-shared key is manually configured into the client utility software and into the access point's firmware. This key is used for all station pairs in an ESS and for many successive sessions before being changed, if ever. Once authenticated, a station pair link is secured using the RC4 or AES encryption algorithms. Once a data link is secured, a protocol analyzer will display "WEP Data" for encrypted data frames.

Some wireless protocol analyzers have the ability to decrypt frames encrypted with pre-shared keys by entering the pre-shared key into the analyzer software. Most analyzers support this function using static WEP keys, and some products are now capable of performing this task with WPA-PSK.

802.1X/EAP

There are numerous EAP types on the market today, and each works slightly different. There are advantages and disadvantages to each, and it is common to see many types in the real world. Besides having to understand the intricacies of each EAP type, the analyst must also understand the differences in implementation of the same EAP type across multiple vendors. Figures 9.15 and 9.16 illustrate two frame captures of 802.1X/EAP-PEAP-MS-CHAPv2. The first capture is an implementation from one vendor, and the second capture is an implementation from another vendor. One vendor completes the PEAP authentication in 58 packets (including ACKs) and gives an anonymous username that can be viewed in clear text with an analyzer as shown in Figure 9.15. This anonymous user name is entered into the client software by the user of the client device. The other vendor completes the PEAP authentication in 38 packets (including ACKs) and allows the real user name to be transmitted in clear text as shown in Figure 9.16.

FIGURE 9.15 802.1X/EAP-PEAP-MS-CHAPv2 Implementation #1

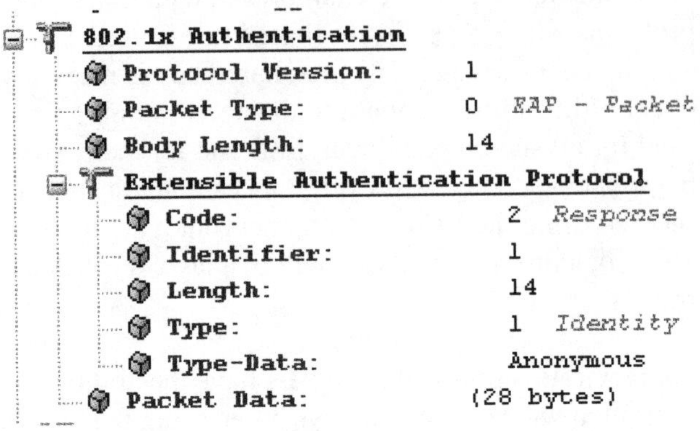

```
802.1x Authentication
  Protocol Version:        1
  Packet Type:             0  EAP - Packet
  Body Length:             14
  Extensible Authentication Protocol
    Code:                    2  Response
    Identifier:              1
    Length:                  14
    Type:                    1  Identity
    Type-Data:               Anonymous
  Packet Data:             (28 bytes)
```

FIGURE 9.16 802.1X/EAP-PEAP-MS-CHAPv2 Implementation #2

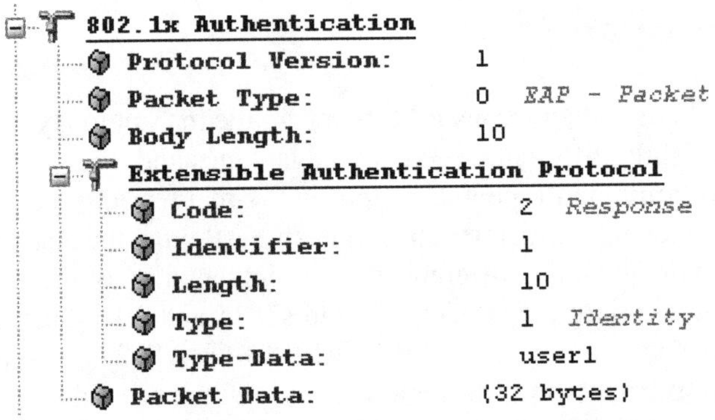

```
802.1x Authentication
  Protocol Version:        1
  Packet Type:             0  EAP - Packet
  Body Length:             10
  Extensible Authentication Protocol
    Code:                    2  Response
    Identifier:              1
    Length:                  10
    Type:                    1  Identity
    Type-Data:               user1
  Packet Data:             (32 bytes)
```

Information for the frame captures above was taken from the first EAP Response frame in the EAP authentication frame exchange.

802.11i / WPA2 / AES

AES options are now available in some wireless LAN products. In order to use AES encryption, the 802.1X supplicant, authenticator, and authentication server must all support AES. Figure 9.17 illustrates an access point that supports 802.1X/EAP authentication with AES encryption. Figure 9.18 illustrates a configuration page from a RADIUS server that supports 802.1X/WPA2 with AES encryption.

FIGURE 9.17 Access Point Configuration - AES

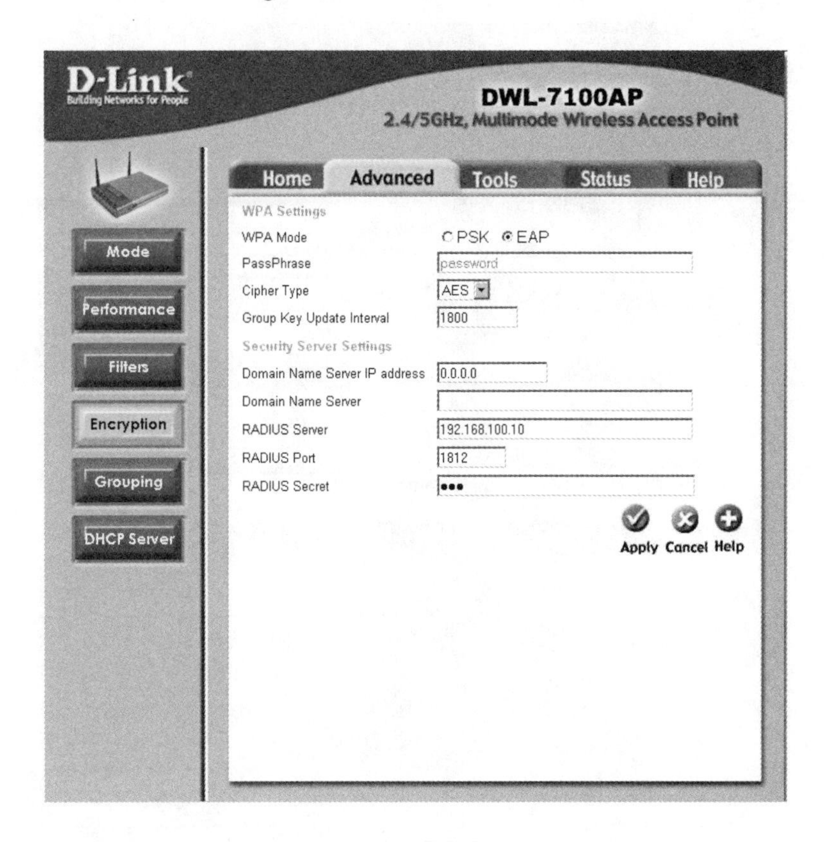

FIGURE 9.18 RADIUS Server Configuration - AES

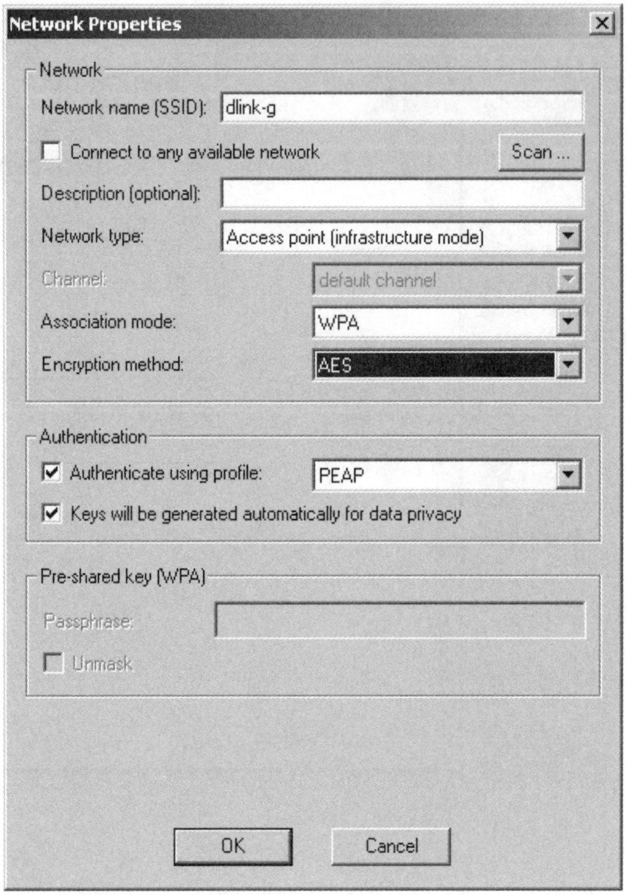

FIGURE 9.19 802.1X/PEAP Authentication Using CCMP/AES

Packet	Address 1	Address 2	Address 3	Channel	Data Rate	Size	Protocol
7	00:0F:3D:E9:05:BF	00:40:96:A1:9A:F9	00:0F:3D:E9:05:BF	6	1.0	34	802.11 Auth
8	00:40:96:A1:9A:F9			6	1.0	14	802.11 Ack
9	00:40:96:A1:9A:F9	00:0F:3D:E9:05:BF	00:0F:3D:E9:05:BF	6	1.0	34	802.11 Auth
10	00:0F:3D:E9:05:BF			6	1.0	14	802.11 Ack
11	00:0F:3D:E9:05:BF	00:40:96:A1:9A:F9	00:0F:3D:E9:05:BF	6	1.0	81	802.11 Assoc Req
12	00:40:96:A1:9A:F9			6	1.0	14	802.11 Ack
13	00:40:96:A1:9A:F9	00:0F:3D:E9:05:BF	00:0F:3D:E9:05:BF	6	1.0	50	802.11 Assoc Rsp
14	00:0F:3D:E9:05:BF			6	1.0	14	802.11 Ack
15	00:40:96:A1:9A:F9	00:0F:3D:E9:05:BF	00:0F:3D:E9:05:BF	6	54.0	45	EAP Request
16	00:0F:3D:E9:05:BF			6	24.0	14	802.11 Ack
17	00:0F:3D:E9:05:BF	00:40:96:A1:9A:F9	00:0F:3D:E9:05:BF	6	54.0	50	EAP Response
18	00:40:96:A1:9A:F9			6	24.0	14	802.11 Ack
19	00:40:96:A1:9A:F9	00:0F:3D:E9:05:BF	00:0F:3D:E9:05:BF	6	54.0	46	EAP Request
20	00:0F:3D:E9:05:BF			6	24.0	14	802.11 Ack
21	00:0F:3D:E9:05:BF	00:40:96:A1:9A:F9	00:0F:3D:E9:05:BF	6	54.0	170	EAP Response
22	00:40:96:A1:9A:F9			6	24.0	14	802.11 Ack
23	00:40:96:A1:9A:F9	00:0F:3D:E9:05:BF	00:0F:3D:E9:05:BF	6	54.0	180	EAP Request
24	00:0F:3D:E9:05:BF			6	24.0	14	802.11 Ack
25	00:0F:3D:E9:05:BF	00:40:96:A1:9A:F9	00:0F:3D:E9:05:BF	6	54.0	101	EAP Response
26	00:40:96:A1:9A:F9			6	24.0	14	802.11 Ack
27	00:40:96:A1:9A:F9	00:0F:3D:E9:05:BF	00:0F:3D:E9:05:BF	6	54.0	83	EAP Request
28	00:0F:3D:E9:05:BF			6	24.0	14	802.11 Ack
29	00:0F:3D:E9:05:BF	00:40:96:A1:9A:F9	00:0F:3D:E9:05:BF	6	54.0	46	EAP Response
30	00:40:96:A1:9A:F9			6	24.0	14	802.11 Ack
31	00:40:96:A1:9A:F9	00:0F:3D:E9:05:BF	00:0F:3D:E9:05:BF	6	54.0	44	EAP Success
32	00:0F:3D:E9:05:BF			6	24.0	14	802.11 Ack
33	00:40:96:A1:9A:F9	00:0F:3D:E9:05:BF	00:0F:3D:E9:05:BF	6	54.0	135	EAPOL-Key
34	00:0F:3D:E9:05:BF			6	24.0	14	802.11 Ack
35	00:0F:3D:E9:05:BF	00:40:96:A1:9A:F9	00:0F:3D:E9:05:BF	6	54.0	159	EAPOL-Key
36	00:40:96:A1:9A:F9			6	24.0	14	802.11 Ack
37	00:40:96:A1:9A:F9	00:0F:3D:E9:05:BF	00:0F:3D:E9:05:BF	6	54.0	159	EAPOL-Key
38	00:0F:3D:E9:05:BF			6	24.0	14	802.11 Ack
39	00:0F:3D:E9:05:BF	00:40:96:A1:9A:F9	00:0F:3D:E9:05:BF	6	54.0	135	EAPOL-Key
40	00:40:96:A1:9A:F9			6	24.0	14	802.11 Ack

Figure 9.19 illustrates an entire 802.1X/PEAP authentication sequence using CCMP/AES. Figure 9.20 illustrates an EAPOL Key frame decode. Use of the CCMP protocol denotes use of AES encryption. In contrast, Figure 9.21 illustrates an EAPOL Key frame that was captured during the same authentication process using TKIP instead of CCMP.

FIGURE 9.20 Using AES with 802.1X/PEAP

```
802.1x Authentication
    Protocol Version:       1
    Packet Type:            3  EAPOL - Key
    Body Length:            119
    EAPOL - Key
        Type:                   254  SSN key descriptor
        Key Information:        %0000000111001010
                                    xxx. .... .... .... Reserved
                                    ...1 .... .... .... Pairwise Key
                                    .... 00.. .... .... Key index 0
                                    .... ..1. .... .... Install/Tx
                                    .... ...1 .... .... Ack
                                    .... .... 1... .... MIC
                                    .... .... .0.. .... Not secure
                                    .... .... ..0. .... No Error
                                    .... .... ...0 .... Not Request
                                    .... .... .... xxxx Reserved
        Key Length:             16
        Replay Counter:         0x0000000000000002
        Key Nonce:              0x40A9F990C36D1AEC8F4DBFED0279406CA5BE
        Key IV:                 0x00000000000000000000000000000000
        Key Sequence Counter:   0x0000000000000000
        Key ID:                 0x0000000000000000
        Key MIC:                0x2512B79DB3A7B5A893B793913A2ABAD5
        Key Data Length:        0x0018
        OUI:                    00-50-F2-01
        Version:                1
        Multicast cipher OUI:   00-50-F2-04  CCMP
        Number of Unicast       1
        Unicast cipher OUI:     00-50-F2-04  CCMP
        Number of Auths         1
        Auth OUI:               00-50-F2-01  SSN
```

FIGURE 9.21 Using TKIP with 802.1X/PEAP

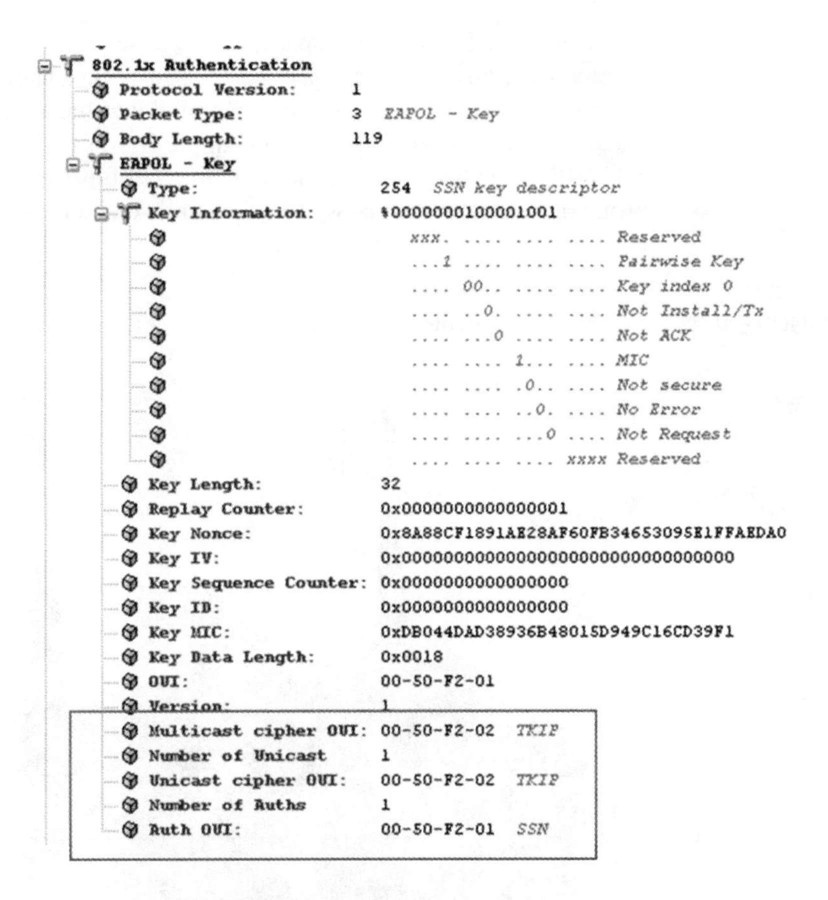

Wireless Router Architecture

By definition, a router transfers packets between logically addressed networks. Each router will choose the best link on which to send packets in order to reach one step closer to the destination. Since routers operate at layer 3 of the OSI model, this functionality may be unseen on wireless LAN links due to layer 2 encryption. If the wireless LAN data link is unencrypted, troubleshooting may, in fact, be simplified. Additionally, many routers support Network Address Translation (NAT), which may add some level of complexity to the troubleshooting process.

WLAN Access Point Router

In the wireless LAN market space, there are several types of routers commonly used. First is the wireless access point router. This is a piece of equipment with a L3 IP router integrated into the access point, which is usually capable of much more than simple IP routing. It is common to find VPN client and server functions, firewall functions, and even routing protocol functions such as RIP. These units come in two types: SOHO class and Enterprise class.

FIGURE 9.22 SOHO Class Wireless Router

FIGURE 9.23 Enterprise Class Wireless Router

Besides the obvious cost difference, these two product classes differ in their feature sets, reliability, security features, manageability, and many other parameters. Introducing IP routing can make design and troubleshooting difficult enough, but adding VPN technology usually complicates the scenario much further.

PPTP VPN Example

The Point-to-Point Tunneling Protocol (PPTP) encapsulates Point-to-Point Protocol (PPP) frames into IP datagrams for transmission over an IP-based internetwork, such as the Internet or a private intranet. PPTP is documented in RFC 2637. PPTP uses a TCP connection known as the PPTP control connection to create, maintain, and terminate the tunnel and a modified version of Generic Routing Encapsulation (GRE) to encapsulate PPP frames as tunneled data. The payloads of the encapsulated PPP frames can be encrypted or compressed or both. PPTP assumes the availability of an IP internetwork between a PPTP client (a VPN client using PPTP) and a PPTP server (a VPN server using the PPTP). The PPTP client might already be attached to an IP internetwork that can reach the PPTP server, or the PPTP client might have to dial into a network access server (NAS) to establish IP connectivity as in the case of dial-up Internet users.

Authentication that occurs during the creation of a PPTP-based VPN connection uses the same authentication mechanisms as PPP connections, such as Extensible Authentication Protocol (EAP), Microsoft Challenge-Handshake Authentication Protocol (MS-CHAP), CHAP, Shiva Password Authentication Protocol (SPAP), and Password Authentication Protocol (PAP).

PTP inherits encryption or compression, or both, of PPP payloads from PPP. For Windows 2000, either EAP-Transport Level Security (EAP-TLS) or MS-CHAP must be used in order for the PPP payloads to be encrypted using Microsoft Point-to-Point Encryption (MPPE). MPPE provides only link encryption, not end-to-end encryption. End-to-end encryption is data encryption between the client application and the server hosting the resource or service being accessed by the client application.

Consider a simple scenario in which Point-to-Point Tunneling Protocol (PPTP) is used as a wireless security mechanism. In this scenario, the clients' VPN tunnel terminates on the access point. Further, assume that roaming while maintaining the VPN connection is necessary to sustain the level of security mandated by corporate policy. Figure 9.24 illustrates this scenario. If only one wireless LAN access point router is used to terminate PPTP tunnels, then each of the other wireless LAN access point routers in the network must route the tunnel traffic to that PPTP server. If each wireless LAN access point router services its own subnet and acts as a stand-alone PPTP server, then each would have to be configured for DHCP, VPN DHCP, upstream routing, and likely a protocol like RIP so that it could trade subnet information with other peer wireless LAN routers for roaming purposes. This configuration could lead to countless subnets, could cause the network design to quickly become very complex, and is not recommended. The most simplistic design that will meet the organization's needs is usually best for keeping troubleshooting time to a minimum when the network has a problem.

FIGURE 9.24 PPTP VPN Network with Roaming

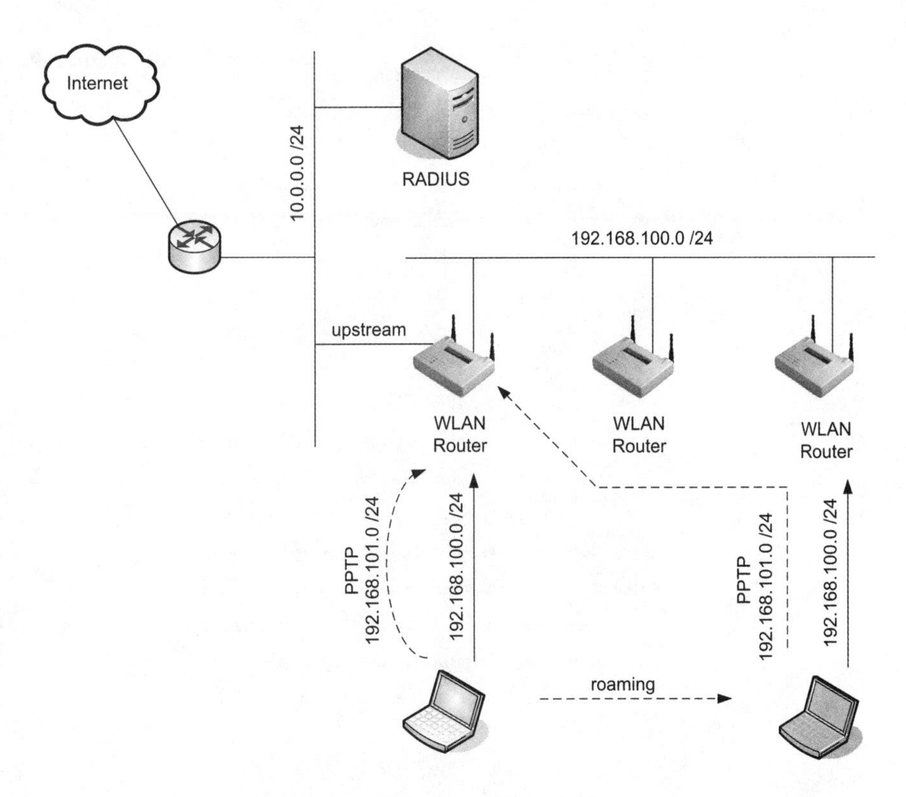

This scenario is only one of many possible wireless LAN access point router solutions that an analyst may find in the real world. Many wireless LAN access point routers support PPTP, L2TP, and IPSec VPN protocols, client and server VPN endpoint connectivity, NAT, and routing protocols such as RIPv1 and RIPv2. Each of these scenarios may make analyzing and troubleshooting a complex task. Figures 9.25 and 9.26 illustrate PPTP Authentication and MPPE-encrypted data frames traversing a wireless LAN. As you can see in Figure 9.26, even with MS-CHAPv2 as the authentication protocol used by PPTP, the username is sent in clear text which makes the PPTP wireless security solution vulnerable to dictionary attacks.

FIGURE 9.25 PPTP Authentication and MPPE Encrypted Data Capture

Address 1	Address 2	Address 3	Flags	Channel	Data Rate	Size	Protocol
00:02:2D:1C:66:63	00:02:2D:1C:6C:25	00:02:2D:1C:66:63		1	11.0	84	PPTP
00:02:2D:1C:6C:25			#	1	2.0	14	802.11 Ack
00:02:2D:1C:6C:25	00:02:2D:1C:66:63	00:02:2D:1C:66:63		1	11.0	84	PPTP
00:02:2D:1C:66:63			#	1	2.0	14	802.11 Ack
00:02:2D:1C:66:63	00:02:2D:1C:6C:25	00:02:2D:1C:66:63		1	11.0	76	PPTP
00:02:2D:1C:6C:25			#	1	2.0	14	802.11 Ack
00:02:2D:1C:66:63	00:02:2D:1C:6C:25	00:02:2D:1C:66:63		1	11.0	232	PPTP
00:02:2D:1C:6C:25			#	1	2.0	14	802.11 Ack
00:02:2D:1C:6C:25	00:02:2D:1C:66:63	00:02:2D:1C:66:63		1	11.0	76	PPTP
00:02:2D:1C:66:63			#	1	2.0	14	802.11 Ack
00:02:2D:1C:6C:25	00:02:2D:1C:66:63	00:02:2D:1C:66:63		1	11.0	232	PPTP
00:02:2D:1C:66:63			#	1	2.0	14	802.11 Ack
00:02:2D:1C:66:63	00:02:2D:1C:6C:25	00:02:2D:1C:66:63		1	11.0	244	PPTP
00:02:2D:1C:6C:25			#	1	2.0	14	802.11 Ack
00:02:2D:1C:6C:25	00:02:2D:1C:66:63	00:02:2D:1C:66:63		1	11.0	76	PPTP
00:02:2D:1C:66:63			#	1	2.0	14	802.11 Ack
00:02:2D:1C:6C:25	00:02:2D:1C:66:63	00:02:2D:1C:66:63		1	11.0	108	PPTP
00:02:2D:1C:66:63			#	1	2.0	14	802.11 Ack
00:02:2D:1C:6C:25	00:02:2D:1C:66:63	00:02:2D:1C:66:63		1	11.0	76	PPTP
00:02:2D:1C:66:63			#	1	2.0	14	802.11 Ack
00:02:2D:1C:66:63	00:02:2D:1C:6C:25	00:02:2D:1C:66:63		1	11.0	116	GRE
00:02:2D:1C:6C:25			#	1	2.0	14	802.11 Ack
00:02:2D:1C:6C:25	00:02:2D:1C:66:63	00:02:2D:1C:66:63		1	11.0	96	GRE
00:02:2D:1C:66:63			#	1	2.0	14	802.11 Ack
00:02:2D:1C:6C:25	00:02:2D:1C:66:63	00:02:2D:1C:66:63		1	11.0	101	GRE
00:02:2D:1C:66:63			#	1	2.0	14	802.11 Ack
00:02:2D:1C:66:63	00:02:2D:1C:6C:25	00:02:2D:1C:66:63		1	11.0	101	GRE
00:02:2D:1C:6C:25			#	1	2.0	14	802.11 Ack
00:02:2D:1C:6C:25	00:02:2D:1C:66:63	00:02:2D:1C:66:63		1	11.0	68	GRE
00:02:2D:1C:66:63			#	1	2.0	14	802.11 Ack
00:02:2D:1C:6C:25	00:02:2D:1C:66:63	00:02:2D:1C:66:63		1	11.0	87	GRE
00:02:2D:1C:66:63			#	1	2.0	14	802.11 Ack
00:02:2D:1C:66:63	00:02:2D:1C:6C:25	00:02:2D:1C:66:63		1	11.0	113	GRE
00:02:2D:1C:6C:25			#	1	2.0	14	802.11 Ack
00:02:2D:1C:6C:25	00:02:2D:1C:66:63	00:02:2D:1C:66:63		1	11.0	113	GRE
00:02:2D:1C:66:63			#	1	2.0	14	802.11 Ack

FIGURE 9.26 PPTP Authentication Using MS-CHAPv2 – Series of Frame Decodes

```
Point-to-Point Protocol
    PPP                        0xC223  Challenge-Handshake Authentication Protocol
    Authentication Code:       2  Response
    Identifier:                0x01
    CHAP Length:               59
    Value Size:                49
    Value:                     0xEF6A64B7EDC4C9D3ADCE2D9956AC06B40000000000000000A063
    Name:                      user1
```

```
Point-to-Point Protocol
    PPP                        0xC223  Challenge-Handshake Authentication Protocol
    Authentication Code:       3  Success
    Identifier:                0x01
    Length:                    46
    Message:                   S=9F6F16014304A8984283DDCF527D66B5A4107D17
```

```
Point-to-Point Protocol
    PPP                        0x80FD  PPP Compression Control Protocol
    Code:                      0x01  Configure-Request
    Identifier:                0x01
    Length:                    10
    Option Type:               18  Microsoft Point-To-Point Compression/Encryption Protocol
        Option Length:             6
        Supported Bits:            0x01000040
                                     .......1 ........ ........ ........ Stateless Mode On
                                     ........ ........ ........ 0....... 56 bit Encryption Off
                                     ........ ........ ........ .1...... 128 bit Encryption On
                                     ........ ........ ........ ..0..... 40 bit Encryption Off
                                     ........ ........ ........ .......0 No Desire to Negotiate MPPC
```

Wireless LAN access point routers may also support layer 2 authentication/encryption types such as WPA, WPA-PSK, and others. It is usually unnecessary to use layer 3 VPN solutions for security when layer 2 mechanisms are in place and doing so introduces added troubleshooting complexity since a wireless protocol analyzer can then only see the layer 2 information.

Enterprise Wireless Gateway (EWG)

A second router type commonly introduced into today's wireless LANs is the Enterprise Wireless Gateway (EWG). These units do not have a wireless interface, but rather are positioned between the wireless segment and the network backbone to provide gateway services such as Role-based Access Control (RBAC), Authentication, VPN services, Firewall, NAT, Captive Portal, and many others. The EWG is representative of any of the individual items that it comprises such as VPN concentrators, firewalls, Ethernet routers, etc. VPN technologies are often used for security on EWGs, and the VPN tunnels terminate on the EWG itself. The downstream access points have no knowledge of the layer 3 security mechanism in place between the client station and the EWG.

FIGURE 9.27 Enterprise Wireless Gateway

On links not encrypted by layer 2 security mechanisms, it may be necessary for an analyst to be familiar with IP routing, various VPN authentication/encryption protocols, role-based access control, and firewall services. Some wireless protocol analyzers can decode and analyze protocols ranging from layer 2 to layer 7 of the OSI model, which is helpful in situations where layer 2 encryption is not in use. Generally, the more information the analyst has to work with, the more accurate and simplified the troubleshooting process can be. It is very common to see layer 2 (WEP, TKIP, CCMP) encryption on wireless links, so the analyst may not often have the opportunity to troubleshoot IP and VPN connectivity over the wireless LAN.

IPSec VPN Example

When maximum strength layer 3 security is required, and it often is, IPSec is the VPN protocol of choice. IPSec has many advantages such as data authentication, non-repudiation, anti-replay prevention, strong encryption, and many others. It is important for an analyst to be familiar with IKE authentication and IPSec data encryption technology because troubleshooting these protocols over wireless LANs is very common with wireless LAN switches, EWGs, and wireless access point routers. Figure 9.28 illustrates an IKE Authentication and IPSec/ESP encrypted data transfer.

FIGURE 9.28 ISAKMP (IKE) Authentication and IPSec/ESP Encrypted Data

Address 1	Address 2	Address 3	Channel	Data Rate	Size	Protocol
00:09:5B:66:E6:09	00:0F:3D:E9:05:BF	00:E0:B8:5C:2E:B5	6	54.0	280	ISAKMP
00:0F:3D:E9:05:BF			6	24.0	14	802.11 Ack
00:0F:3D:E9:05:BF	00:09:5B:66:E6:09	00:E0:B8:5C:2E:B5	6	54.0	172	ISAKMP
00:09:5B:66:E6:09			6	24.0	14	802.11 Ack
00:09:5B:66:E6:09	00:0F:3D:E9:05:BF	00:E0:B8:5C:2E:B5	6	54.0	248	ISAKMP
00:0F:3D:E9:05:BF			6	24.0	14	802.11 Ack
00:0F:3D:E9:05:BF	00:09:5B:66:E6:09	00:E0:B8:5C:2E:B5	6	54.0	248	ISAKMP
00:09:5B:66:E6:09			6	24.0	14	802.11 Ack
00:09:5B:66:E6:09	00:0F:3D:E9:05:BF	00:E0:B8:5C:2E:B5	6	54.0	132	ISAKMP
00:0F:3D:E9:05:BF			6	24.0	14	802.11 Ack
00:0F:3D:E9:05:BF	00:09:5B:66:E6:09	00:E0:B8:5C:2E:B5	6	54.0	132	ISAKMP
00:09:5B:66:E6:09			6	24.0	14	802.11 Ack
00:09:5B:66:E6:09	00:0F:3D:E9:05:BF	00:E0:B8:5C:2E:B5	6	54.0	348	ISAKMP
00:0F:3D:E9:05:BF			6	24.0	14	802.11 Ack
00:0F:3D:E9:05:BF	00:09:5B:66:E6:09	00:E0:B8:5C:2E:B5	6	54.0	228	ISAKMP
00:09:5B:66:E6:09			6	24.0	14	802.11 Ack
00:09:5B:66:E6:09	00:0F:3D:E9:05:BF	00:E0:B8:5C:2E:B5	6	54.0	116	ISAKMP
00:0F:3D:E9:05:BF			6	24.0	14	802.11 Ack
00:0F:3D:E9:05:BF	00:09:5B:66:E6:09	00:E0:B8:5C:2E:B5	6	54.0	148	ISAKMP
00:09:5B:66:E6:09			6	24.0	14	802.11 Ack
00:09:5B:66:E6:09	00:0F:3D:E9:05:BF	00:E0:B8:5C:2E:B5	6	54.0	116	ESP
00:0F:3D:E9:05:BF			6	24.0	14	802.11 Ack
00:0F:3D:E9:05:BF	00:09:5B:66:E6:09	00:E0:B8:5C:2E:B5	6	54.0	116	ESP
00:09:5B:66:E6:09			6	24.0	14	802.11 Ack
00:09:5B:66:E6:09	00:0F:3D:E9:05:BF	00:E0:B8:5C:2E:B5	6	54.0	108	ESP
00:0F:3D:E9:05:BF			6	24.0	14	802.11 Ack
00:0F:3D:E9:05:BF	00:09:5B:66:E6:09	00:E0:B8:5C:2E:B5	6	48.0	148	ESP
00:09:5B:66:E6:09			6	24.0	14	802.11 Ack
00:09:5B:66:E6:09	00:0F:3D:E9:05:BF	00:E0:B8:5C:2E:B5	6	54.0	108	ESP
00:0F:3D:E9:05:BF			6	24.0	14	802.11 Ack

Figure 9.29 illustrates use of the ESP protocol within the IPSec framework to encrypt data traffic.

FIGURE 9.29 ESP Data Frame Decode

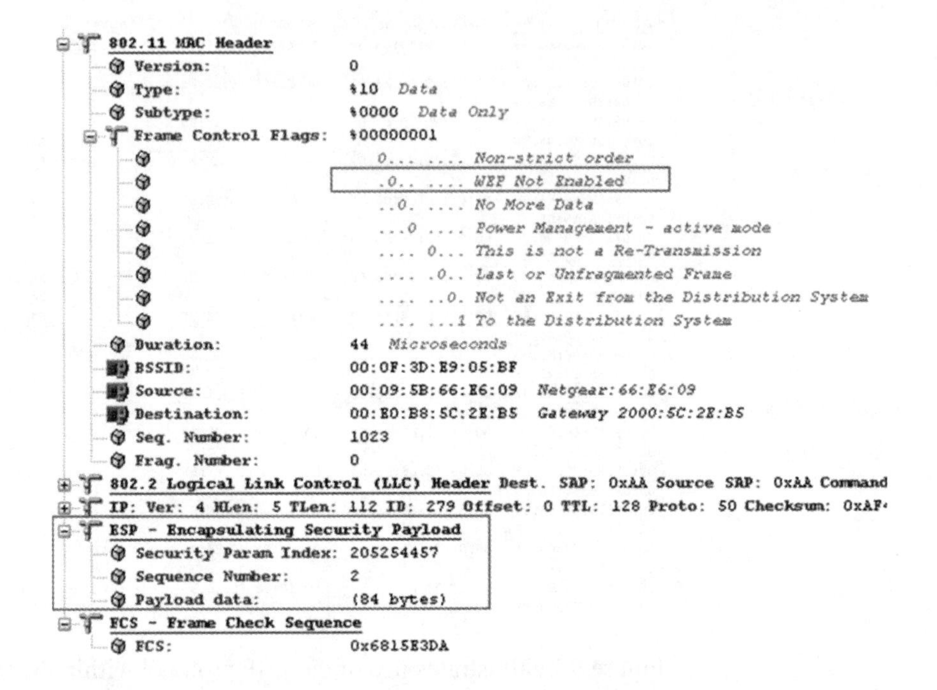

Introducing a wireless gateway sometimes means introduction of a layer 3 boundary between the wireless LAN segment and the backbone segment of the network. This layer 3 boundary may introduce a level of complexity with the introduction of NAT (or PAT) and roaming across access points downstream from different gateways. This scenario sometimes introduces subnet roaming, which every vendor implements differently. Some vendors have constructed ways for a client device to maintain its IP address and VPN connection across EWG boundaries. Figure 9.30 illustrates an example of layer 3 roaming.

FIGURE 9.30 Roaming Across EWGs

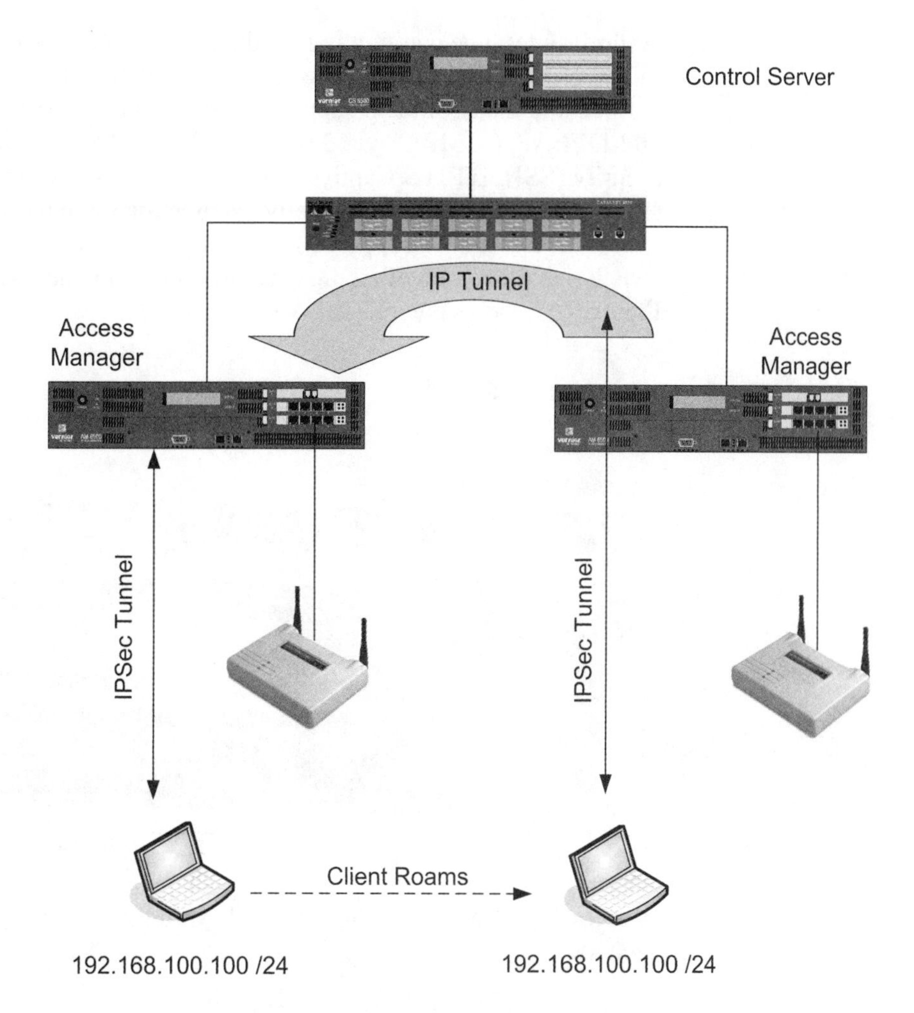

Scenarios such as these are reasonable to troubleshoot using a protocol analyzer capable of layer 2-7 analysis, but can become difficult when layer 2 encryption, such as TKIP, is introduced at the access points for additional security. Encryption of all layer 3 information in TKIP-encrypted frames makes end-to-end connectivity troubleshooting with a protocol analyzer quite an ordeal.

Wireless LAN Switch

A third router type commonly introduced into today's wireless LANs is the wireless LAN switch. Wireless LAN switches are common in the market, and many of them support layer 2 security (WEP, TKIP, 802.1X/EAP, CCMP), layer 3 security (PPTP, IPSec, L2TP), and layer 7 security (SSH, HTTPS/Captive Portal). The leading edge wireless LAN switches and EWGs have basically become the same product with the exception that EWGs support use of "fat access points" and wireless LAN switch manufacturers often provide their own "thin access points." Figure 9.31 shows a wireless LAN switch.

FIGURE 9.31 Wireless LAN Switch

Wireless LAN switches support a wide variety of security mechanisms such as:

- 802.1q VLAN tagging
- multiple SSIDs per VLAN
- per-VLAN security policies
- inter-VLAN routing
- per-user firewalls
- per-user or per-group security policies
- SSID suppression

Each of these features is performed in software within the switch, and thus very complex wireless LANs can be constructed using a single switch or set of switches.

Wireless LAN Switch Scenario #1

As an example of how these features might affect protocol analysis, consider the simple scenario presented in Figure 9.32.

FIGURE 9.32 Wireless LAN Switch Scenario

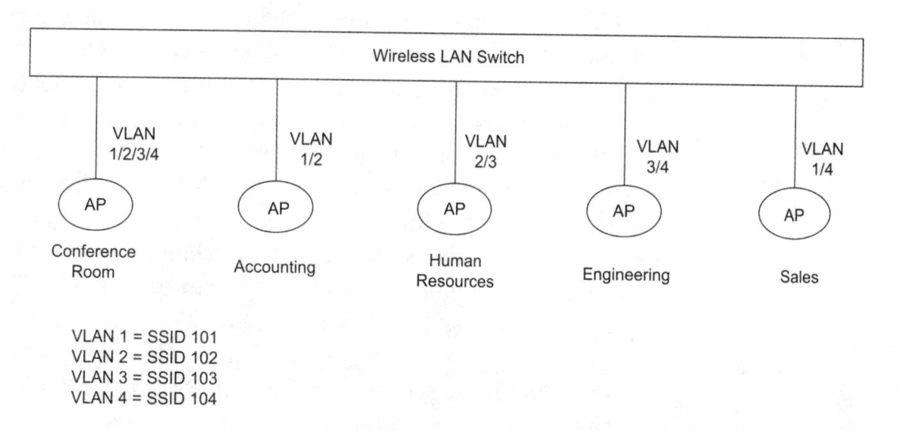

In this example, each VLAN/SSID pair in the wireless LAN switch may have a different security mechanism assigned. For example, SSID 101 may use EAP-PEAP-MS-CHAPv2, and SSID 102 may use TKIP. A client authenticates using SSID 104 and is connected to VLAN-4. The client is now able to connect to the access points in Engineering, Sales, and the Conference Room only. If the client roams into the Accounting department and the computer tries to associate to the access point, it will fail authentication. Additionally, if the client tries to use the Conference Room access point after 6pm each day, it will fail authentication due to corporate policy implemented on the wireless LAN switch. In both of these cases, there is no clear indication of why the authentication fails by viewing a frame capture. Knowledge of the policy implementation in the switch is also required to effectively troubleshoot this scenario.

There are many security configurations using a variety of layer 2-7 features that can be implemented in a wireless LAN switch. Each new security feature adds an additional layer of complexity to analysis, but many times troubleshooting boils down to the security feature that resides lowest in the OSI model. As a wireless network analyst, it is not enough to know only protocol analysis, but you must also be very familiar with wireless LAN security design practices.

Wireless LAN Switch Scenario #2

As an example of how security protocols may affect protocol analysis, consider the 802.1X/EAP-TTLS layer 2 wireless security protocol. EAP-TTLS constructs a TLS tunnel using a server-side digital certificate for authentication. Once the TLS tunnel is constructed, a number of authentication protocols may be used for client authentication inside the TLS tunnel. None of the authentication credentials (username, password, certificate, etc.) used inside the encrypted TLS tunnel will be seen by a wireless protocol analyzer. EAP-TTLS uses an "anonymous" username when establishing the TLS tunnel. Someone unfamiliar with EAP-TTLS might assume this username is valid when in reality it is arbitrary and meaningless. Figure 9.33 illustrates part of an EAP-TTLS response frame decode. Wireless LAN switches, "thick" or "fat" access points, and some software solutions support EAP-TTLS. This situation illustrates that knowledge of the security protocol in use is essential to effective wireless design, security, and troubleshooting.

FIGURE 9.33 EAP-TTLS Capture Showing Use of Anonymous User Name

```
802.1x Authentication
    Protocol Version:          1
    Packet Type:               0  EAP - Packet
    Body Length:               14
    Extensible Authentication Protocol
        Code:                  2  Response
        Identifier:            1
        Length:                14
        Type:                  1  Identity
        Type-Data:             anonymous
    Packet Data:               (28 bytes)
```

Wireless Mesh Architecture

Wireless LAN mesh networks typically come in two types: infrastructure-only and infrastructure-client. In the infrastructure-only model, only infrastructure devices such as purpose-built mesh routers are used to move client data across the network. In the infrastructure-client model, both mesh routers and client devices forward 802.11 frames toward their destination. Both types of mesh networks have their advantages, but we will only cover the infrastructure-only design in this text. Figures 9.34 and 9.35 illustrate an indoor wireless mesh router, which physically looks almost identical to any normal access point on the market with the exception of multiple Ethernet interfaces.

FIGURE 9.34 Indoor Wireless Mesh Router (front)

FIGURE 9.35 Indoor Wireless Mesh Router (rear)

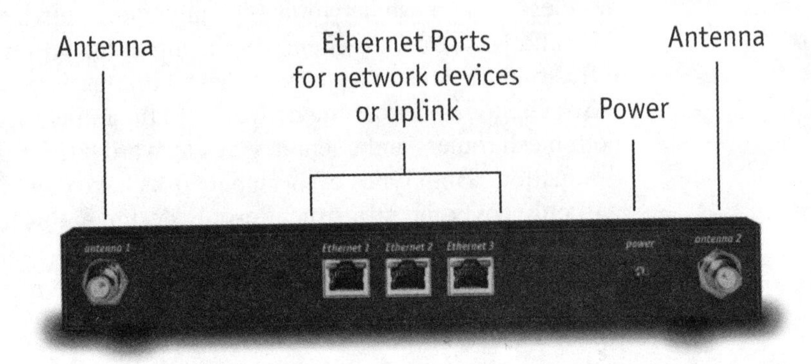

With wireless infrastructure mesh networks, the mesh routers act as a distributed Ethernet switch where the ports are connected by radios rather than by an internal CPU. Obviously there is the disadvantage of low throughput across data links such as 802.11b in comparison to the 100Tx or 1000Tx wired Ethernet networks users have become accustomed to, but the advantage of having a self-forming, resilient, scalable infrastructure outweigh the disadvantages in some scenarios.

Network Design

Conceptually, understanding the intended use of wireless mesh routers is simple, as shown in Figures 9.36 and 9.37.

FIGURE 9.36 Indoor Wireless Mesh

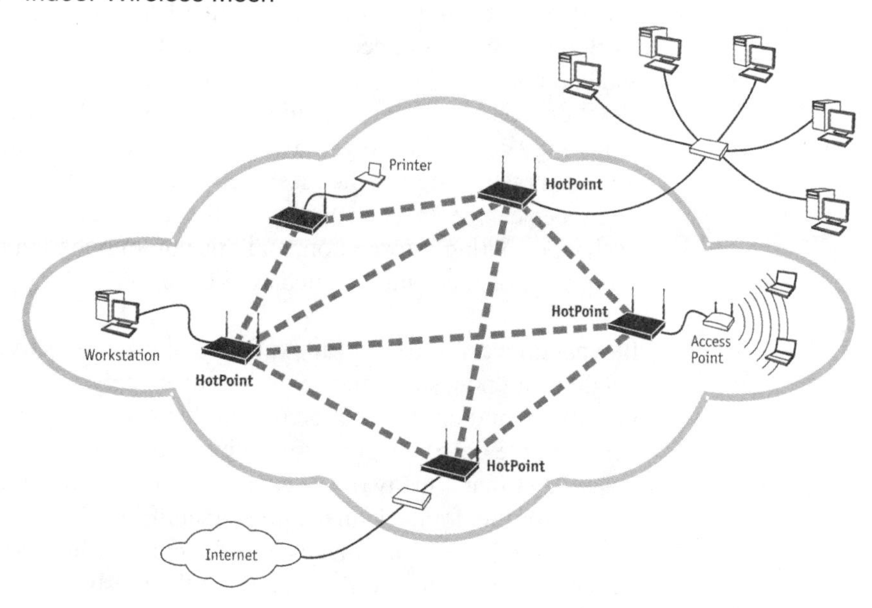

FIGURE 9.37 Metro Wireless Mesh

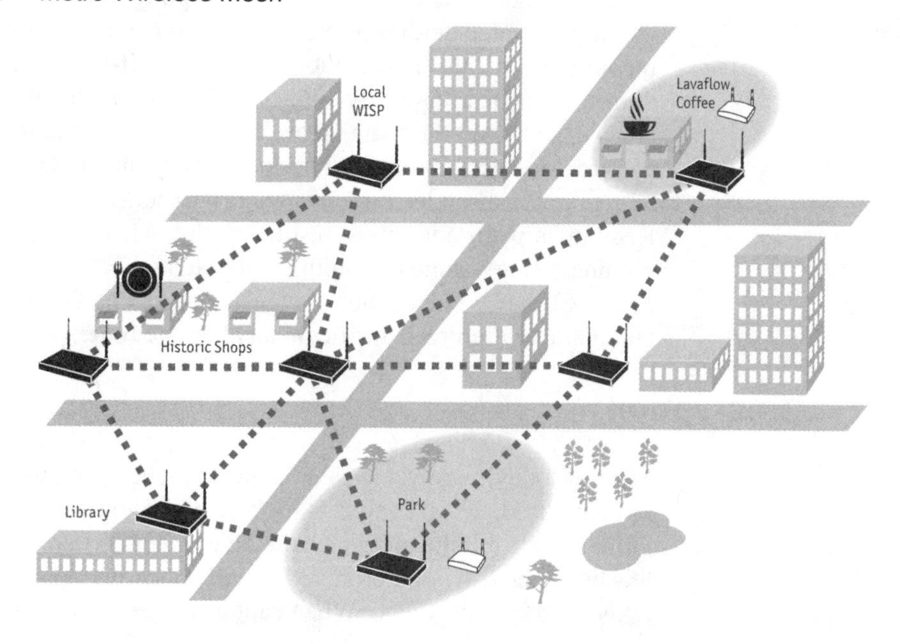

You can see from these illustrations that Ethernet connections on each mesh router allow connectivity to an Internet connection, local printers and computers, Ethernet switches, and various other devices. Wireless client devices do not connect to the wireless mesh routers directly, but rather may connect through an access point that is connected to a mesh router. Traffic entering the mesh network through one of the Ethernet interfaces can be encrypted, often with AES. The mesh routers must exchange routing information, and this communication may be secured using various mechanisms such as WEP or AES.

In what appears to be a paradox, the complex connectivity among neighboring nodes can make wireless mesh networks very simple to implement and operate. Overcoming this apparent paradox requires a considerable development effort to make the mesh as self-managing as possible. In fact, many of the advantages of wireless mesh networking derive from its four self-managing capabilities. First, the mesh is self-configuring, which also makes it self-*re*configuring. New nodes automatically become full members of the mesh topology shortly after booting up. Adding, moving, or removing nodes and their attached Ethernet devices (clients, servers, access points, surveillance cameras, gateways, routers, etc) is as easy as it is immediate. Intelligent self-reconfiguration also makes the entire mesh self-tuning end-to-end, allowing traffic to move dynamically along optimal paths. As those paths change, so too do the route tables that direct traffic on the shortest, fastest, and least-congested routes. The self-configuring and self-tuning abilities help give the mesh its third advantage as a self-healing network. Redundant paths add robust resiliency and, when properly arranged, eliminate single points of failure and potential bottlenecks within the mesh. Should a link become congested or a node fail, the mesh automatically redirects traffic on an alternate route.

Encryption

Most of today's analyzers cannot decrypt AES, even when pre-shared keys are used. Most analyzers look at the Protected Frame bit in the Frame Control field, and make the determination that WEP is enabled if this bit is set to 1. Therefore, AES will appear as WEP in most wireless LAN protocol analyzers. When capturing frames across a mesh infrastructure directly between two mesh routers, it may appear as illustrated by Figure 9.38.

FIGURE 9.38 Wireless Mesh Analysis

Packet	Transmitter	Receiver	Flags	Channel	Data Rate	Size	Protocol
217	00:02:6F:08:0C:AD	00:02:6F:08:0C:C1	CW	6	11.0	1683	802.11 WEP Data
218			#C	6	11.0	30	802.11 Control
219	00:02:6F:08:0C:AD	00:02:6F:08:0C:C1	CW	6	11.0	811	802.11 WEP Data
220			#C	6	11.0	30	802.11 Control
221	00:02:6F:08:0C:AD	00:02:6F:08:0C:C1	CW	6	11.0	1683	802.11 WEP Data
222			#C	6	11.0	30	802.11 Control
223	00:02:6F:08:0C:AD	00:02:6F:08:0C:C1	CW	6	11.0	811	802.11 WEP Data
224			#C	6	11.0	30	802.11 Control
225	00:02:6F:08:0C:AD	00:02:6F:08:0C:C1	CW	6	11.0	1683	802.11 WEP Data
226			#C	6	11.0	30	802.11 Control
227	00:02:6F:08:0C:AD	00:02:6F:08:0C:C1	CW	6	11.0	811	802.11 WEP Data
228			#C	6	11.0	30	802.11 Control
229	00:02:6F:08:0C:C1	00:02:6F:08:0C:AD	CW	6	11.0	229	802.11 WEP Data
230			#C	6	11.0	30	802.11 Control
231	00:02:6F:08:0C:C1	00:02:6F:08:0C:AD	CW	6	11.0	229	802.11 WEP Data
232			#C	6	11.0	30	802.11 Control
233	00:02:6F:08:0C:C1	00:02:6F:08:0C:AD	CW	6	11.0	229	802.11 WEP Data
234			#C	6	11.0	30	802.11 Control
235	00:02:6F:08:0C:AD	00:02:6F:08:0C:C1	CW	6	11.0	1683	802.11 WEP Data
236			#C	6	11.0	30	802.11 Control
237	00:02:6F:08:0C:AD	00:02:6F:08:0C:C1	CW	6	11.0	811	802.11 WEP Data
238			#C	6	11.0	30	802.11 Control
239	00:02:6F:08:0C:C1	00:02:6F:08:0C:AD	CW	6	11.0	229	802.11 WEP Data
240			#C	6	11.0	30	802.11 Control
241	00:02:6F:08:0C:AD	00:02:6F:08:0C:C1	CW	6	11.0	1683	802.11 WEP Data
242			#C	6	11.0	30	802.11 Control
243	00:02:6F:08:0C:AD	00:02:6F:08:0C:C1	CW	6	11.0	811	802.11 WEP Data
244			#C	6	11.0	30	802.11 Control
245	00:02:6F:08:0C:C1	00:02:6F:08:0C:AD	CW	6	11.0	229	802.11 WEP Data
246			#C	6	11.0	30	802.11 Control
247	00:02:6F:08:0C:AD	00:02:6F:08:0C:C1	CW	6	11.0	1683	802.11 WEP Data
248			#C	6	11.0	30	802.11 Control
249	00:02:6F:08:0C:AD	00:02:6F:08:0C:C1	CW	6	11.0	811	802.11 WEP Data
250			#C	6	11.0	30	802.11 Control
251	00:02:6F:08:0C:AD	00:02:6F:08:0C:C1	CW	6	11.0	1683	802.11 WEP Data
252			#C	6	11.0	30	802.11 Control
253	00:02:6F:08:0C:C1	00:02:6F:08:0C:AD	CW	6	11.0	229	802.11 WEP Data
254			#C	6	11.0	30	802.11 Control
255	00:02:6F:08:0C:AD	00:02:6F:08:0C:C1	CW	6	11.0	811	802.11 WEP Data
256			#C	6	11.0	30	802.11 Control

Notice the flags column shows that every frame is corrupt. The reason for this occurrence is that AES-encrypted frames are in use and the analyzer cannot decode them properly. Also, the 802.11 Control frames that obviously should be 14 bytes are 30 bytes. This additional bytes demonstrate that 16 bytes of encryption overhead, likely CCMP, is being used even in management frames which carry no payload.

Analysis

Typically, troubleshooting of wireless mesh networks is done by checking connections and monitoring the mesh network management application included with the hardware. Since data may travel one way through the mesh at one point in time and another way through the mesh at another point in time, troubleshooting and baselining is almost impossible. Depending on the vendor's implementation, traffic may flow between end nodes asymmetrically across the mesh. This means that the traffic from station-1 going to station-2 may take one route, and traffic returning to station-1 from station-2 may take another route.

Latency greatly affects mobile clients such as VoIP phones, and latency is partially dependent on system load, the number of hops, the data rate of each link, symmetric data flow, and use of available QoS features. Measuring real-time load statistics and data paths across the mesh are features of the management application bundled with the mesh networking product because making such measurements via the RF medium using an analyzer would be very difficult, if not impossible. If the mesh routers themselves support SNMP, then management applications could pull statistics directly from the mesh routers, but this feature is often not available.

For a real-world look at the throughput an analyst should expect from a mesh network, consider a data transfer across a 3-router mesh network. A 3-router mesh network forms the shape of a triangle. Therefore, transferring data between two wired devices across the mesh will be directly between two mesh routers if the shortest path is chosen.

FIGURE 9.39 Three Router Wireless Mesh Network

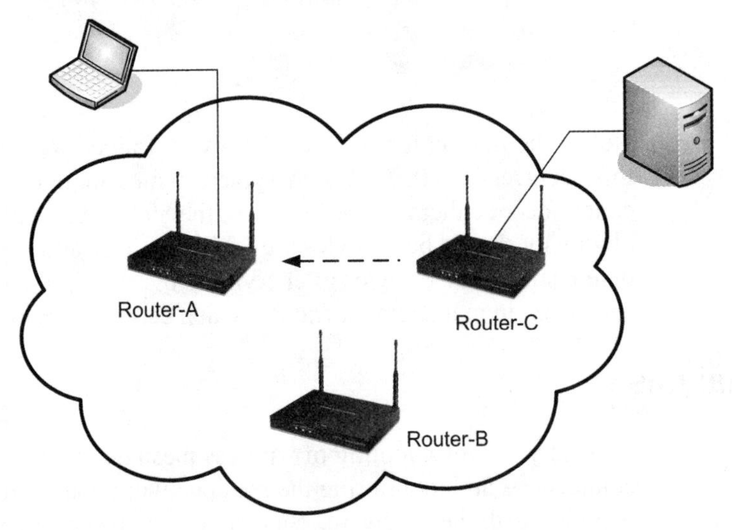

Whether the shortest path is chosen or not will depend on the routing algorithm used by the vendor. Most of today's mesh products have layer 2 routing protocols that are based primarily on hop count, but as the products mature, more sophisticated protocols will be used. Since the routers can be placed at different distances from each other, it is possible that the A/B router link could be at 11 Mbps, the B/C link could be at 5.5 Mbps, and the A/C link could be at 1 Mbps. The routing algorithm might then decide to route the data through router B as the "best case" scenario. Knowing the data rates on each link, the path the traffic is likely to take, and being able to manually manipulate the routing algorithm are all important to optimizing throughput across the mesh. If multiple hops are taken, the mesh routers act as wireless repeaters, having to repeat each frame on the same channel on which it was received. This cuts throughput in half (or perhaps less). Figure 9.40 illustrates the typical throughput seen across a single router-to-router mesh link.

FIGURE 9.40 Wireless Mesh Network Throughput Example

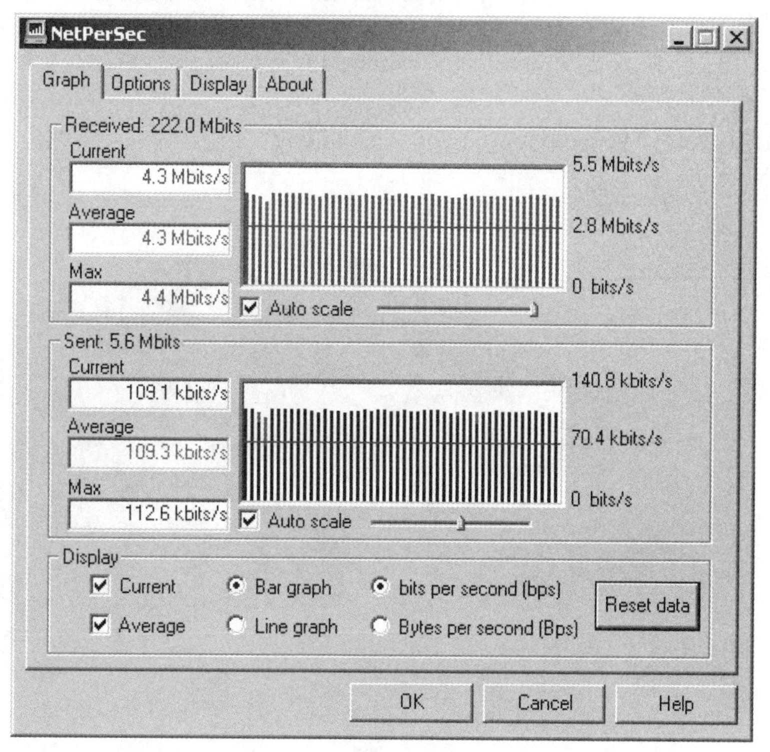

Trying to perform a baseline performance analysis or perform connectivity troubleshooting across a network that may only be partially meshed and on which routing is performed automatically may be close to impossible.

Enterprise Encryption Gateways (EEG)

Enterprise Encryption Gateways are commonly used for maximum strength layer 2 wireless security due to their use of AES-based encryption and compression. Typical EEG architecture has the wireless network on one side of the gateway, and the wired network on the other side of the gateway. This design is typical of all gateways, but the uniqueness of using an EEG is its use of layer 2 encryption of the wireless link. Client software is loaded onto each client device, which adds a proprietary layer 2 protocol. The client and gateway use this protocol for secure, compressed communication. Advanced encryption typically brings throughput down, but using data compression on the link typically offsets any loss due to encryption. Figure 9.41 shows a common EEG, and Figure 9.42 illustrates the typical EEG network design.

FIGURE 9.41 Enterprise Encryption Gateway

FIGURE 9.42 Typical EEG Network Architecture

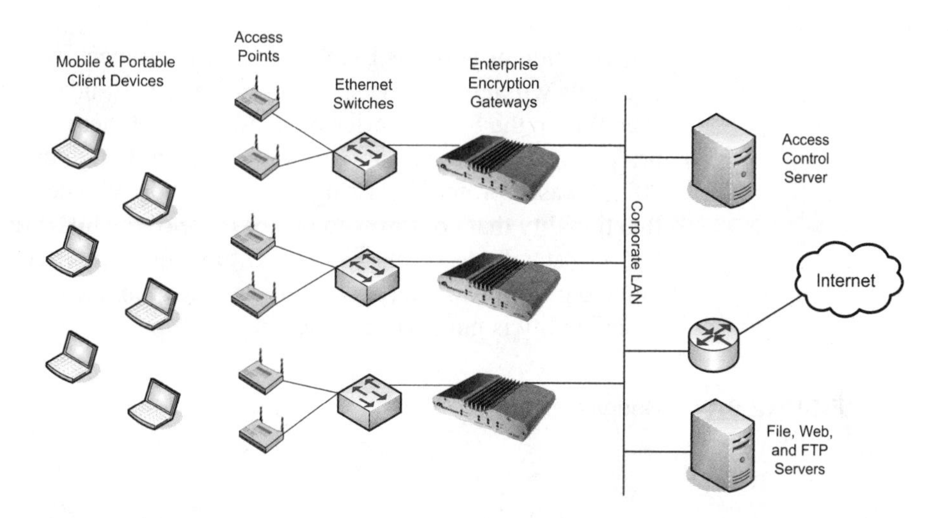

Notice in Figure 9.43 that standard Open System authentication is used on the access point, and then a protocol that is not understood by the analyzer is shown. The analyzer is displaying LLC protocol type SNAP 8895 in this capture. The entire MAC header of data frames (the only MAC frame type that is encrypted) is still intact. It is the data being placed into the MAC frame's frame body that is encrypted with an encryption type that is not understood by the analyzer. The encryption only happens between the client software and the EEG device. Data upstream from the EEG is not encrypted.

FIGURE 9.43 EEG Frame Capture Example

Packet	Address 1	Address 2	Address 3	Channel	Data Rate	Size	Protocol
18	00:0D:ED:A5:51:70	00:0A:8A:47:BC:1A	00:0D:ED:A5:51:70	1	1.0	34	802.11 Auth
19	00:0A:8A:47:BC:1A			1	1.0	14	802.11 Ack
20	00:0A:8A:47:BC:1A	00:0D:ED:A5:51:70	00:0D:ED:A5:51:70	1	11.0	34	802.11 Auth
21	00:0D:ED:A5:51:70			1	11.0	14	802.11 Ack
22	00:0D:ED:A5:51:70	00:0A:8A:47:BC:1A	00:0D:ED:A5:51:70	1	1.0	88	802.11 Assoc Req
23	00:0A:8A:47:BC:1A			1	1.0	14	802.11 Ack
24	00:0A:8A:47:BC:1A	00:0D:ED:A5:51:70	00:0D:ED:A5:51:70	1	11.0	84	802.11 Assoc Rsp
25	00:0D:ED:A5:51:70			1	11.0	14	802.11 Ack
26	00:0A:8A:47:BC:1A	00:0D:ED:A5:51:70	00:0D:65:C9:32:76	1	11.0	86	SNAP-00-40-96-00-00
27	00:0D:ED:A5:51:70			1	11.0	14	802.11 Ack
28	00:0D:ED:A5:51:70	00:0A:8A:47:BC:1A	00:0D:65:C9:32:76	1	1.0	110	SNAP-00-40-96-00-00
29	00:0A:8A:47:BC:1A			1	1.0	14	802.11 Ack
30	00:0D:ED:A5:51:70	00:0A:8A:47:BC:1A	FF:FF:FF:FF:FF:FF	1	11.0	270	SNAP-00-00-00-88-95
31	00:0A:8A:47:BC:1A			1	11.0	14	802.11 Ack
32	00:0D:ED:A5:51:70	00:0A:8A:47:BC:1A	FF:FF:FF:FF:FF:FF	1	11.0	126	SNAP-00-00-00-88-95
33	00:0A:8A:47:BC:1A			1	11.0	14	802.11 Ack
34	FF:FF:FF:FF:FF:FF	00:0D:ED:A5:51:70	00:0A:8A:47:BC:1A	1	11.0	270	SNAP-00-00-00-88-95
35	FF:FF:FF:FF:FF:FF	00:0D:ED:A5:51:70	00:0A:8A:47:BC:1A	1	11.0	126	SNAP-00-00-00-88-95
36	00:0A:8A:47:BC:1A	00:0D:ED:A5:51:70	00:E0:F4:11:F3:B2	1	11.0	286	SNAP-00-00-00-88-95

Bridging

The function of wireless LAN bridges is to connect two wired network segments wirelessly. The specifics of bridging functionality are not specified in the 802.11 series of standards, and therefore wireless LAN bridge implementations are proprietary, even though they may use 802.11-based protocols. Bridges are available with access point functionality that conforms to one or more of the 802.11a, 802.11b, and 802.11g standards. This access point functionality may allow the bridge units to pass Wi-Fi interoperability certification testing, but the bridging functionality is not part of the testing.

FIGURE 9.44 Bridging Scenario

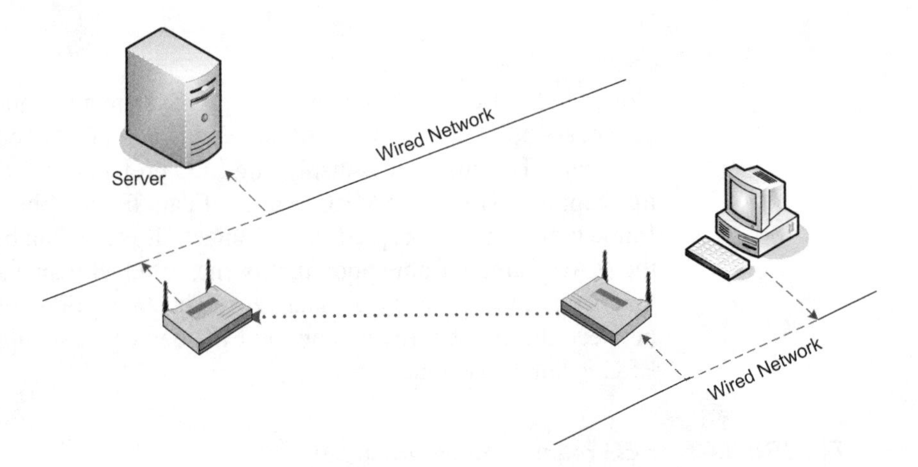

Figures 9.45 and 9.46 illustrate that wireless bridges use a WDS-like addressing scheme. All four address slots in the MAC frame are used, and the ToDS and FromDS bits are both set to 1 for frames transmitted between bridges.

FIGURE 9.45 Bridging Frame Capture

Packet	Address 1	Address 2	Address 3	Address 4	Channel	Data Rate	Size	Protocol
4	00:40:96:53:CF:EB	00:40:96:56:23:EF	00:0B:CD:7B:3A:CE	00:0B:CD:A6:CF:0F	6	11.0	102	PING Req
5	00:40:96:56:23:EF				6	11.0	14	802.11 Ack
6	00:40:96:56:23:EF	00:40:96:53:CF:EB	00:0B:CD:A6:CF:0F	00:0B:CD:7B:3A:CE	6	11.0	102	PING Reply
7	00:40:96:53:CF:EB				6	11.0	14	802.11 Ack
8	00:40:96:53:CF:EB	00:40:96:56:23:EF	00:0B:CD:7B:3A:CE	00:0B:CD:A6:CF:0F	6	11.0	102	PING Req
9	00:40:96:56:23:EF				6	11.0	14	802.11 Ack
10	00:40:96:56:23:EF	00:40:96:53:CF:EB	00:0B:CD:A6:CF:0F	00:0B:CD:7B:3A:CE	6	11.0	102	PING Reply
11	00:40:96:53:CF:EB				6	11.0	14	802.11 Ack

FIGURE 9.46 Bridging Frame Decode

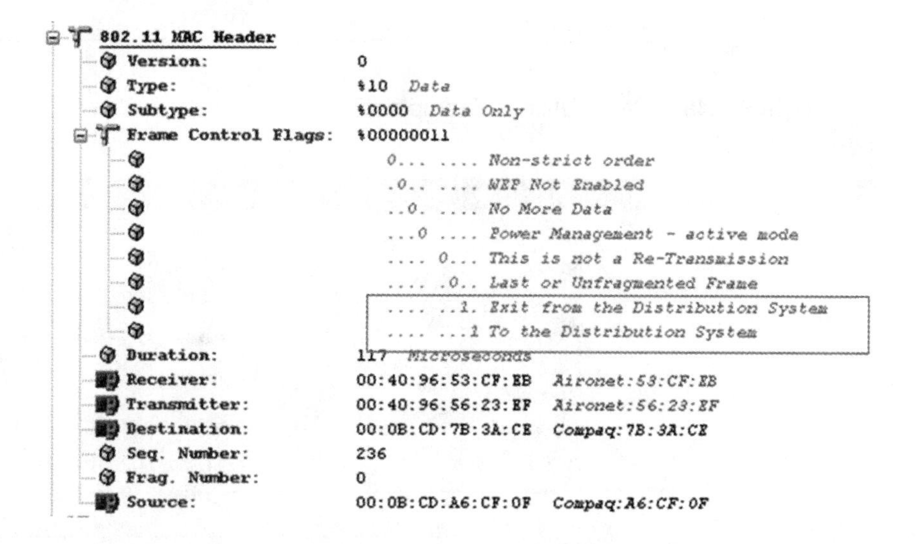

Client Configuration Options

Client utilities have varied widely since the introduction of the wireless LAN PC card. There are a wide range of features available across all of the vendors, some of which allow the user or administrator to configure MAC layer protocol settings such as RTS/CTS Threshold and Fragmentation Threshold. Figures 9.47 and 9.48 illustrate these settings from two different client utilities.

FIGURE 9.47 Client Utilities Example #1

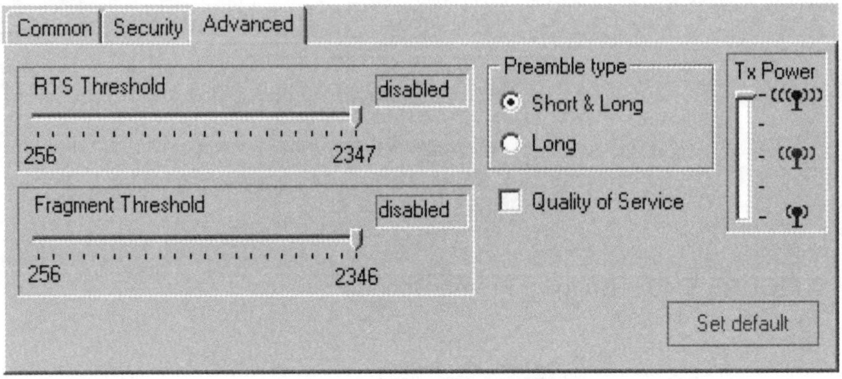

FIGURE 9.48 Client Utilities Example #2

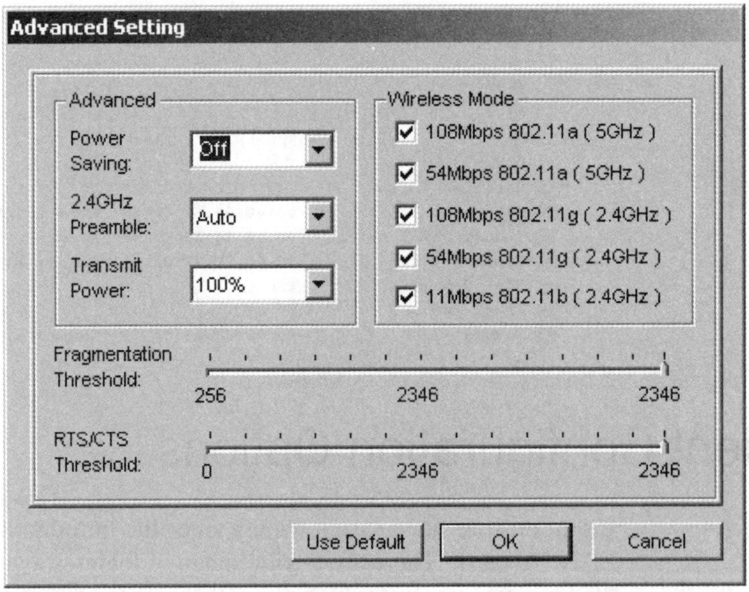

Settings such as these allow the user to optimize wireless LAN connectivity for any given physical environment. Keep in mind that these settings strictly affect transmissions from this client device, and every client device and access point is configured independent of any other. A wireless-to-wireless data transmission within the same BSS may be fragmented in two different ways – once at the transmitting station and again at the access point.

Authentication

Understanding authentication protocols is a large part of wireless LAN protocol analysis, and there are many pieces to the authentication puzzle. RADIUS is the most common authentication service used by wireless LAN infrastructure devices. In some cases the native (internal) RADIUS user database is used for user authentication, and in other cases, the RADIUS server only serves as a proxy to point to some other user database such as Microsoft's Active Directory, LDAP, or Novell's eDirectory. Access points, EWGs, wireless access point routers, wireless LAN switches, and many other wireless LAN infrastructure devices use RADIUS as the authentication protocol of choice.

There are many versions of the RADIUS protocol, each customized for the vendor's specific purposes. Figure 9.49 and 9.50 illustrate two different RADIUS servers' support for multiple versions of the RADIUS protocol. Figures 9.49 and 9.50 each illustrate where the administrator may choose the Network Access Server (NAS; access point or wireless access point router in this case).

FIGURE 9.49 RADIUS NAS Selection – RADIUS Server #1

FIGURE 9.50 RADIUS NAS Selection – RADIUS Server #2

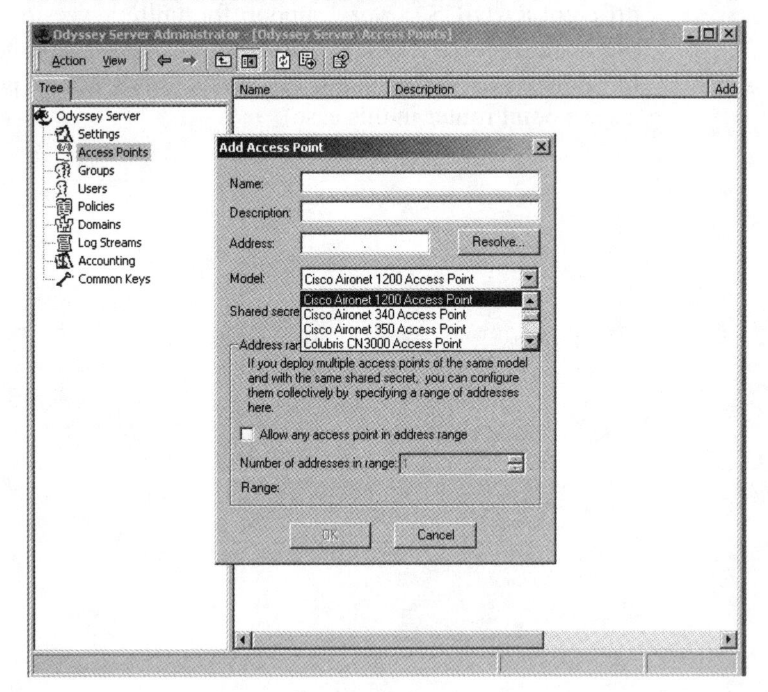

Figure 9.51 shows a wireless LAN authentication frame exchange between an access point and a RADIUS server across the wired network. In this example, Cisco's LEAP protocol was used. The unicast encryption key is passed from the RADIUS server to the access point as an encrypted RADIUS attribute as shown in Figure 9.52. This encrypted attribute is carried in the Access Accept frame.

FIGURE 9.51 RADIUS Authentication Exchange with Access Point

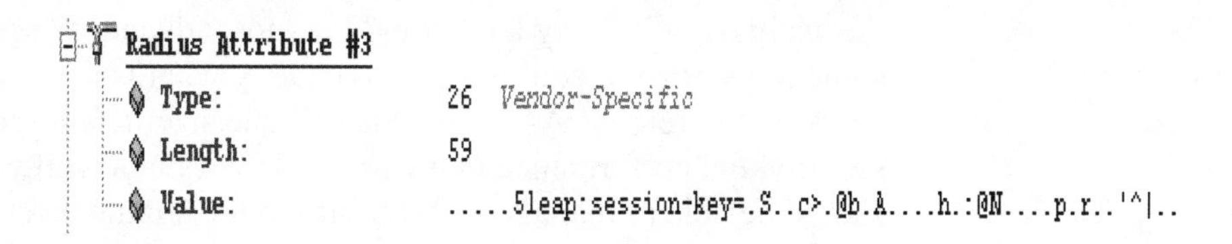

Packet	Source	Destination	Size	Absolute Time	Protocol	Summary
3	IP-192.168.100.1	IP-192.168.100.10	165	18:16:05.210399	RADIUS	C Access Request User:user1 NASPort:263
4	IP-192.168.100.10	IP-192.168.100.1	126	18:16:05.210807	RADIUS	C Access Challenge
5	IP-192.168.100.1	IP-192.168.100.10	205	18:16:05.213784	RADIUS	C Access Request User:user1 NASPort:263
6	IP-192.168.100.10	IP-192.168.100.1	109	18:16:05.214065	RADIUS	C Access Challenge
7	IP-192.168.100.1	IP-192.168.100.10	189	18:16:05.217014	RADIUS	C Access Request User:user1 NASPort:263
8	IP-192.168.100.10	IP-192.168.100.1	237	18:16:05.217629	RADIUS	R Access Accept

FIGURE 9.52 Encrypted RADIUS Attribute Carrying Unicast Encryption Key

Radius Attribute #3

Type:	26	*Vendor-Specific*
Length:	59	
Value:	.....5leap:session-key=.S..c>.@b.A....h.:@N....p.r..'^|..	

Summary

Each scenario an analyst may encounter in the real-world will use a different combination of network design, security protocols, and network equipment. Many vendors implement proprietary features which are often not supported in protocol analyzers. Understanding the basics of protocol analysis, security protocols, expected performance, wireless LAN topologies, proprietary protocol types, and equipment types will arm the analyst with the tools to make informed decisions. It is important for the wireless LAN analyst to have a thorough understanding of all the possible pieces of wireless LAN equipment that might be encountered during the actual analysis, and the expected behavior and capabilities of each piece of equipment.

The experienced analyst will more quickly get to the root of a performance or security problem by understanding that, by its very nature, the analyzer will likely lag behind the most advanced security protocol, sometimes causing confusion as to exactly what is being transmitted across the wireless LAN. There is no "one size fits all" solution for security and performance in a wireless LAN, but a well-equipped analyst should be able to manage and troubleshoot even the most complex wireless LAN.

Key Terms

Before taking the exam, you should be familiar with the following terms:

802.1h

Advanced Encryption Standard (AES)

auto-channel selection

frame forwarding

frame translation

Internet Protocol Security (IPSec)

Layer2 Tunneling Protocol (L2TP)

Point-to-Point Tunneling Protocol (PPTP)

RADIUS Attribute

RFC 1042

Role-based Access Control (RBAC)

wireless LAN bridging

wireless LAN mesh router

wireless repeater

Voice over IP (VoIP)

Voice over WLAN (VoWLAN)

Review Questions

1. Name one reason that an administrator may choose to configure an access point not to respond to probe request frames that contain a null SSID field.

2. AES encryption appears as ______ in most of today's wireless LAN protocol analyzers.

3. If a layer 2 security protocol such as TKIP and a layer 3 security protocol such as IPSec are used simultaneously, what security protocol will the wireless LAN protocol analyzer see?

4. Name two specific access point enhancements not specified by the 802.11 series of standards.

5. Name one advantage and one disadvantage to using a wireless LAN repeater.

6. Name two frame translation standards commonly used by 802.11 compliant access points as a portal between 802.11 and 802.3 networks.

7. In an 802.11a or 802.11g network operating in "turbo" mode, what is one system requirement?

8. In addition to the information obtained through a wireless LAN protocol analyzer, what other information may be required to troubleshoot a wireless LAN?

9. How is a baseline analysis performed in an infrastructure mesh wireless LAN?

10. What troubleshooting approach to client IPSec VPN connectivity problems should an analyst take when TKIP is in place between clients and the access point?

CHAPTER

10

802.11 Protocol Analyzers

CWAP Exam Objectives Covered:

- Demonstrate appropriate application of an 802.11a/b/g protocol analyzer
  - Troubleshooting
  - Performance testing
  - Security analysis
  - Intrusion analysis
  - Distributed analysis
- Apply generic features common to most 802.11a/b/g protocol analyzers
  - Protocol decodes
  - Peer map functions
  - Conversation analysis
  - Expert functions

In This Chapter

WLAN Analysis Dynamics

There are many factors that influence an analyzer's ability to properly capture and decode 802.11 frames. While there are far too many influences to cover in one book, we will cover an important subset of these influences in this chapter. Most wireless LAN analyzers share a common set of features, and have additional higher-end features that make them unique. All wireless LAN analyzers fall prey to the same limitations such as PHY support, encryption support, and their physical environment. This chapter will discuss the scope of wireless LAN analysis, analyzer types and features, protocol parameter impacts, mobility effects, and environmental impacts. Three abbreviated case studies are included to give the analyst a feel for common scenarios experienced in the wireless LAN analysis field.

MAC vs. PHY

Most information presented by a wireless LAN protocol analyzer will be MAC sublayer information, but there are certain pieces of physical layer information that are presented by the analyzer. Some analyzers present more physical layer information than others, but there are many items common to most analyzers. Physical layer parameters such as signal and noise levels, data rate, the channel on which a frame was captured, frame length, and errors experienced are commonly interpreted by an analyzer. Illustrations of what an analyzer might present are shown in Figures 10.1 and 10.2.

FIGURE 10.1 Wireless LAN Analyzer's PHY Parameter Interpretation

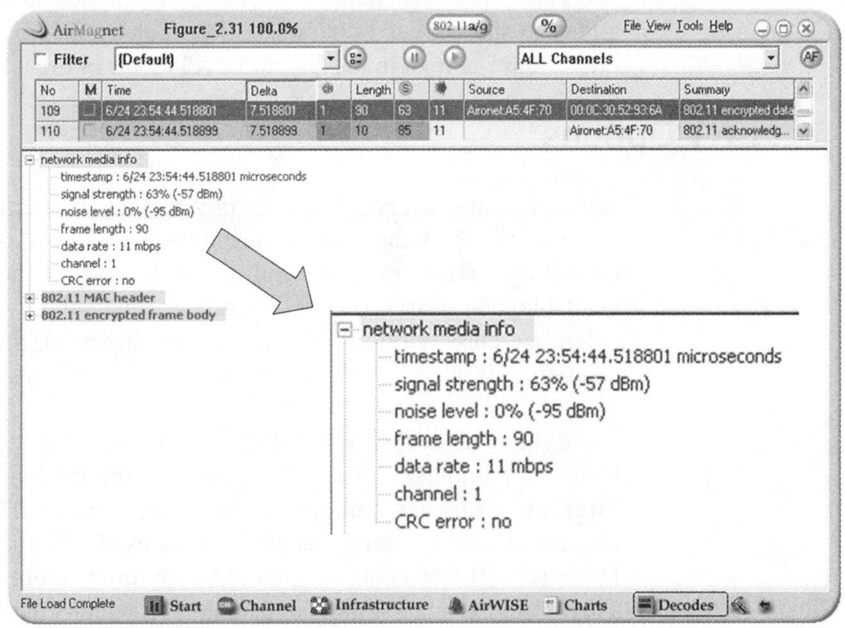

FIGURE 10.2 Wireless LAN Analyzer Displaying RF Characteristics

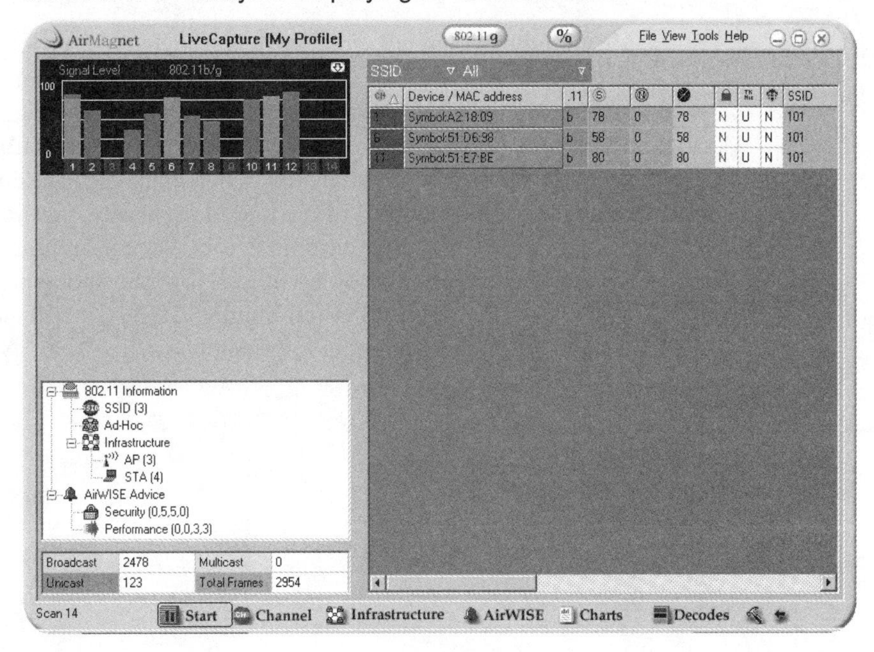

Figure 10.2 illustrates a wireless LAN analyzer displaying three DSSS access points using channels 1, 6, and 11. Some analyzers have the ability to show layer 1 characteristics, such as the cross-channel interference shown in the Signal Level meter in the left pane.

PHY Options

Analyzers must support the same physical layer parameters (PHY) as the network on which they are capturing frames. Due mostly to a quickly dwindling market share of installed FHSS systems, there are no readily available, enterprise-class FHSS protocol analyzers. All of today's enterprise-class wireless LAN analyzers support DSSS and OFDM technologies in the 2.4 GHz ISM and 5 GHz UNII bands.

Stand-alone analyzers, which typically run as software on laptops or handheld computers, often use PCMCIA, CF, or MiniPCI RF radios. These analyzers can only tune to and capture on one DSSS or OFDM channel at a time, which means that wireless LAN traffic on all 2.4 GHz DSSS and OFDM channels cannot be captured simultaneously. Therefore, the analyst must understand on which channel(s) to capture the traffic he or she is seeking at any point in time.

Most analyzers have a scanning feature whereby a user can program the analyzer to capture on a given list of channels for a certain amount of time per channel. This feature is useful for detecting wireless networks (access points and Ad Hoc networks) and analyzing a small subset of the wireless LAN traffic on a set of channels. Each analyzer has settings for both the channels to be scanned and the rate of scanning. Finding wireless LANs is important for security administrators when searching for rogue devices, for users when looking for a wireless LAN hotspot, and for site surveyors when verifying an installation. Figures 10.3 - 10.5 illustrate the scanning features in three different enterprise-class wireless LAN analyzers.

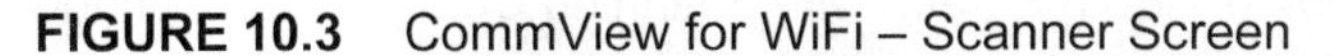

FIGURE 10.3 CommView for WiFi – Scanner Screen

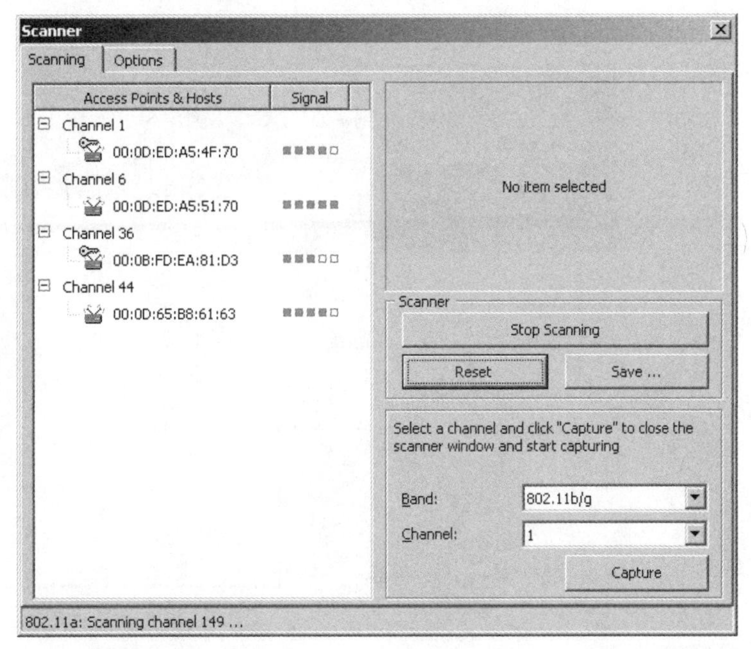

FIGURE 10.4 Network Instruments Observer – Site Survey Screen with Scanning Configuration Window

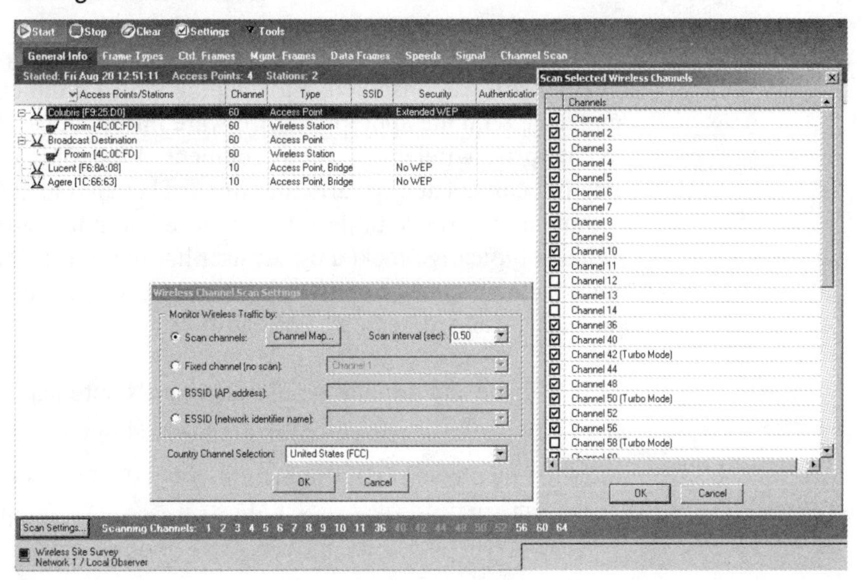

FIGURE 10.5 AirMagnet Laptop – Start Screen with Scanning Configuration Window

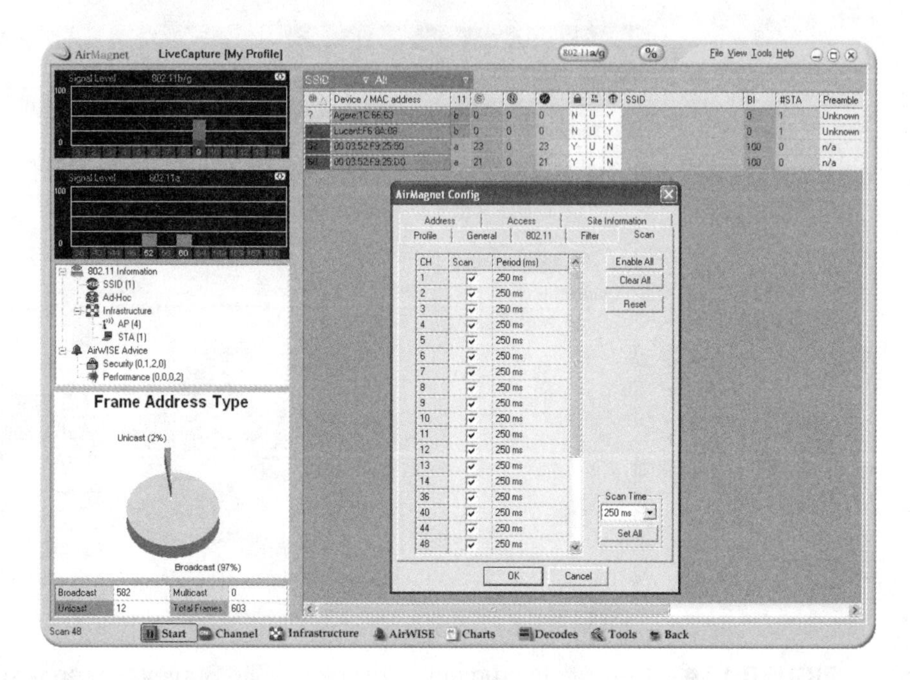

Laptop vs. Distributed

The analyst should realize that a single wireless LAN protocol analyzer, whether software installed onto a laptop with a radio card or a stand-alone appliance with the appropriate firmware installed, has many limitations in an enterprise wireless LAN environment. First, only one physical area may be scanned at any particular time. That area is the physical area within hearing range of the antenna installed in the radio card. Part of this signal is typically blocked by the user holding the analyzer, and many other factors (some of which are listed below) may also affect what the analyzer is able to perceive correctly.

- Type and orientation of radio card's antenna
- Receive sensitivity of the radio card in use
- The physical environment in the radio card's immediate vicinity
- The RF environment in the radio card's immediate vicinity

Since the analyzer can only listen in one area at a time, problems that are occurring network-wide or across a large physical area are not easily tracked down.

FIGURE 10.6 Laptop Analyzer Scenario

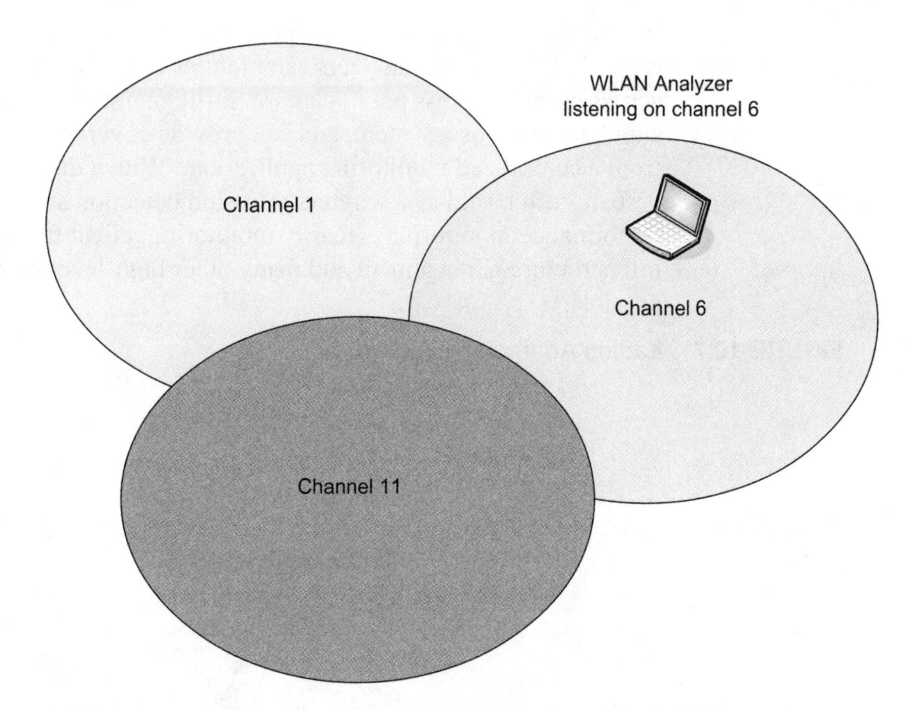

Wireless LAN analyzers are typically used on one channel at a time, but there are analyzer products available that can analyze 802.11a and 802.11b/g channels simultaneously using a PC card with multiple internal radios. Having the ability to scan an 802.11a and 802.11g channel simultaneously may allow the user to look for 802.11a rogue access points while troubleshooting a client connectivity problem on an 802.11b channel. This scenario is not very probable since it is not likely that you would do troubleshooting on one network while doing a security audit on another, but it is possible using multiple capture windows within the same analyzer. The ability to scan on one channel per radio limits a laptop analyzer's use in an enterprise environment where many channels are in use simultaneously and clients are constantly roaming between access points on different channels. If a mobile client is having roaming problems, for instance, the analyzer could only capture traffic for that client before it roams. After that particular client has roamed, the analyzer loses track of it. Also, there is often no way to accurately predict to which access point the client will roam, so monitoring the client both before and after it roams is difficult.

It is because of these and other limitations that distributed analyzers were created. Distributed analyzers can monitor clients as they move from access point to access point, capture traffic on multiple channels simultaneously as a system, and can provide coverage of large premises from a centralized monitoring application. With a distributed analyzer system - often sold as a wireless intrusion detection system (WIDS) - performance monitoring, security monitoring, client tracking, infrastructure management and many other high-level functions are all possible.

FIGURE 10.7 Laptop Analyzer Scenario

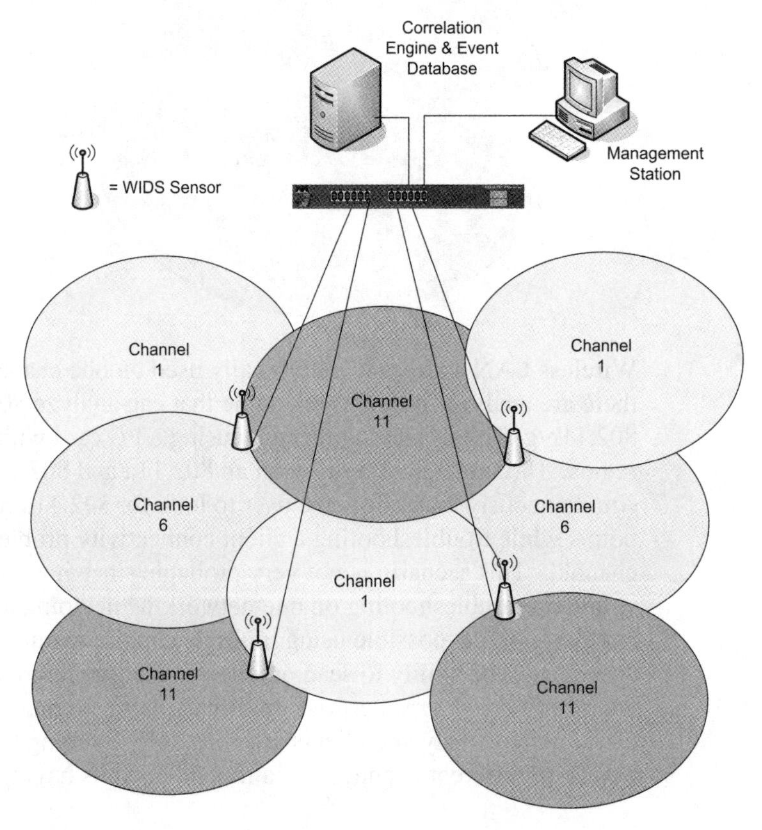

WIDS systems are generally comprised of a central correlation engine, event database, and distributed sensors. The sensors stream 802.11 frames to the correlation engine, and interpreted events are logged into the database. Sensors may be remotely manageable and may be able to provide 802.11 frames to a protocol analyzer console in real time for manual analysis.

WIDS Sensors

All of the WIDS on the market today have 802.11a/b/g radios, but one fact that some manufacturers tout as important is that their sensors have separate 802.11a and 802.11b/g radios in each sensor. This feature allows for faster scanning through the many 802.11a and 802.11b/g channels and it allows for simultaneous scanning in both the 2.4 GHz and 5 GHz bands.

Another common feature of WIDS sensors is the ability to connect to the sensor from within a laptop analyzer. This feature allows the laptop analyzer to use the radio in the WIDS sensor to capture frames at the sensor's location. The captured frames are encapsulated in Ethernet frames and sent across the wired network to the laptop analyzer.

Analyzer/Sensor Placement

WIDS sensor placement and the number of sensors deployed depend on the intended purpose of the system. More deployed sensors equates to better event correlation across the system as a whole. The WIDS correlation engine can give more detailed analysis when more frames on more channels are captured simultaneously. Each frame is time stamped, and the sensor knows which frame was captured on which channel and in what order.

Most engineers that have worked in the wireless LAN market for a while understand the importance of a site survey. The analyst should also understand the importance of and procedures involved in a WIDS sensor site survey. Most WIDS manufacturers provide product training, and it is important to learn how to perform a sensor site survey for optimal coverage and analysis detail levels.

Encryption Impact

If you have been working in protocol analysis for years, you are no doubt accustomed to Ethernet and other layer 2 protocols having unencrypted traffic in a LAN environment. On a wireless LAN, data links between clients and access points are almost always encrypted in corporate environments. This level of data protection may be required by government regulations in some industries. There are several different types of data link layer encryption protocols available.

A small percentage of today's wired LANs are protected by layer 2 or layer 3 authentication and layer 3 encryption, but use is on the rise due to the security panic sparked by wireless LAN technology. It is no longer acceptable in the eyes of many IT managers to encrypt only wireless traffic. Since layer 2 authentication and encryption are the most popular security mechanisms used by wireless LAN devices, analysts must focus closely on layer 2 header information for troubleshooting purposes. This focus is due to information inside the layer 2 frames being encrypted.

It is not considered overkill in some organizations to perform layer 2, 3, and 7 data encryption. The layer 3 and 7 encrypted data will not be seen by the protocol analyzer when layer 2 encryption is in use, though use of multiple layers of encryption might have an adverse affect on data throughput. Enterprise-class layer 2 encryption schemes that use rotating encryption keys such as WPA-TKIP and 802.1X/EAP-TTLS are not supported in protocol analyzers because each wireless data link is unique. Wireless LAN protocol analyzers often support shared key encryption such as static WEP, but support for more complex encryption techniques like TKIP, used in WPA-PSK, is currently limited to only a few analyzers. Figures 10.8 and 10.9 illustrate common scenarios where layer 2, 3, and 7 data encryption is in use.

FIGURE 10.8 Triple Data Encryption Scenario #1

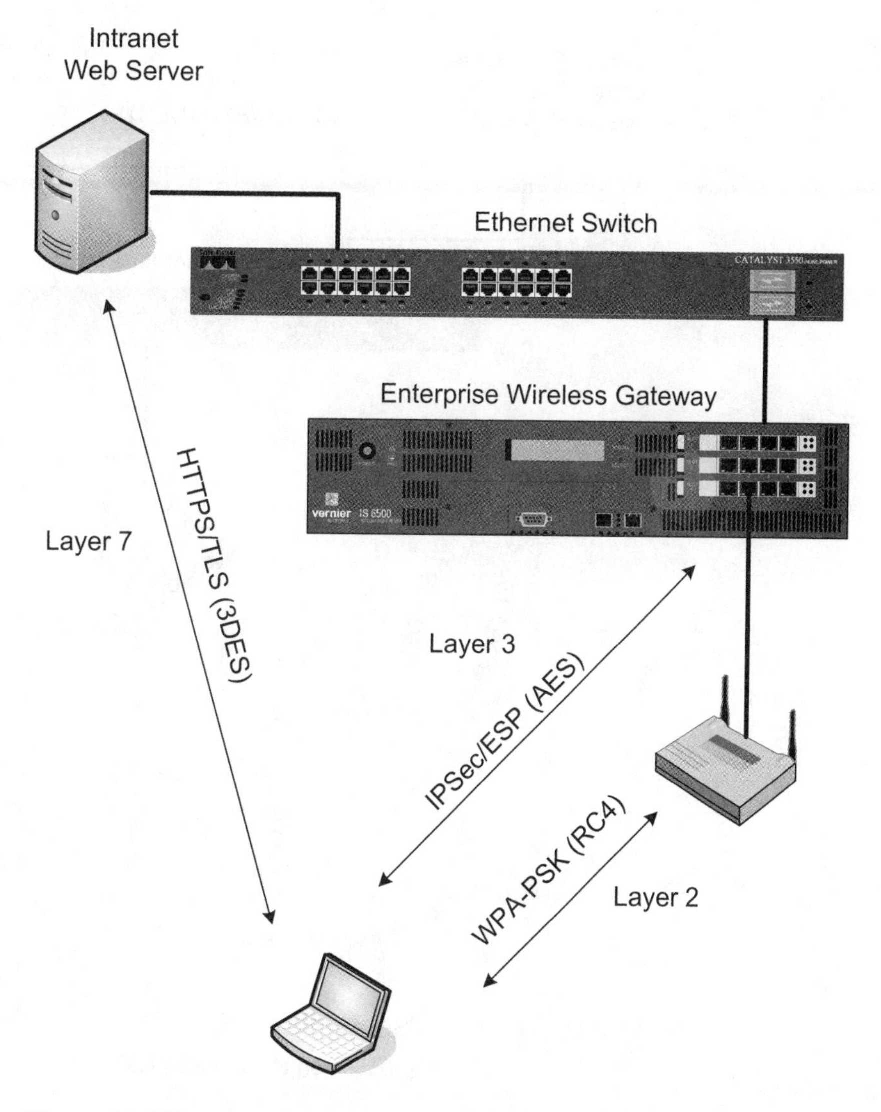

Figure 10.8 illustrates layer 2 encryption (WPA-PSK) between the client and access point. An IPSec VPN tunnel is built to an EWG, and then a secure application (HTTPS) is used to connect to the Intranet Web Server.

FIGURE 10.9 Triple Data Encryption Scenario #2

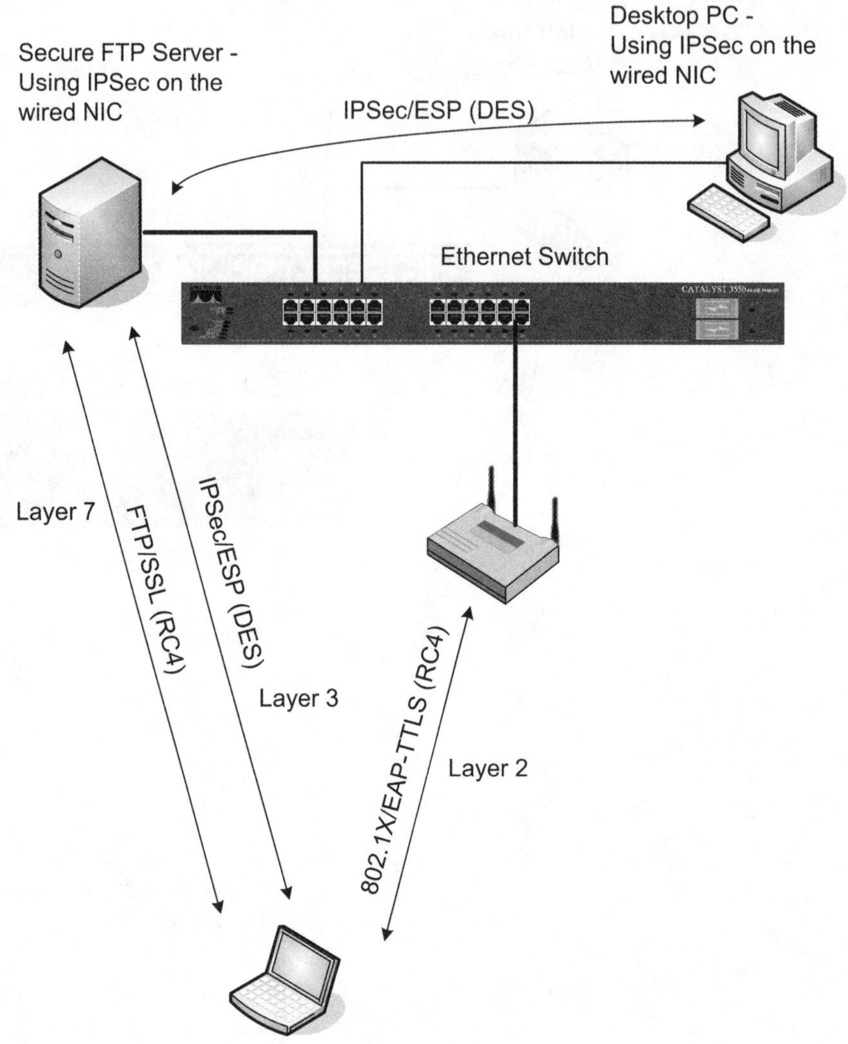

Figure 10.9 illustrates a scenario in which IPSec/ESP is being used on the wired and wireless segments of the network. All nodes on this network must use IPSec/ESP in order to connect to other nodes. EAP-TTLS is being used to secure the wireless link, and FTP over SSL is being used to secure the data transfer at the application layer.

Remember that as data moves down the OSI stack, it is encapsulated at each lower layer. The encrypted layer 7 data is encapsulated into layer 3 packets, which then get encrypted and passed to layer 2 for further encapsulation and encryption. This multiple encapsulation means that analysts will only have layer 1 and 2 information available for analysis regardless of layer 2-7 protocols and encryption being used on the wireless link. Layer 3-7 encryption will affect troubleshooting on the wired side of the end-to-end link.

Common Features

Most wireless LAN protocol analyzers on the market today are enterprise-class analyzers. All wireless LAN protocol analyzers have a common base of features which allow them to perform manual analysis, and each analyzer has unique features that make it particularly attractive to certain user groups. Some of the areas that should be considered when purchasing a wireless LAN protocol analyzer are (in no particular order):

- User interface (ease of use and discernable features)
- Cost
- Security analysis features
- Performance analysis features
- Expert features (automated analysis features)
- Stand-alone vs. distributed options
- Supported wireless LAN cards
- Site survey features
- Pre- and post-capture filtering features
- Networking tools
- Integration with other software products from the same vendor
- Frame capture import/export features
- Graphical node maps (also called peer maps)
- Real-time vs. Non Real-time frame decoding
- Layers of the OSI model that can be analyzed

Most manufacturers publicly release user manuals, offer downloadable demo versions of their products, and some even have online simulators of their products. These offerings allow the analyst to become familiar with their analyzer and to compare the aforementioned features between products. There are many cases where an analyst may need to purchase two different analyzers to successfully perform his duties as an analyst.

Analyzer Types and Features

There are two categories of wireless LAN protocol analyzers: portable and distributed. Portable analyzers come in the form of software that can be installed on a laptop or handheld computer and in the form of a portable hardware appliance with pre-loaded firmware. The most common analyzer today is the software analyzer that is installed on laptop computers. Distributed analyzers are made up of multiple software applications and often hardware sensors that coordinate the capture, recording, and correlation of frames and network events on the wireless medium. All distributed analyzers have a centralized database and event correlation engine application, and the main differences lie in whether a software or hardware sensor is used. The degree of scalability of distributed systems also varies.

Laptop and Handheld Analyzers

Wireless LAN protocol analyzers began as hardware appliances or software that were installed on a portable computer such as a laptop, tablet PC, or handheld. Handheld analyzers offer great portability required by site surveyors and some security professionals. Laptop and tablet PCs are great for thorough security and performance analysis because of their large display, powerful CPU, large amount of memory, and portability.

Software analyzers can be installed on a number of operating systems such as MAC OS X, Microsoft Windows, and a variety of Linux distributions. Figure 10.10 shows and example of both a laptop and a handheld analyzer.

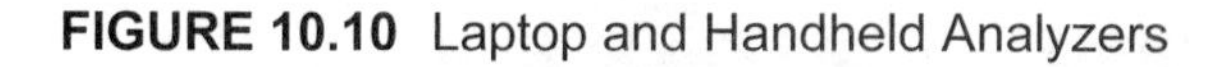
FIGURE 10.10 Laptop and Handheld Analyzers

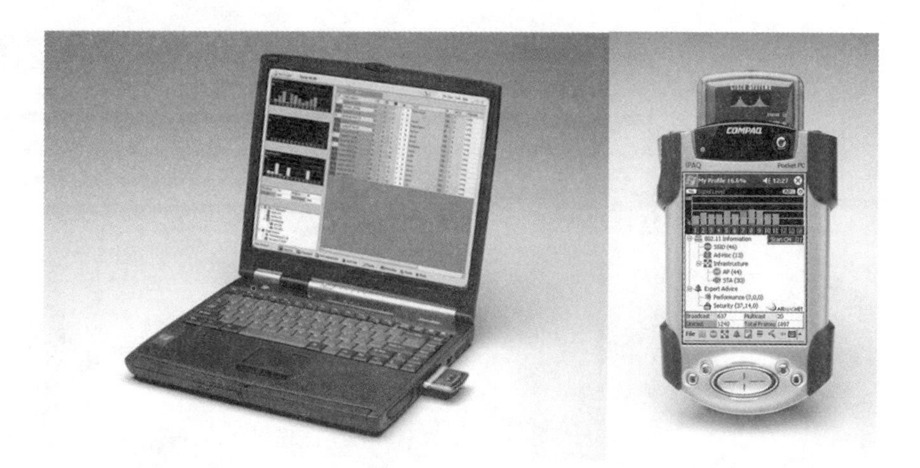

Frame Decoding

The most common use of a protocol analyzer is capturing and decoding individual frames or frame exchanges. The analyzer interprets and sorts the layer 2 through 7 header fields so that the analyst can understand what is being transmitted across the wireless medium. Figures 10.11 and 10.12 illustrate two analyzers performing decoding functions. Some analyzers perform decoding while they are still capturing frames. This feature is often referred to as real-time decoding. Some analyzers require that you stop the capture in order to decode the frames. Some analyzers focus their wireless analysis strictly at the MAC layer while others can fully decode and interpret all portions of a data frame from the MAC layer to the Application layer. Some analyzers can display limited physical layer information such as found in Figure 10.1, but this information is not decoded from the PPDU preamble, header, or payload.

FIGURE 10.11 Frame Decoding - Observer

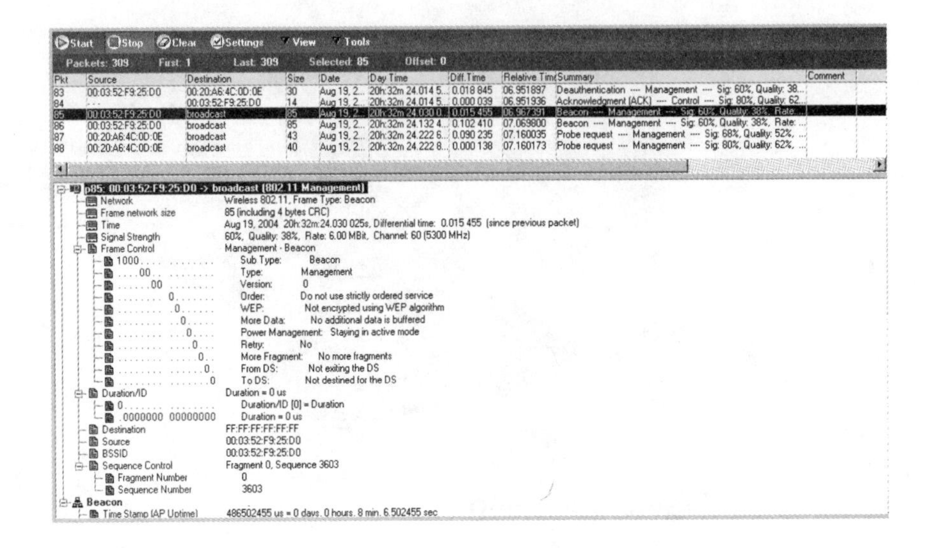

FIGURE 10.12 Frame Decoding – AirMagnet Laptop

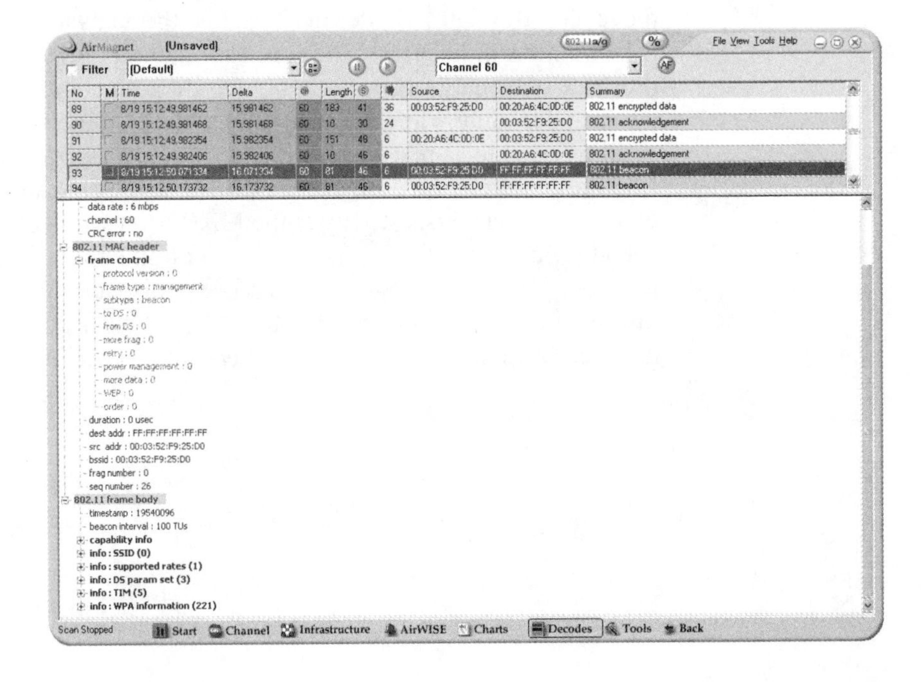

Each line in the packet (upper) window represents a single frame that the analyzer captured and that is now buffered for this capture window. The packet window has a number of columns of information. Depending on the analyzer, each column can typically be enabled or disabled, moved into a different order, and made wider or narrower. Each column shows different information about the packet, and colors may be associated with certain aspects of the protocol or conversation such as source or destination address, protocol, etc. Most analyzers have a quick way of changing the columns such as the context menu shown in Figure 10.13 or by right-clicking the column headers themselves.

FIGURE 10.13 Frame Decoding – Changing Information Columns

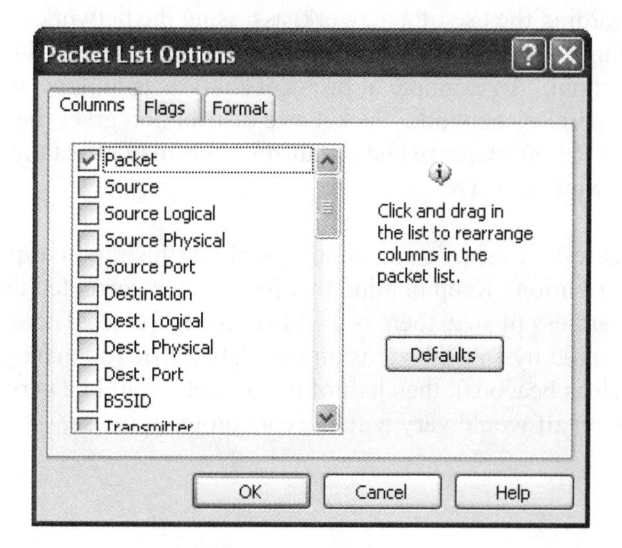

Protocol Statistics

The objectives of using protocol statistics analysis are to understand the general interaction between network protocols and to identify potential protocol problems before they impact the end-user community. An analyst might apply this method of analysis when concerned about network or server performance degradation or when multiple networks have been connected together to form a larger infrastructure. Protocol statistics analysis would be used to confirm proper operation. The scenario of combining network infrastructures is common considering many wireless networks are built as separate entities from the regular network and then later merged.

An analyst would characterize network traffic to determine which protocols are operating and what applications are being used as well as establishing or confirming expectations regarding the use, loading, operation, and problem status of a network. When working with a vendor to solve equipment problems, the analyst may need to identify the types of traffic and applications that are using the vendor's equipment. This ability to identify the types of traffic is especially necessary because vendors are constantly creating new features and new protocols that behave in unexpected ways. In order to understand the "big picture" regarding the use of a network, assessing the network as part of an on-going program of network management, monitoring, and analysis is essential. An example of protocol analysis is looking at packet sizes on the wireless network. Packet size distribution gives the analyst an idea of how RF interference and retransmissions may affect the performance of the wireless LAN.

Figure 10.14 shows a common pie chart illustration of packet size distribution. Keep in mind that for every uncorrupted unicast data frame, regardless of size, there is a 14 byte ACK frame. Therefore, if all traffic captured by an analyzer is unicast data traffic (no management traffic such as beacons), then half of the packets would be very small while the other half would vary with the data protocol in use.

FIGURE 10.14 Protocol Statistics – Packet Size Distribution – AiroPeek NX

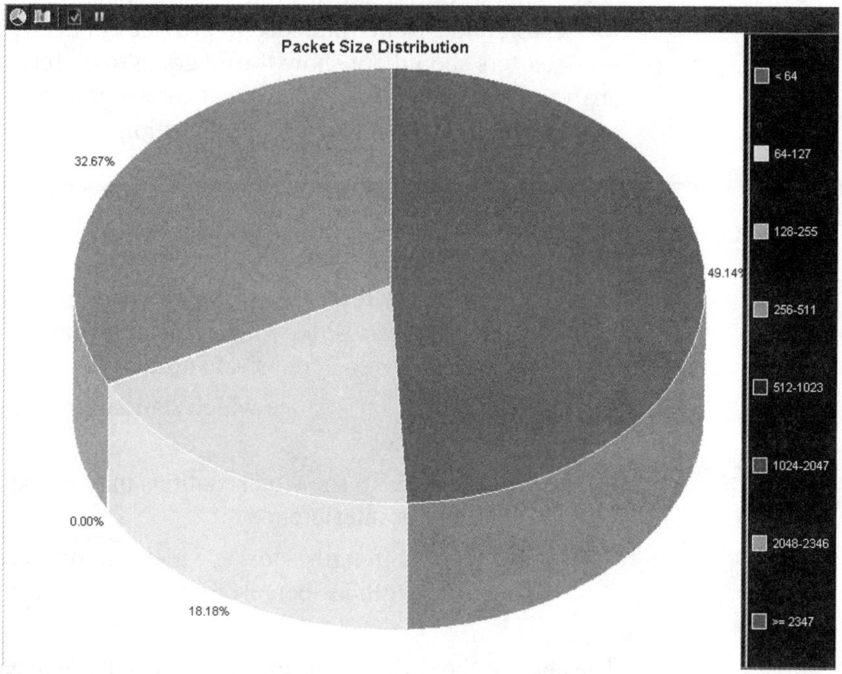

Node Statistics

The objectives of analyzing node statistics are:

- to determine the most active communicators on a wireless network,
- to evaluate the types of protocols being used by individual stations, and
- to identify potential problem stations or protocol characteristics.

An analyst may want to apply this type of analysis when initially exploring a specific end-user complaint or determining the status of a particular device prior to upgrade.

Access points will typically show up as the most active nodes on the wireless network because several stations may be talking to each access point. End-user stations that show very high utilization may be abusing the "acceptable use" policy for the wireless LAN. The analyst should keep in mind relative traffic patterns. Most data is typically read *from* a file server, and a router intended to provide connectivity to corporate headquarters should not show traffic going to an Internet site. Most analyzers have detailed statistics that can be used by the analyst to evaluate the protocols in use by each station and which stations are communicating with each other.

Node statistics can be sorted in a variety of manners such as:

- By MAC address – to see which vendors' radios are in use and special scenarios like routers that may have multiple logical addresses for each physical address
- By Encryption – to see which stations are using encryption and which are not
- By Retries – to see which stations may be experiencing signal blockage or interference
- By Signal Strength – to see which stations have the weakest signal strength (as perceived by the analyzer)

The analyst should look for unexpected results such as stations sending more data than they should be sending, stations talking to servers that they should not be talking to, and stations using unexpected protocols. Packet size and utilization should be reasonable for applications being used (e.g. terminal access should produce smaller frames; file transfers should produce larger frames, etc.). Utilization is especially important on an 802.11b network, a half-duplex environment in which bandwidth is at a premium. Figure 10.15 illustrates a snapshot of typical node statistics.

FIGURE 10.15 Node Statistics – AiroPeek NX

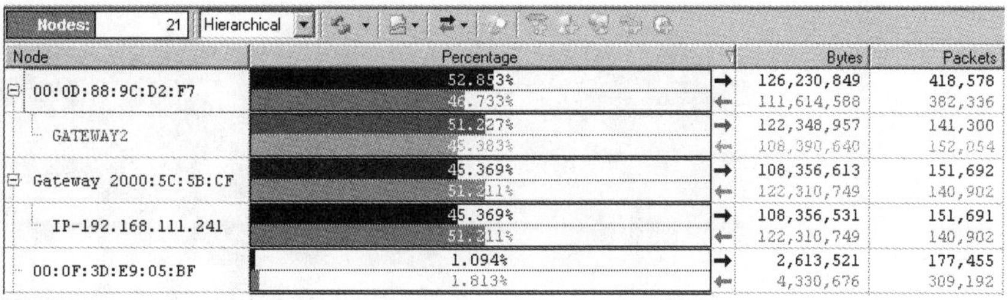

Nodes: 21 Hierarchical

Node	Percentage		Bytes	Packets
00:0D:88:9C:D2:F7	52.853%	→	126,230,849	418,578
	46.733%	←	111,614,588	382,336
GATEWAY2	51.227%	→	122,348,957	141,300
	45.383%	←	108,390,640	152,054
Gateway 2000:5C:5B:CF	45.369%	→	108,356,613	151,692
	51.211%	←	122,310,749	140,902
IP-192.168.111.241	45.369%	→	108,356,531	151,691
	51.211%	←	122,310,749	140,902
00:0F:3D:E9:05:BF	1.094%	→	2,613,521	177,455
	1.813%	←	4,330,676	309,192

Channel Analysis

Some analyzers give extreme amounts of detail about the channel on which they are capturing. As you can see in Figure 10.16, this analyzer can see signal strength, CRC errors, noise, data rates of each frame, every frame type sent, throughput, and more. This detailed information comes from the analyzer capturing the frames, interpreting the information in the frames, and then doing analysis against the frames to arrive at the statistics shown below. Additional information may be gathered when information is passed by the PHY to the MAC with the octets of the frames or otherwise reported in MIB variables.

FIGURE 10.16 Channel Analysis – AirMagnet Laptop

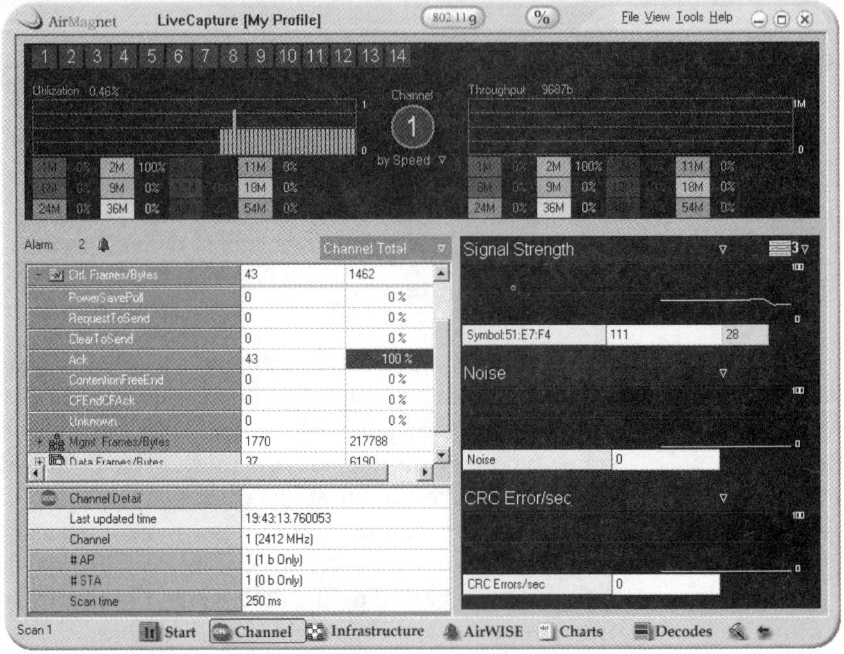

Conversation Analysis

The objectives of conversation analysis are to understand protocol behavior between two communicating stations, to recognize common conversation-related problems, and to differentiate between traffic categories. Differentiating between traffic categories such as client/server, terminal/host, and peer-to-peer allows the analyst to recognize acceptable types of activities on the wireless LAN.

An analyst may want to apply conversation analysis if he or she is troubleshooting a problem involving a specific node or protocol. When analyzing conversations, it is helpful to determine whether the conversation involves user or background traffic. User traffic is traffic that directly carries out user requests, such as downloading email or transferring a file. Background traffic is traffic that supports user traffic, such as 802.11 management traffic or looking up the address of a server using WINS or DNS.

Background traffic may include:

- Network management traffic such as SNMP
- Router traffic such as RIP, OSPF, EIGRP
- Spanning Tree Protocol traffic such as BPDUs
- IP connection management traffic such as ARP, DHCP, DNS
- 802.11 data-link management traffic such as association, acknowledgement, and beacon frames

All user traffic and all background traffic have certain characteristics in common, making it easier to troubleshoot one user or background protocol if you already understand another. An example of conversation analysis in wireless networks is repetitive loss of signal with the access point due to excessive distance or poor atmospheric conditions.

Some analyzers can track and illustrate conversations step by step as shown in Figure 10.17. This level of detail allows the analyst to troubleshoot not only connections between hosts, but also the proper operation of applications. In this illustration, a TCP session is set up and torn down.

FIGURE 10.17 Session Analysis – Network Chemistry Packetyzer

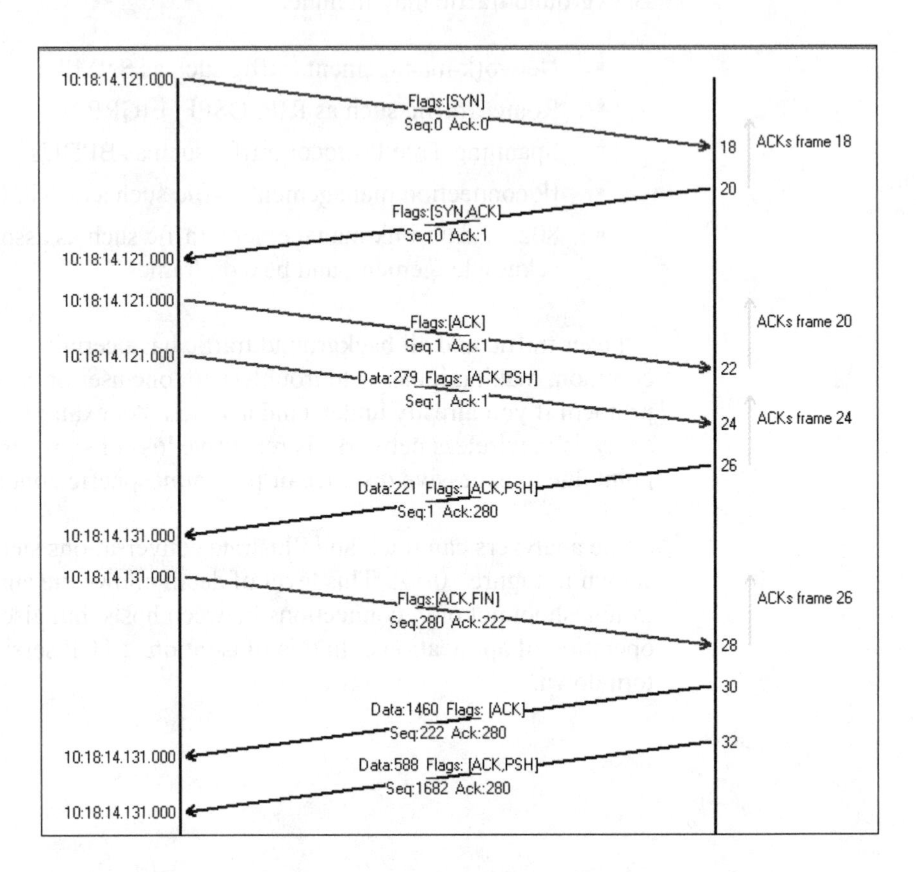

Analyzers have the ability to associate frames that are part of a conversation based on knowledge of the protocol, sequence numbers, and other factors. The analyst can configure the analyzer to show only the frames related to a specific conversation and to hide the other frames.

The TFTP protocol is a good example of when this feature might be very useful. TFTP negotiates a connection using port 69, but subsequently chooses random ports on both sides of the connection for data transfer. Port 69 is only used to establish the connection. Figures 10.18 and 10.19 illustrate an analyzer associating the port 69 traffic with the subsequent random-port traffic.

FIGURE 10.18 TFTP Conversation

```
User Datagram Protocol, Src Port: 1085 (1085), Dst Port: tftp (69)
    Source port: 1085 (1085)        ---> Initial Source at Client
    Destination port: tftp (69)     ---> Initial Destination at Server
    Length: 24
    Checksum: 0xa7eb (correct)
Trivial File Transfer Protocol
    Opcode: Read Request (1)
    Source File: big.txt
    Type: octet

User Datagram Protocol, Src Port: 1105 (1105), Dst Port: 1085 (1085)
    Source port: 1105 (1105)        ---> Initial Source at Server
    Destination port: 1085 (1085)   ---> Returned to Initial Source at Client
    Length: 524
    Checksum: 0x8e49 (correct)
Trivial File Transfer Protocol
    Opcode: Data Packet (3)
    Block: 1
    Data (512 bytes)

User Datagram Protocol, Src Port: 1085 (1085), Dst Port: 1105 (1105)
    Source port: 1085 (1085)        ---> Conversation now between 1085
    Destination port: 1105 (1105)        at Client and 1105 at Server
    Length: 12
    Checksum: 0xadcd (correct)
Trivial File Transfer Protocol
    Opcode: Acknowledgement (4)
    Block: 1
```

FIGURE 10.19 Watch Protocol – Network Chemistry Packetyzer

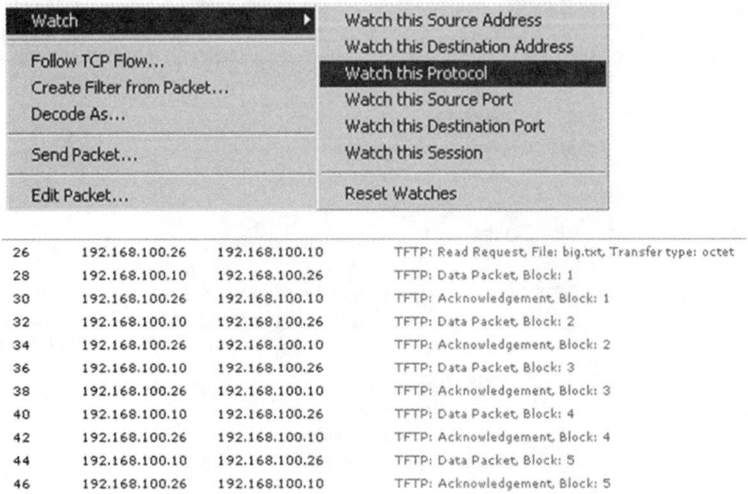

Watch
Follow TCP Flow...
Create Filter from Packet...
Decode As...
Send Packet...
Edit Packet...

Watch this Source Address
Watch this Destination Address
Watch this Protocol
Watch this Source Port
Watch this Destination Port
Watch this Session
Reset Watches

26	192.168.100.26	192.168.100.10	TFTP: Read Request, File: big.txt, Transfer type: octet
28	192.168.100.10	192.168.100.26	TFTP: Data Packet, Block: 1
30	192.168.100.26	192.168.100.10	TFTP: Acknowledgement, Block: 1
32	192.168.100.10	192.168.100.26	TFTP: Data Packet, Block: 2
34	192.168.100.26	192.168.100.10	TFTP: Acknowledgement, Block: 2
36	192.168.100.10	192.168.100.26	TFTP: Data Packet, Block: 3
38	192.168.100.26	192.168.100.10	TFTP: Acknowledgement, Block: 3
40	192.168.100.10	192.168.100.26	TFTP: Data Packet, Block: 4
42	192.168.100.26	192.168.100.10	TFTP: Acknowledgement, Block: 4
44	192.168.100.10	192.168.100.26	TFTP: Data Packet, Block: 5
46	192.168.100.26	192.168.100.10	TFTP: Acknowledgement, Block: 5

Expert Feature

The "Expert" feature, an automated analysis tool found in most enterprise-class analyzers, may detect problems such as:

- Overloaded RF medium or server
- Inefficient client
- Low server to client throughput
- Low client to server throughput
- One-way traffic
- Slow server response time
- Failing radio hardware
- Unanswered requests

Some analyzers are designed to be mostly an "Expert" analyzer, meaning that frame decoding is their secondary function. Figure 10.20 and 10.21 illustrate two different analyzers' Expert feature.

FIGURE 10.20 Expert Feature Example #1 – AiroPeek NX

Conversations analyzed: 8
Problems detected: 3,732
(Using MyExpertProfile.exp)

Net Node 1 (Client)	Net Node 2	Problems	Data Rate	Signal	Packets	Bytes	Duration	Avg Delay	TCP Status
COMPUTER4	COMPUTER-2	3,587			19,266	14,603,998		1.729 ms	
TCP/Port 1032<->ftp			11	78%	3	262	00:00:00.150	2.516 ms	Open
TCP/Port 1042<->ftp-data		1,348	11	68%	15,315	14,040,004	00:04:23.554	0.942 ms	Closed
TCP/Port 1024<--ftp-data		2	11	51%	1	1,536	00:02:37.550		Open
TCP/Port 1042-->ftp-data		180	11	70%	170	13,940	00:00:10.244		Closed
TCP/Port 1042<--ftp-data		1,964	11	62%	2,851	233,782	00:02:17.161		Closed
Wireless Transmission Retry		3							
IP Packet with CRC Frame Error		4							
TCP Invalid Checksum		3							
TCP Reset Inactive Connection		1							
Wireless Too Many Retries		288							
Wireless Data Rate Change		6							
One-Way Traffic		1							

Problem Summary | Problem Log | Node Details

Total: 3,732

Layer	Event	Count
Physical	Wireless Access Point - Broadcasting SSID	14
Physical	Wireless Access Point - WEP Not Required	14
Physical	Wireless Access Point - Rogue	28
Physical	Wireless Channel Overlap	25
Physical	Wireless Access Point - Weak Signal	7
Physical	Wireless Client - Rogue	6
Physical	Wireless Access Point - Restarted	3
Transport	TCP Invalid Checksum	62
Network	IP Packet with CRC Frame Error	67
Physical	Wireless Transmission Retry	884
Physical	Too Many Physical Errors	28
Physical	Wireless Client - Physical Errors	13
Physical	Wireless Access Point - Physical Errors	7

Packets | Nodes | Protocols | Summary | Graphs | Channels | Signal | Log | Expert | Peer Map | Filters

FIGURE 10.21 Expert Feature Example #2 - Observer

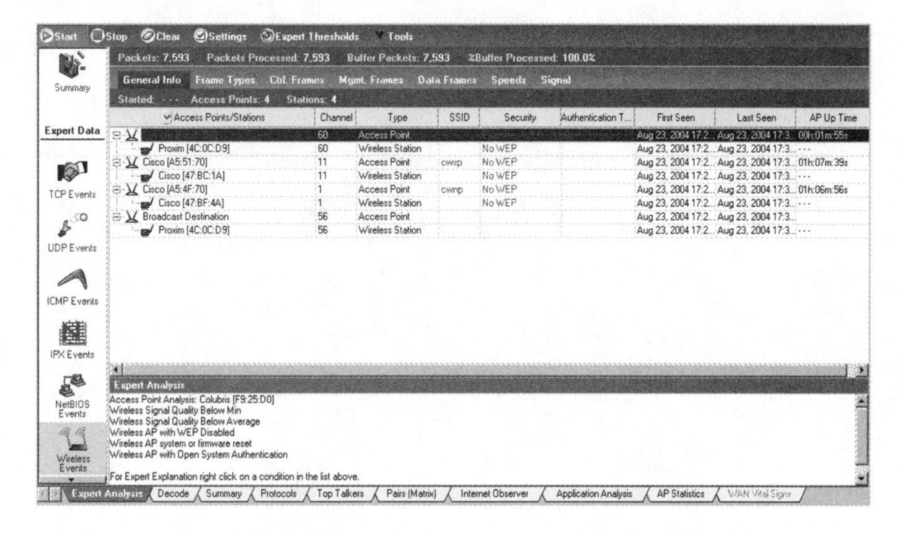

The Expert feature is essential when problematic events are occurring rapidly at random intervals on the wireless medium. An analyst could

never hope to see and interpret events such as rapid rate changes due to RF interference or intermittent radio failure, but an analyzer can easily see that this behavior matches one of its pre-configured performance problem signatures. Figure 10.22 illustrates an intermittently failing radio card due to faulty firmware in a wireless router that is transmitting a continuous stream of frames. Without an analyzer, troubleshooting this problem would take a significant amount of time.

FIGURE 10.22 Intermittently Failing Radio

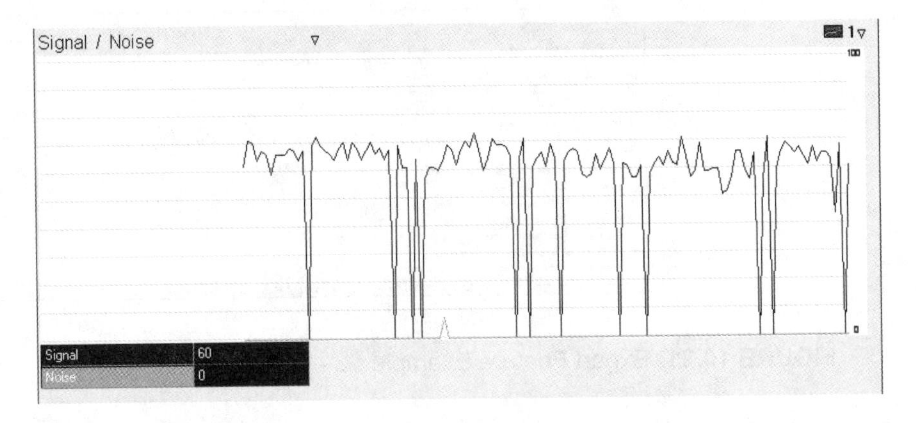

A normal radio transmission should look like the illustration in Figure 10.23.

FIGURE 10.23 Normal Radio Operation

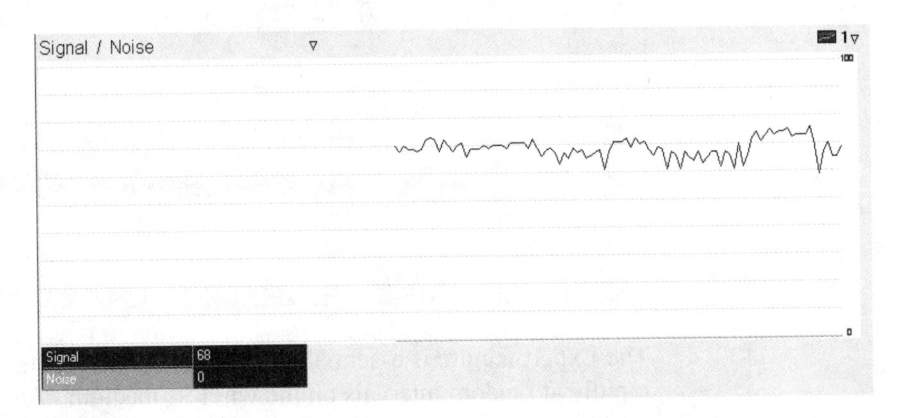

Peer Maps

A peer map, also called a pair matrix, gives the analyst an illustration of all of the wireless conversations on the network. Peers may be immediate radio link partners or bridged data-link partners with an access point invisibly in between. In the later case it may not be obvious which peers are on the wired link and which are on the wireless link. Peer maps may illustrate higher traffic quantities with thicker lines, bigger node dots, etc. MAC addresses or IP addresses may be displayed. Peer maps give the analyst a quick and easy way to identify which nodes are broadcasting heavily, which nodes are access points, which nodes are communicating with each other, and many other important information items. Figures 10.24 and 10.25 illustrate typical peer maps.

FIGURE 10.24 Peer Map – AiroPeek NX

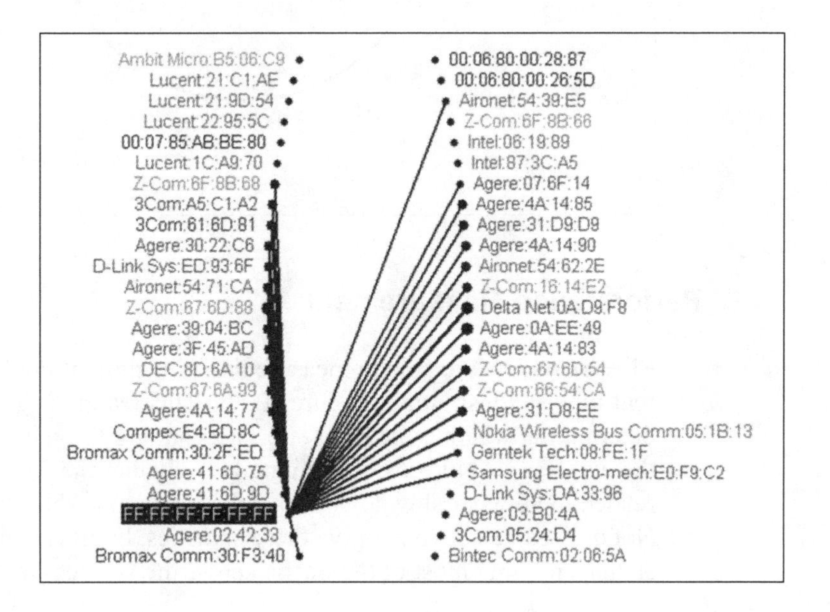

FIGURE 10.25 Pair Matrix - Observer

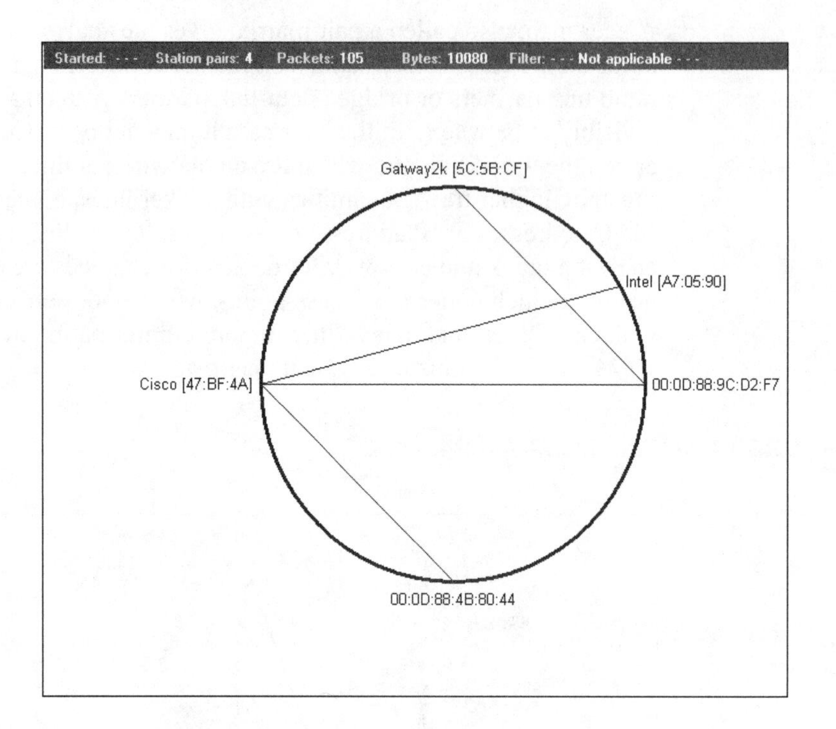

Performance Measurement

Most analyzers are able to measure the utilization of the RF medium in real-time as illustrated in Figure 10.26. This feature is useful for determining whether the right 802.11 technology is being used, whether the wireless LAN needs to be upgraded, whether the RF medium is related to network slowdowns, or if the wireless LAN is being abused. Not only can an analyzer give these statistics, but it can also point out the stations moving most of the traffic across the wireless medium.

FIGURE 10.26 Performance Measurement – TamoSoft CommView for WiFi

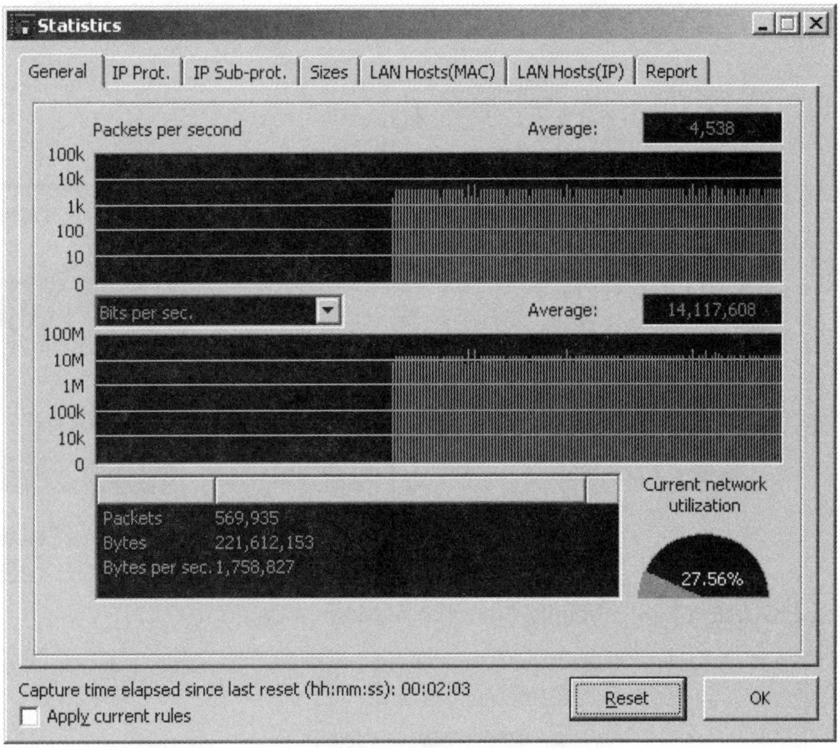

Alarms

There are many performance-related issues that could become problematic on a wireless LAN. An analyst cannot be expected to catch or interpret them all, so analyzers do the job instead. Many analyzers have a pre-programmed set of performance and security related signatures that trigger alarms when their criteria are met by captured frames. Figure 10.27 illustrates an analyzer showing multiple performance-related alarms in the top-left pane with an interpretation of what each alarm means in the top-right pane. Figure 10.28 illustrates the same alarm scenario for security alarms. Using these alarms, the analyst can begin troubleshooting each problem one by one.

FIGURE 10.27 Performance Alarms – AirMagnet Laptop

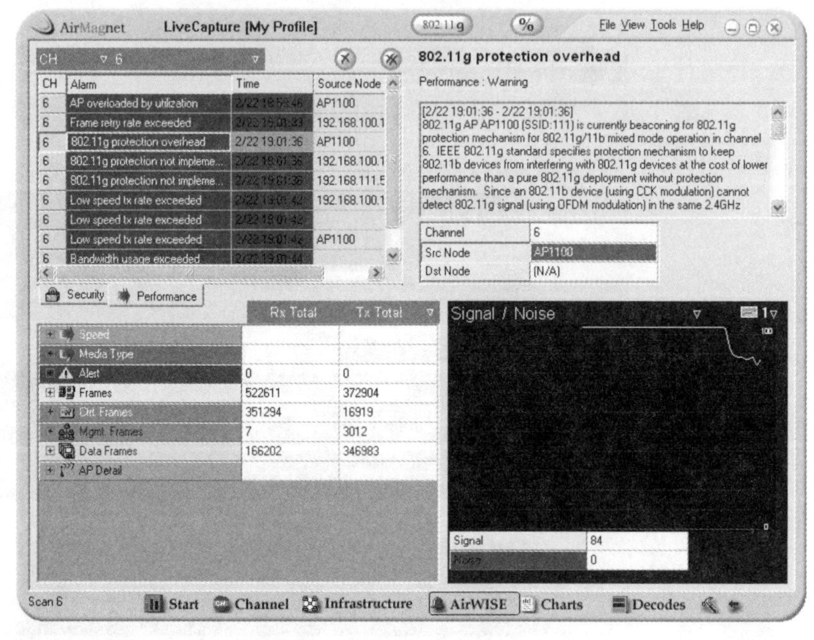

FIGURE 10.28 Security Alarms – AirMagnet Laptop

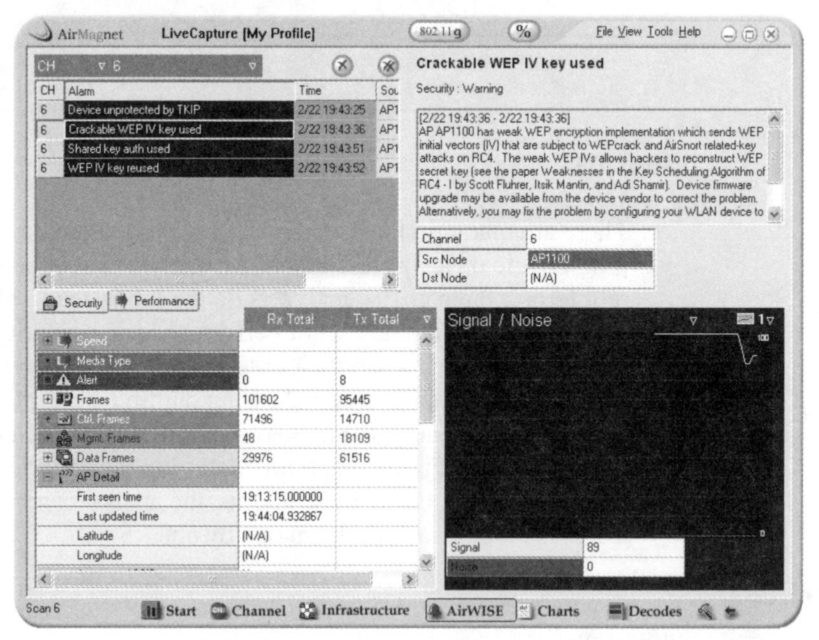

Some analyzers have security and performance alarms integrated into a single interface. When alarms are triggered by a series of captured frames matching a signature, notifications may also be triggered. Notifications may include sounds, logging, display messages, email, etc. Some analyzers allow the analyst to configure custom signatures as shown in Figure 10.29.

FIGURE 10.29 Event Alarms – CommView for WiFi

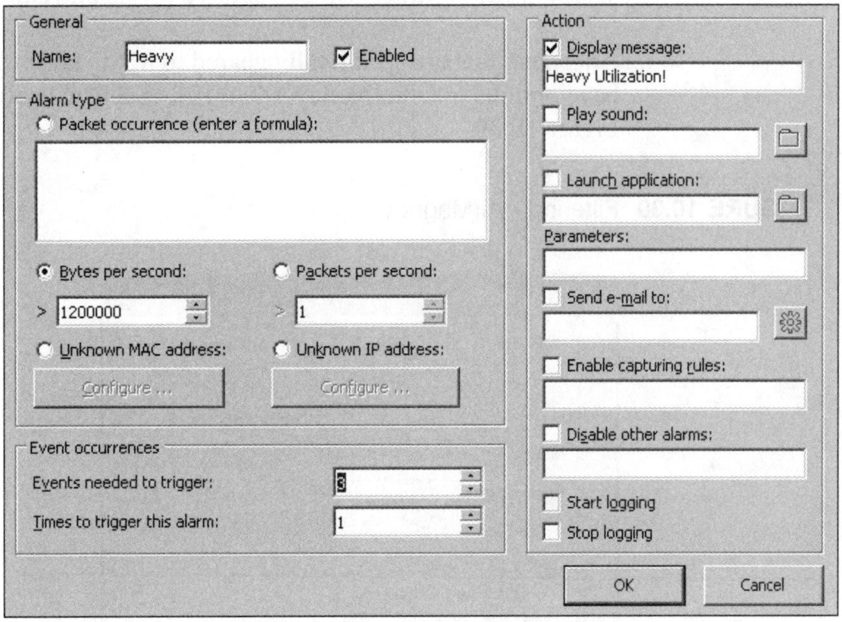

Filtering

Filtering is one of the most important features in a wireless LAN analyzer. Implementation of filtering features varies widely between analyzers, but most enterprise-class analyzers have a thorough filter feature set. One of the most simple and useful filters is the 'no beacons' filter. Some analyzers have this particular filter pre-configured, and some simply allow you to build it yourself.

There are two times that filtering can be accomplished: during a *live capture* and after a capture has been accomplished, called *post capture*. There are two methods of filtering: *capture filtering* and *display filtering*. If the filter is applied to a live capture, then only the frames accepted by the filter will be captured. If the filter is applied to a previously captured set of frames, whether from a buffer or a file, then only the frames accepted by the filter will be displayed. There are two types of filters: *reject matching* and *accept matching*. Frames that match a *reject matching* filter will not be captured (or displayed). Frames that match an a*ccept matching* filter will be captured (or displayed).

Analyzers that are specifically geared to MAC layer captures will only have options for filtering layer 2 traffic. Such an analyzer is shown in Figure 10.30.

FIGURE 10.30 Filtering - AirMagnet

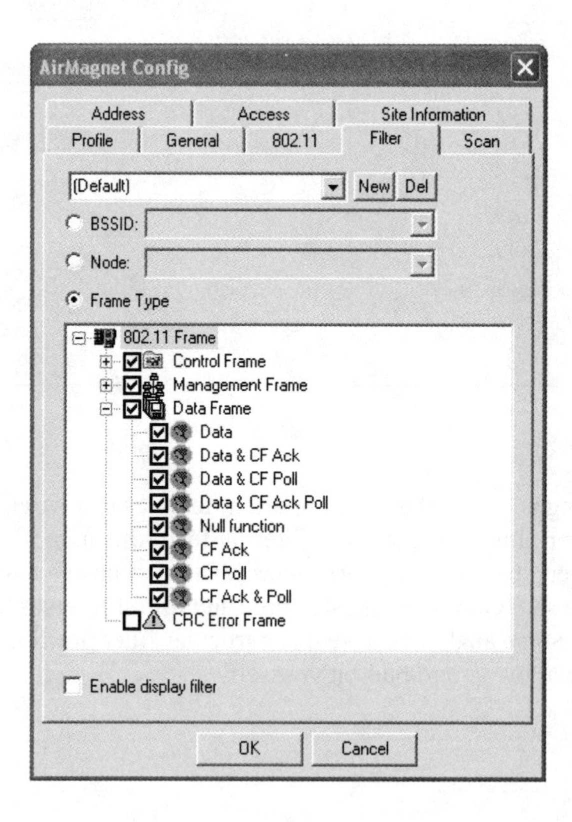

Analyzers that can analyze and decode layer 2-7 traffic will have filters that related to protocols at various layers of the OSI model. Wireless filters may be only one of many filtering options. Such an analyzer is shown in Figure 10.31.

FIGURE 10.31 Filtering - Observer

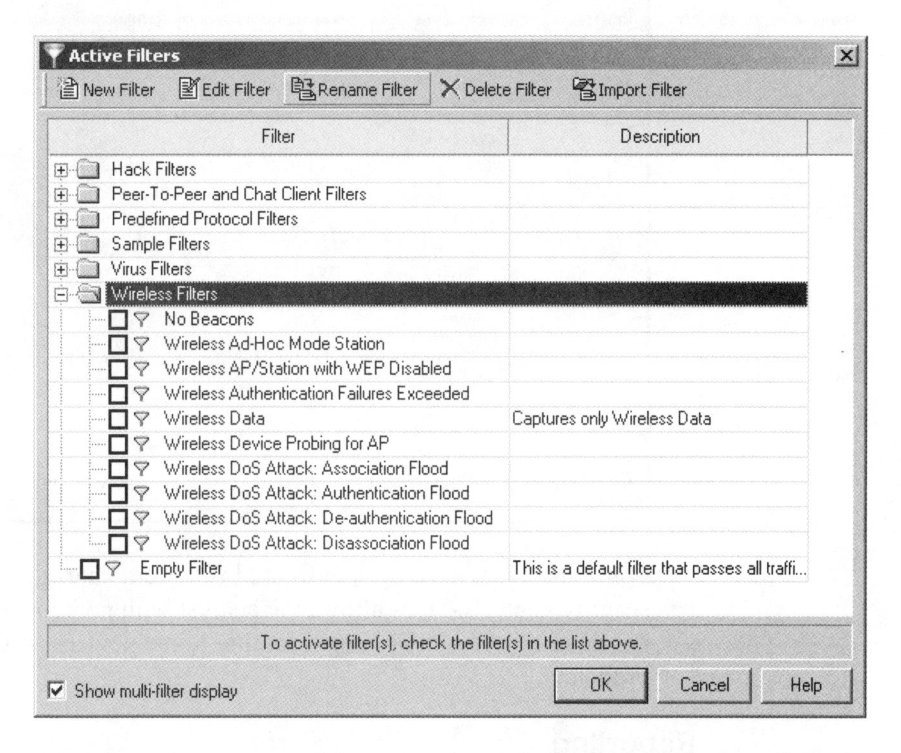

Some analyzers come with a large set of easily-used, pre-configured filters, allowing the analyst to click a couple of check boxes to apply filters that are used on a regular basis with no knowledge of filter configuration. Such a product is shown in Figure 10.32.

FIGURE 10.32 Filtering - AiroPeek

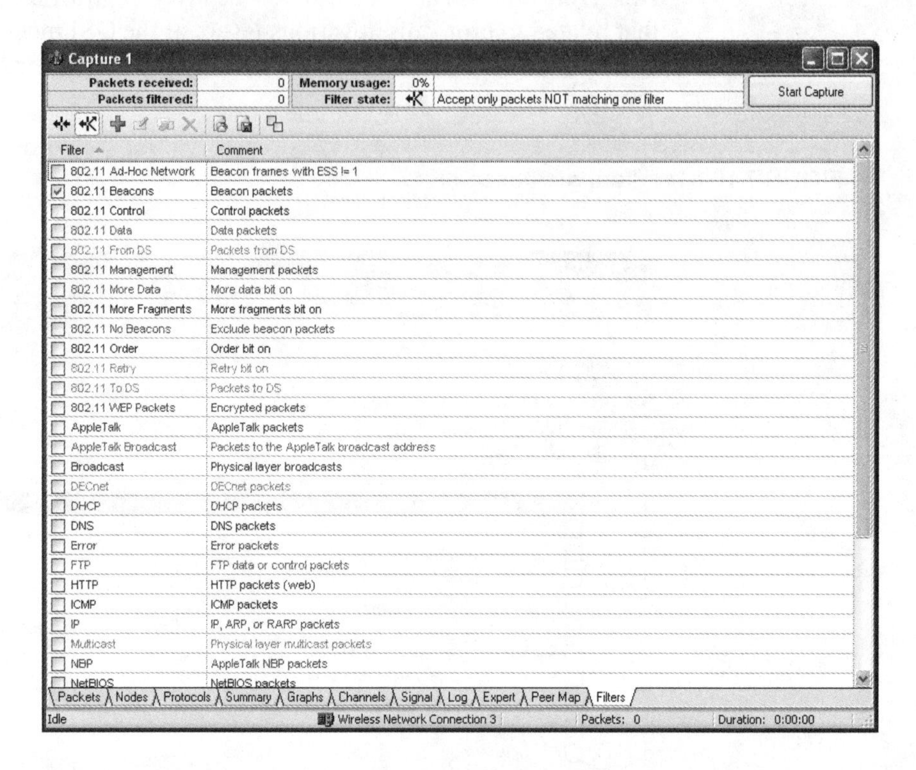

Some analyzers have pre-configured filter templates geared toward security analysis. This allows the analyst with no knowledge of filter configuration to quickly begin finding rogue devices and other security problems.

Reporting

Some analyzers have integrated reporting features, but they are generally minimal. It is becoming more common to see separate reporting applications from analyzer manufacturers that are tightly integrated with the analyzer. These reporting applications can give detailed statistics of information gathered by the analyzer in a useable format. Information is exported from the analyzer, and then imported into the reporting tool. Figure 10.33 illustrates an analyzer reporting tool.

FIGURE 10.33 Reporting – AirMagnet Reporter

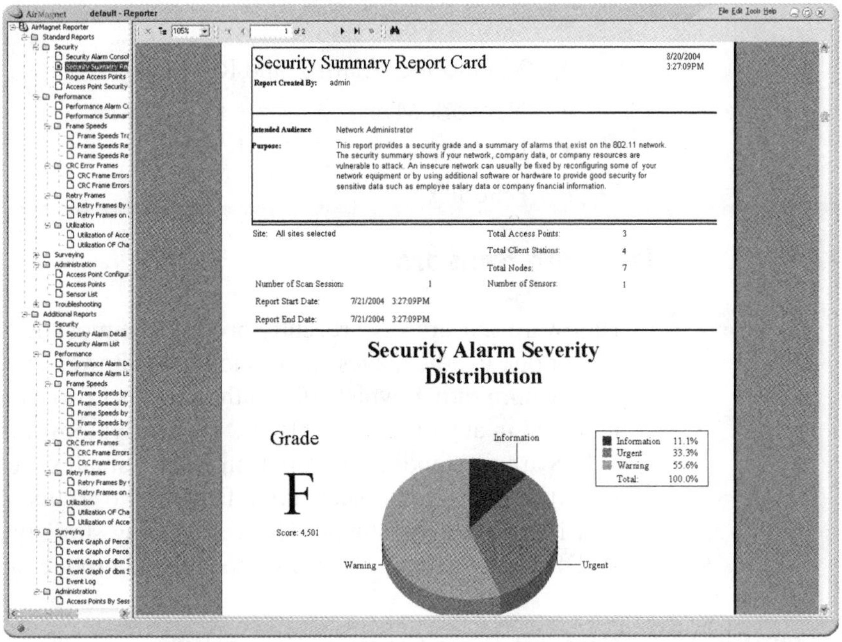

Distributed Analyzers

As previously stated, distributed analyzers are marketed and sold as wireless intrusion detection systems (WIDS). WIDS consist of a central database, a correlation engine, and sensors. It is important to perform a sensor site survey in the same way that you would perform a wireless LAN site survey in order to get the best results. WIDS have a large number of complex features that may be understood in a short time, but it may take a significant amount of time to master use of the WIDS as a whole. WIDS are capable of everything a portable analyzer is capable of with many times the number of features. These additional features deal with correlation of information gathered from distributed sensors on various channels around the network. Not only do events have to be understood by the analyzer, but the relationships between large numbers of events on the network must each be understood. Following is a list of the features of WIDS, each of which is explained in the sections that follow:

- Hardware Sensors
- Software Sensors
- Dashboards
- Monitoring, Alarms, and Reporting
- Network Maps
- Live Views
- Filtering

Hardware Sensors

Hardware sensors have recently become the most popular type of WIDS sensor due to advantages such as low cost, Power-over-Ethernet (PoE), and plenum rating, which allows the sensors (also called "probes") to be mounted in any number of places. Currently, most manufacturers of WIDS are re-branding sensors from only a small set of sensor manufacturers and, as such, are reliant on those subcontractors to provide new leading-edge features. This re-branding strategy means that most WIDS manufacturers release sensor-related product updates at the same time.

FIGURE 10.34 Hardware Probe Example - Wildpackets

FIGURE 10.35 Hardware Probe Example - AirMagnet

Software Sensors

Some WIDS manufacturers offer only software sensors, some offer only hardware sensors and others offer both. Software sensors are software applications that may be installed on any desktop or portable computer with an appropriate wireless LAN PC card installed. Some software sensors allow for multiple radio cards to be installed and monitored simultaneously. An advantage of software sensors is that they can make use of obsolete hardware that is no longer appropriate for desktop applications. The processing requirements of a software sensor are minimal. A disadvantage of software sensors is that they must be installed on a laptop or desktop computer, which cannot be physically placed in many areas such as above a drop ceiling. Figure 10.36 illustrates the main screen from a popular software sensor.

FIGURE 10.36 Distributed Software Sensor – Observer Advanced Multiprobe

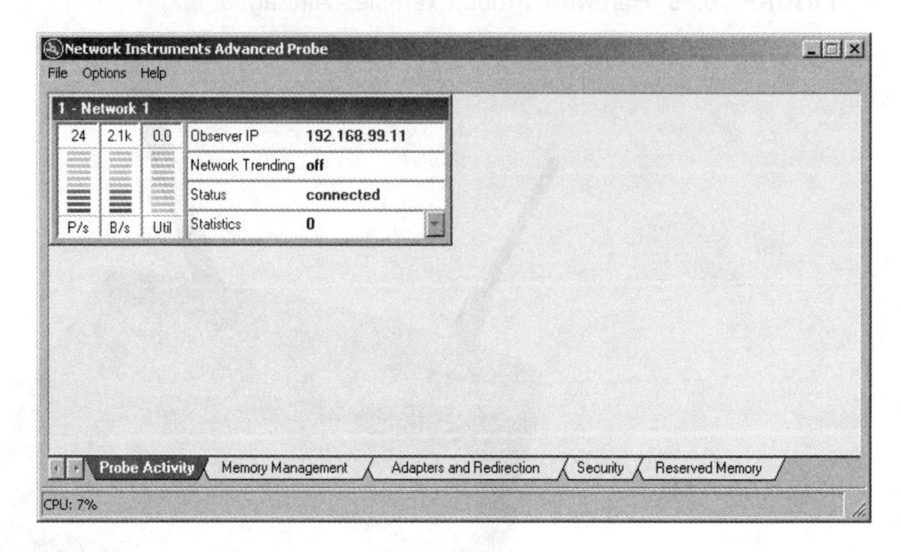

Dashboards

WIDS have a main screen that summarizes network-wide events and provides the administrator with a quick way to drill down to pertinent information. This main screen is often called the "dashboard" and should be simple to use and unambiguous at first glance. Figures 10.37 – 10.39 illustrate dashboards from three different WIDS manufacturers. Events should be listed last to first, prioritized by color, and offer a quick way to identify when and where the event happened.

FIGURE 10.37 WIDS Dashboard Example #1 – Network Chemistry

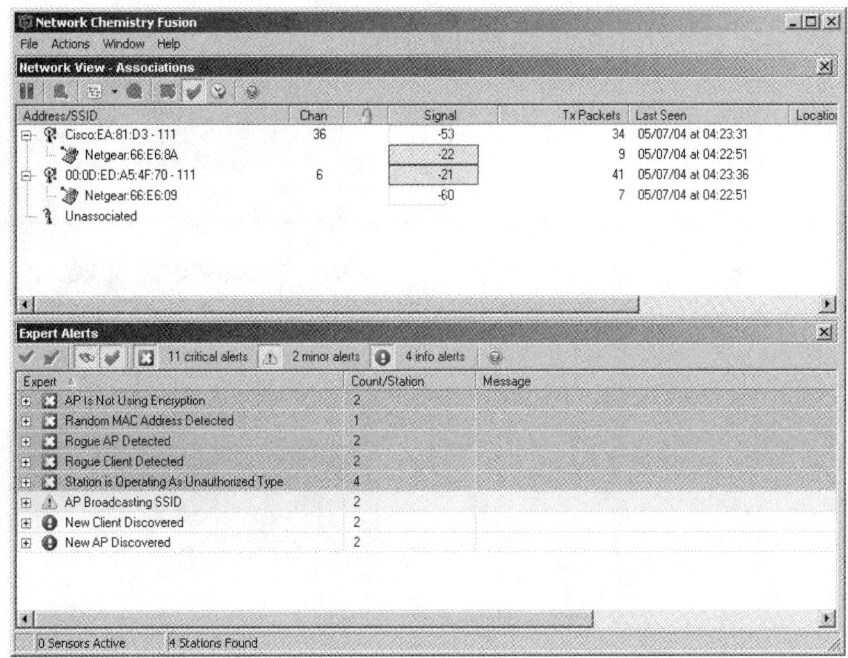

FIGURE 10.38 WIDS Dashboard Example #2 - AirDefense

FIGURE 10.39 WIDS Dashboard Example #3 – AirMagnet Distributed

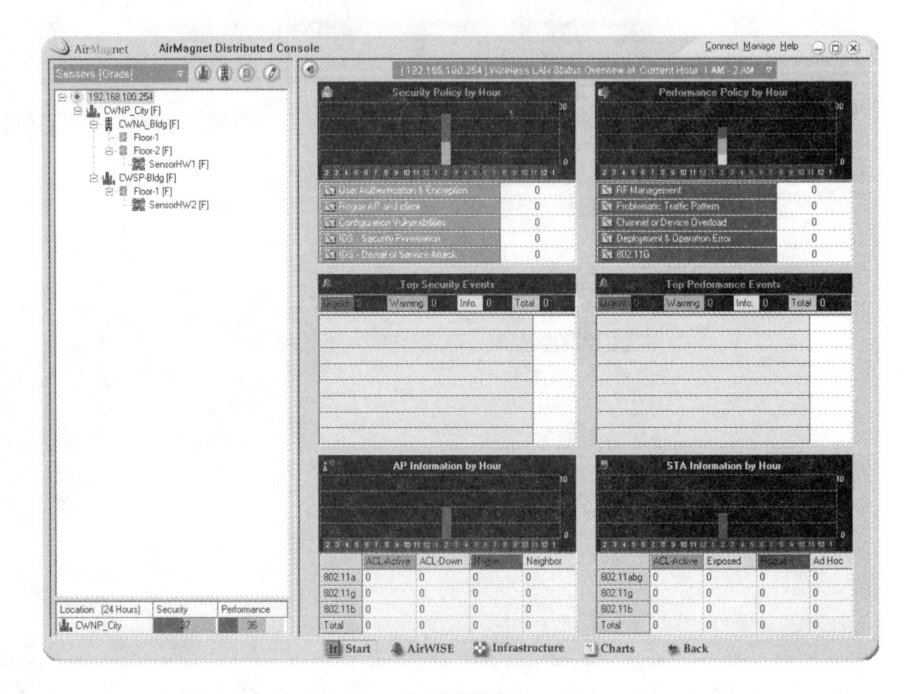

Monitoring, Alarms, and Reporting

One of the most important features of a WIDS is its ability to monitor the network 24x7x365 and inform administrators with an alarm when there is a problem. Availability of this feature enables the organization to make better use of its engineering resources, rather than having to employ several engineers to constantly monitor the system for intrusions, performance and security problems, etc. The WIDS can send notification through email, a paging system, pop-up screen alarms, and other methods. At the end of a reporting period, such as a week, month, or quarter, WIDS are fully capable of generating reports on what types of events have been seen, how often they are experienced, and how quickly they were handled. These reporting features are highly valuable to IT and security executives in any organization.

Some analyzers have integrated databases of instructions on how to generically handle certain network events, whether security or performance related. Figure 10.40 illustrates this feature.

FIGURE 10.40 Monitoring & Reporting Example – AirMagnet Distributed

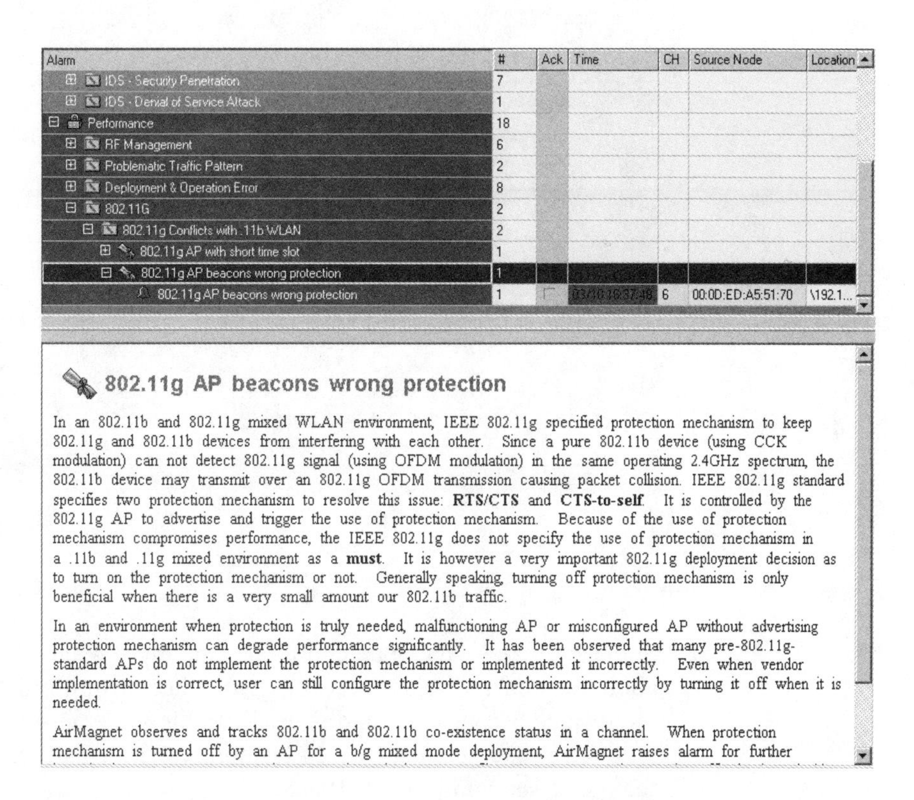

Some WIDS have integrated search functions capable of categorizing, prioritizing, and reporting on security and performance alarms. Such functions allow the administrator to deal efficiently with large amounts of accumulated alarms. Figure 10.41 illustrates this feature.

FIGURE 10.41 Alarms Example - AirDefense

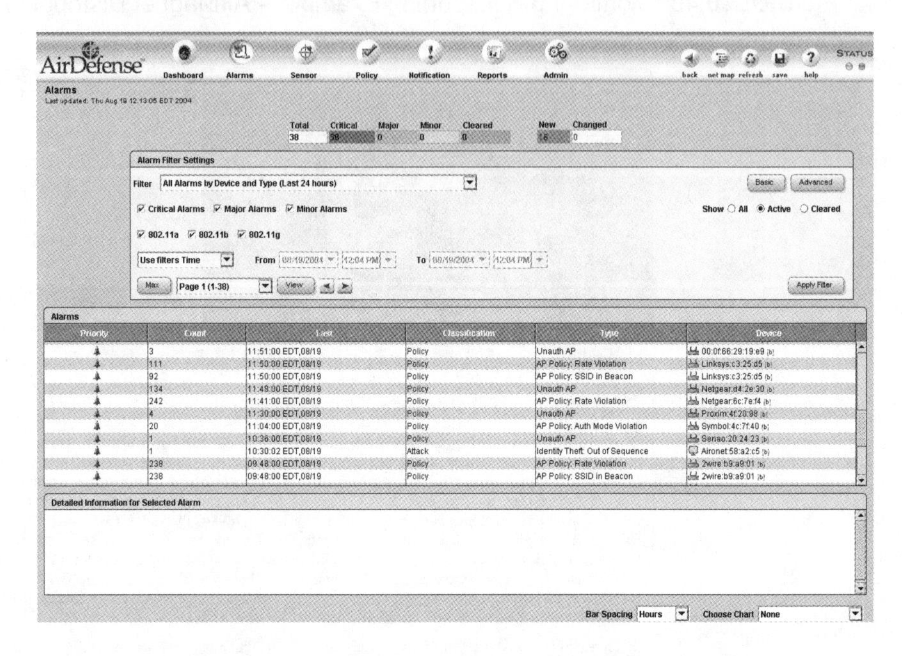

Some analyzers catalog an enormous amount of detailed information about the entire wireless network. Not all of this information is seen on the surface because it would overwhelm the system user. In order to render this data in an organized fashion, WIDS usually have a reporting feature as shown in Figure 10.42.

FIGURE 10.42 Reporting Example - AirDefense

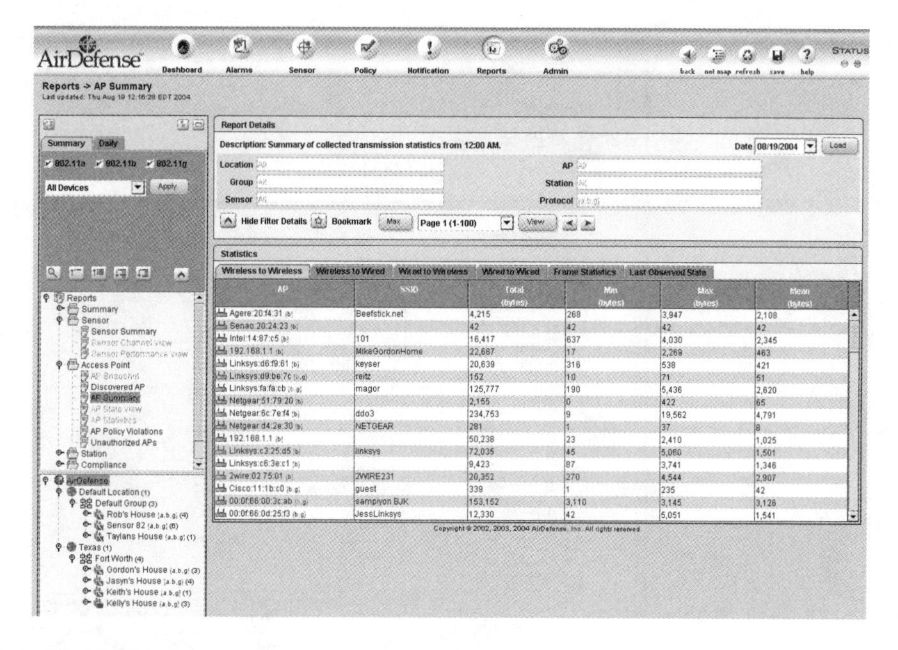

Network Maps

When installed after a proper sensor site survey, a distributed analyzer should be able to hear all infrastructure and client devices as they operate on the wireless LAN. The WIDS knows through which sensor it hears (or has heard) each device. Most of the time, access points and clients can be heard through multiple sensors. The more sensors that can hear a particular device, the better the WIDS can calculate the location of the device. By watching all of the frames on the wireless medium, a WIDS can determine which clients are associated to which access points, which devices are rogues, and even postulate problems such as an access point having too much output power. By analyzing all of this information, the WIDS can form a map of the network as shown in Figure 10.43.

FIGURE 10.43 Network Map – AirDefense

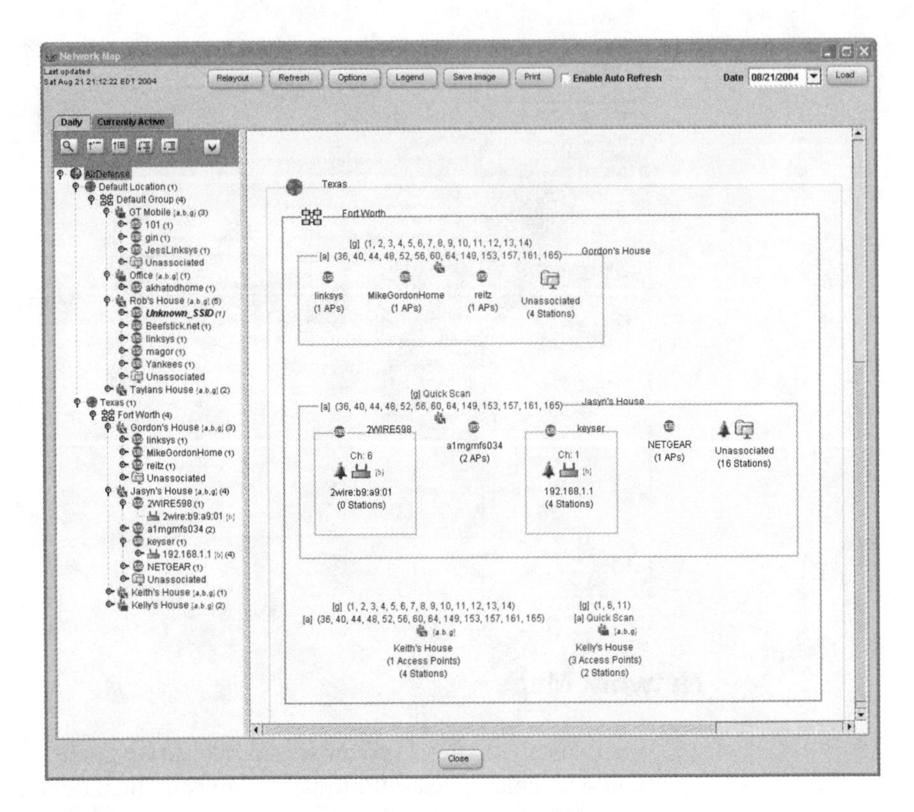

Live Views

Manufacturers that make both portable and distributed analyzers often have features in the portable analyzer to connect to distributed analyzer sensors individually for analysis. WIDS manufacturers have features that allow the administrator to view live frames or statistics from a particular sensor instead of the network as a whole. This feature is particularly useful when the administrator wants to focus on details of a particular area and channel. Figures 10.44 and 10.45 illustrate live views from remote sensors.

FIGURE 10.44 Live View - AirDefense

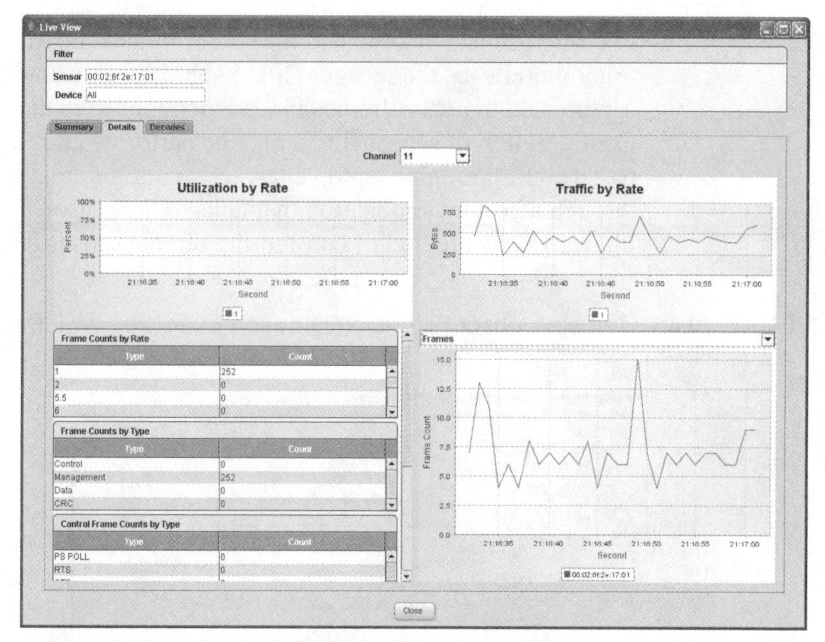

FIGURE 10.45 Live View – AirMagnet Distributed

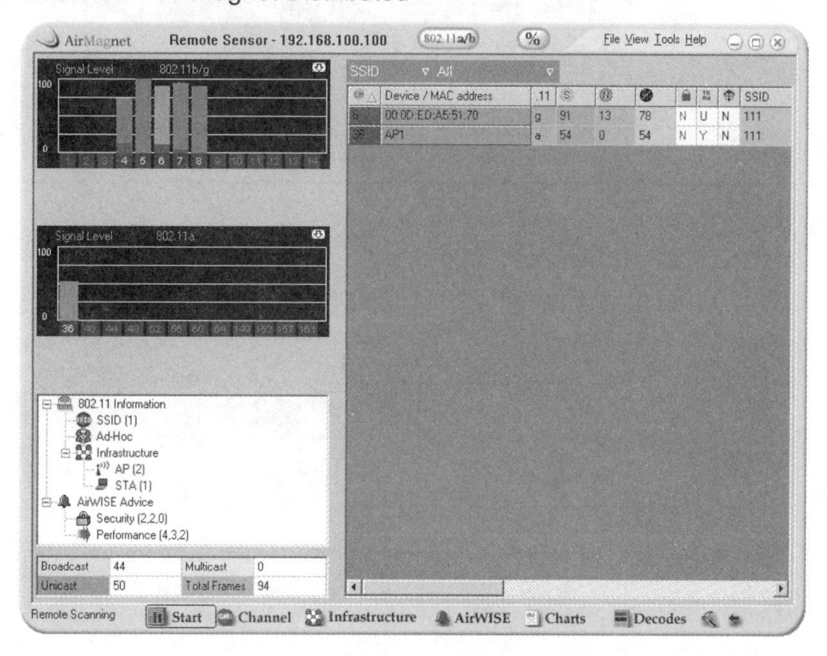

Filtering

Filtering can be performed at the sensor in most distributed analyzers. Filtering relieves the meager CPU in hardware sensors from having to capture and transport frames that are not needed to the central application. Configuration of sensor filters may be performed through CLI, HTTP-based GUI, through a laptop analyzer's configuration interface, or through a custom application that talks to the sensor. Figure 10.46 illustrates one method of configuring remote sensor filters.

FIGURE 10.46 Remote Sensor Filter Configuration Example – WildPackets RF Grabber

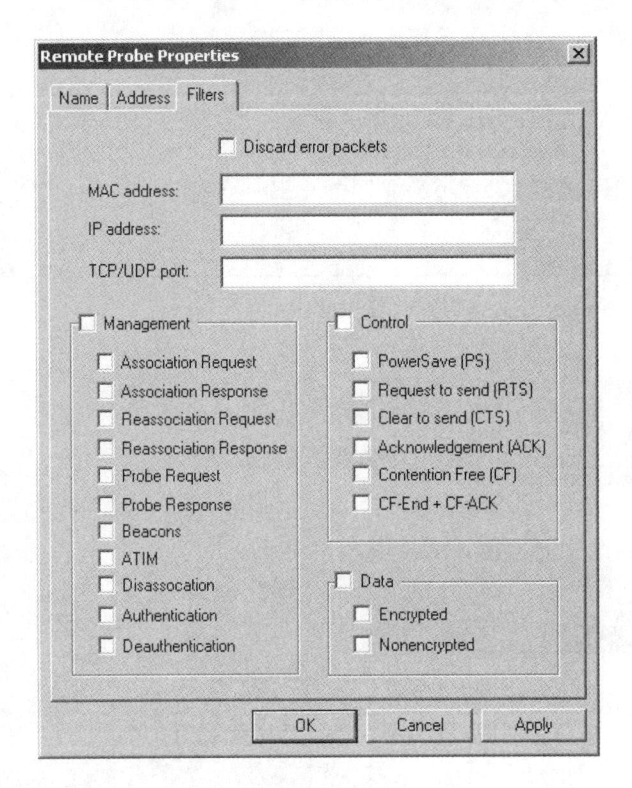

Summary

An in-depth understanding of what information a protocol analyzer can capture, what types of analyzers are currently available on the market, and the differences between analyzers is essential to successful wireless LAN protocol analysis. The introduction of new technologies into the wireless LAN market is making laptop analyzers increasingly obsolete in the wireless LAN market. Mobile, rather than portable, stations are becoming more common, increasing the need for distributed analysis. Knowing how environmental factors affect analysis and wireless LAN performance is crucial in using an analyzer effectively, especially in less than optimum conditions. There is no substitute for gaining hands-on experience with both laptop and distributed analyzers. Fortunately, a complex network design with expensive wireless LAN equipment is not required for the analyst to familiarize himself with the 802.11 standards and compliant analyzers.

Key Terms

Before taking the exam, you should be familiar with the following terms:

alarms

channel analysis

conversation analysis

dashboard

distributed analyzer

expert feature

filtering

frame decoding

laptop analyzer

live view

mobility

network map

node statistics

peer maps

portability

protocol statistics

reporting

scanning

WIDS hardware sensor

WIDS software sensor

Wireless Intrusion Detection System (WIDS)

Review Questions

1. Name two physical layer parameters that can be presented to the analyst by the analyzer about an 802.11 frame.

2. In order to locate active wireless LAN networks in a particular area, which feature of a laptop wireless LAN protocol analyzer would you use?

3. Why is a laptop wireless LAN analyzer unable to track a client device as it roams between access points?

4. A wireless intrusion detection system (WIDS) is generally comprised of what three components?

5. Before deployment of WIDS sensors, what task should be completed?

6. If beacons are filtered out during a live capture, how can the analyst later review the decoded beacons?

7. Where in the beacon frame should the analyst look to see what channel a beacon frame was transmitted on?

8. Configuring specific filters can be tedious and time-consuming. To reuse custom filter configurations at a later time, what should the analyst do?

9. Name two notification methods a WIDS might use to inform an administrator of an intrusion.

CHAPTER

11

802.11 Performance Variables

CWAP Exam Objectives Covered:

- Interpret 802.11a/b/g protocol traces
  - Troubleshooting
  - Performance testing
  - Security analysis
  - Intrusion analysis

In This Chapter

This section is aimed at informing the network analyst of the impact of changes in protocol settings on infrastructure and client devices, changes in the RF environment, and changes from various uses of wireless client devices. In a constantly changing wireless LAN environment, there are configurations that the administrator can adjust manually and that the infrastructure devices can adjust automatically. Each of these adjustments may be intended to improve the performance of the network. It is the analyst's responsibility to determine which changes are most effective and to understand why other changes may be detrimental to the network's efficiency.

Protocol Impacts

Changes to configuration settings in infrastructure and client devices may have an undesirable result if used improperly. Some settings may produce bad results unless optimized. An example of such a setting is the fragmentation threshold setting.

Fragmentation Threshold

When enabled, the fragmentation threshold setting can be configured from 256 to 2346 bytes, and sets the maximum allowable size for an MPDU, before encryption. If the threshold is set too small, then the RF medium utilization will become very busy with small frames and acknowledgements. Data throughput will be reduced due to the additional overhead of MAC headers and ACKs for each fragment. If the threshold is set to large, then large frames may have to be retransmitted often, significantly reducing throughput.

Beacon Rate

Beacons are typically transmitted onto the wireless medium approximately 10 times per second by each access point by default. This rate is fairly common in the industry. Changing the beacon rate will impact how quickly client devices locate access points and how quickly client devices will request, and subsequently receive, buffered unicast, broadcast, and multicast traffic when operating in power save mode.

DTIM Rate

Changing the DTIM rate will impact receipt of broadcast and multicast traffic for stations operating in power save mode. The buffered broadcast/multicast traffic is transmitted immediately after the beacon containing the DTIM. If the DTIM interval is normally set to 2 (meaning every second TIM is a DTIM), and now it is set to 4, buffered broadcast/multicast traffic will be sent only half as often. This setting could cause quality problems for multicast audio or video streaming in particular.

Manual and Automatic Protection Mechanism Activation

RTS/CTS thresholds can be manually configured at each client and at the access point, depending on whether or not the equipment vendor provides the configuration settings. The RTS/CTS functionality was originally used as a protection mechanism in 802.11 for problems such as hidden node due to distance or RF signal blockage. RTS/CTS and CTS-to-Self are now used as protection mechanisms for the same problem, but nodes can now be hidden due to the transmission technology they are using instead of only distance or RF signal blockage. Access points that are compliant with the 802.11g standard are required to enable automatic use of RTS/CTS and/or CTS-to-Self when non-ERP stations associate to the BSS so that 802.11g and 802.11b stations can co-exist in the same BSS without a significant amount of corrupt frames. Figures 11.1 – 11.4 illustrate throughput both in a pure 802.11g environment and in a mixed 802.11b/g environment.

FIGURE 11.1 Normal Throughput When No Protection is specified by the access point

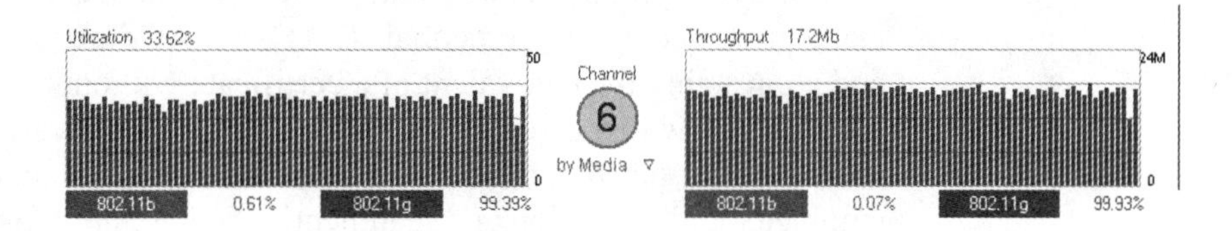

FIGURE 11.2 RF Medium Throughput Reduction When AP Mandates Protection

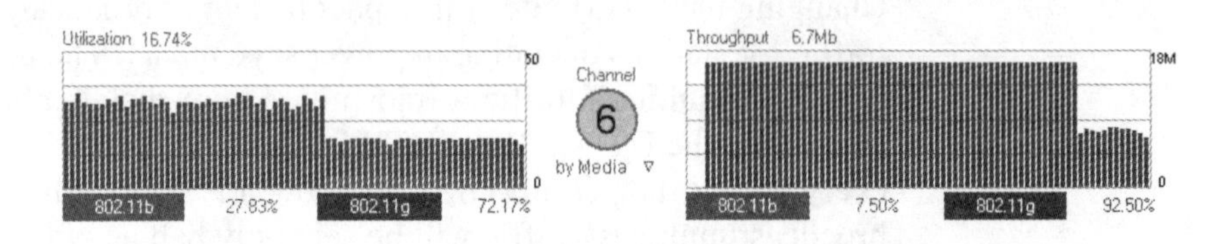

FIGURE 11.3 Mixed Mode 802.11b/g Typical Throughput

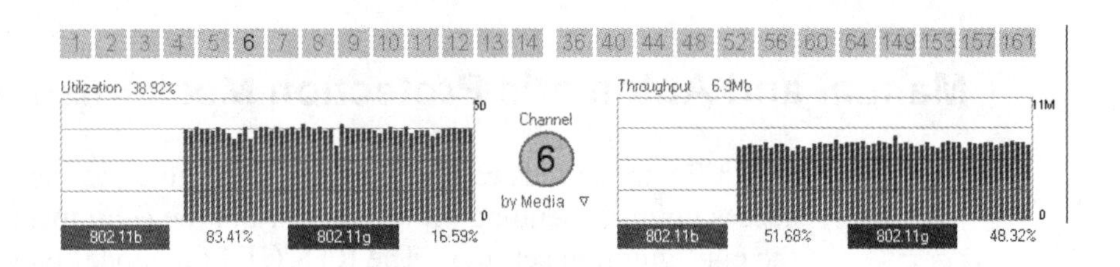

FIGURE 11.4 When 802.11b is removed, and only 802.11g remains

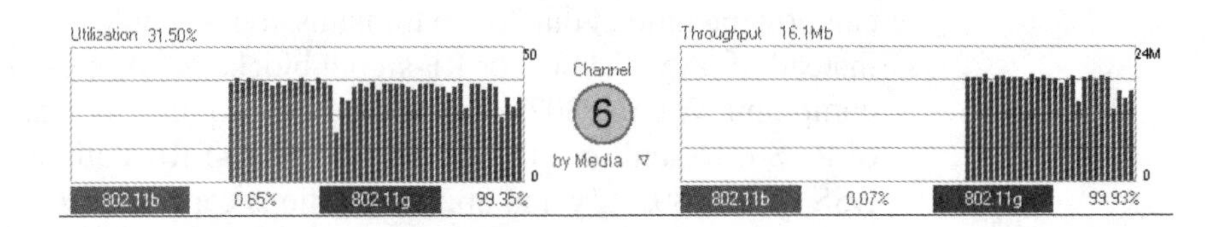

Security Protocols and Encryption

Wireless security protocols based on 802.1X/EAP using an RC4 stream cipher add a minimal amount of encryption overhead. Authentication methods vary greatly, some using passwords, some certificates, and others using protected access credentials (PACs). PACs are part of EAP-FAST and are roughly the equivalent of a certificate without the need for connectivity with a certificate authority. Though the various EAP processes are substantially different and take different amounts of time to complete, authentication and reauthentication are not typically performed on short intervals so their affect on throughput is negligible.

Layer 3 VPN protocols vary greatly in their implementation, and while authentication and reauthentication follow the same rules as layer 2 security methods, layer 3 encryption overhead can range from paltry to profuse. Figures 11.5 – 11.8 illustrate expected throughput for an 802.11b network using no encryption and using an array of security protocols. To take these screenshots, the station and the access point were the same in every scenario, and the VPN concentrator and the FTP server were both able to handle many times the throughput of the wireless client device.

FIGURE 11.5 802.11b File Transfer Throughput – No Encryption

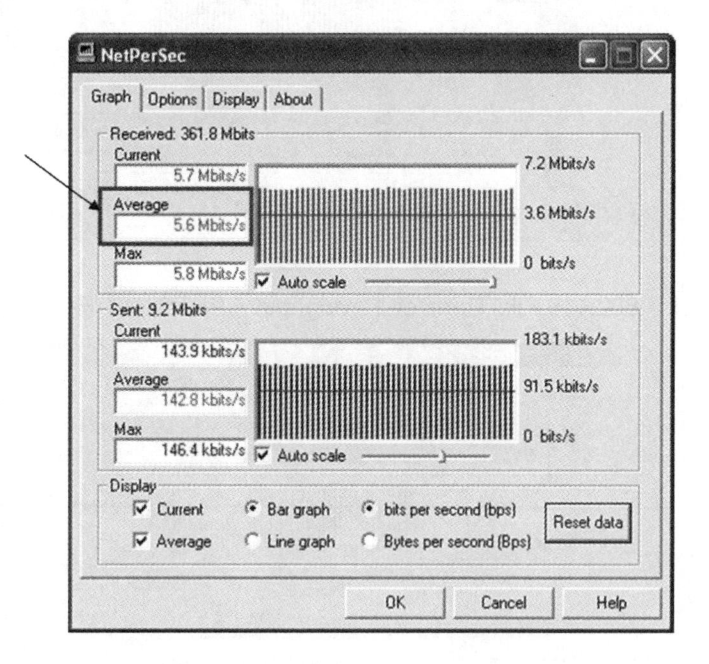

FIGURE 11.6 802.11b File Transfer Throughput – 802.1X/PEAP/WPA

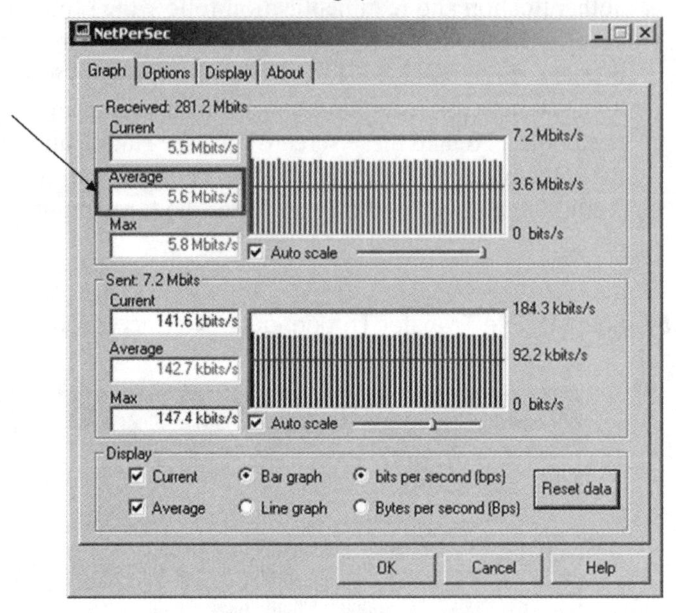

FIGURE 11.7 802.11b File Transfer Throughput – PPTP/MPPE-128

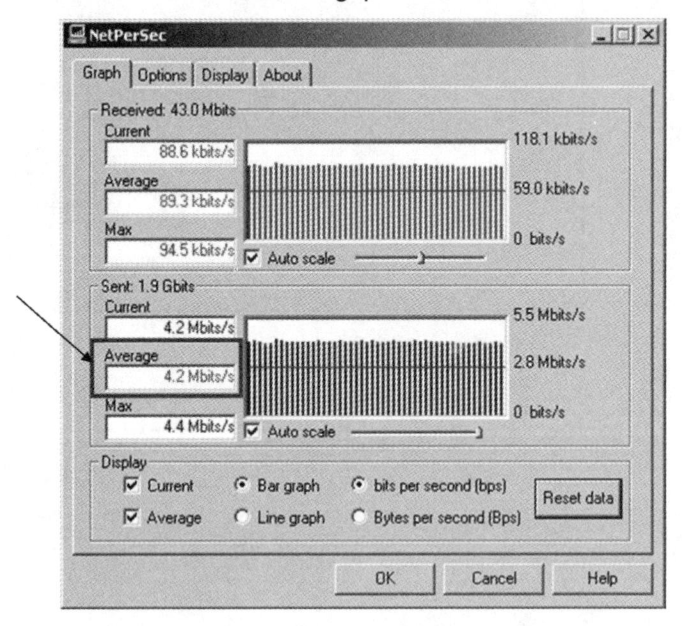

FIGURE 11.8 802.11b File Transfer Throughput – IPSec/ESP/Tunnel/3DES

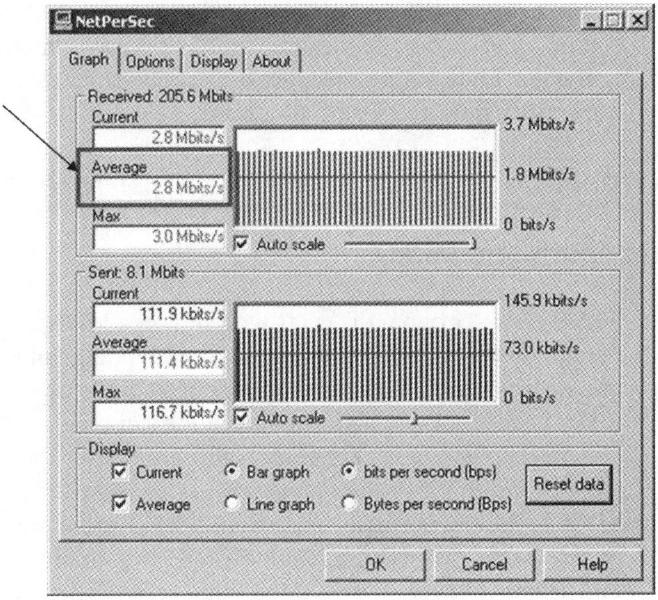

Environmental Impacts

Environmental impacts can affect what frames may be captured by an analyzer. There are many environmental variables to consider, and multiple variables may impact protocol analysis simultaneously. A partial list of environmental factors includes:

- Wireless node proximity
- Output power and antenna types of clients and access points
- Multipath – reflective surfaces in the area
- RF Signal blockage
- RF Interference sources – wideband or narrowband
- Co-channel or Adjacent Channel Interference with significantly overlapping Basic Service Areas (BSAs)

Each of these factors will be described individually in the following sections. Most of these factors will affect wireless LAN performance in a similar way – excessive retransmissions – but each may affect what an analyst sees in a wireless LAN protocol analyzer in a number of ways. When a reasonable quality RF signal is not received by the access point or station, the received frames may not be properly interpreted. The receiver will not acknowledge corrupted or partial frames, and therefore the transmitter must retransmit the frame.

Multipath, Hidden Nodes, and RF Interference

Multipath and MIMO

802.11 Task Group n (TGn) is working on a Multiple Input / Multiple Output (MIMO) based solution to take advantage of multipath propagation instead of multipath being a source of frame corruption. MIMO runs multiple data streams in the same radio channel using "smart antenna" technology. The results are throughput speeds comparable to those of wired Fast Ethernet, as well as increased access point range. MIMO uses the normally superfluous signals that are a byproduct of RF transmission. MIMO uses multiple antennas to break a single fast signal into several slower signals at the same time. The slower signals are sent using a different antenna using the same frequency channel. The receiver then reassembles the signals using complex mathematical algorithms to sort out the jumbled radio waves so that they can be read easily.

FIGURE 11.9 MIMO Wireless Router - Belkin

Hidden Nodes

There are often situations in which one side of the link can hear transmissions and the other side cannot because of a signal blockage or distance between stations. Access points usually have much better antennas than client devices, so the access point is usually the device that can both transmit and receive well. Laptop analyzers use standard client wireless LAN PC cards and therefore have the same antennas. This means that the analyzer itself could potentially be one of the hidden nodes, capturing only part of a conversation. Seeing only a portion of the expected frames may, at first, be confusing to the analyst. Figure 11.10 illustrates a situation in which only a portion of the expected frames are captured by the analyzer. The analyzers proximity to the transmitter and receiver are relative to the frames it can successfully capture.

FIGURE 11.10 Analyzer Captures Part of a Conversation

Packet	Address 1	Address 2	Address 3	Channel	Data Rate	Size	Protocol
20	00:09:5B:66:E5:F1	00:0D:ED:A5:51:70	00:0D:65:C9:32:76	6	11.0	96	PING Reply
21	00:09:5B:66:E5:F1			6	11.0	14	802.11 Ack
22	00:09:5B:66:E5:F1			6	11.0	14	802.11 Ack
23	00:09:5B:66:E5:F1	00:0D:ED:A5:51:70	00:0D:65:C9:32:76	6	11.0	96	PING Reply
24	00:09:5B:66:E5:F1			6	11.0	14	802.11 Ack
25	00:09:5B:66:E5:F1			6	11.0	14	802.11 Ack
26	00:09:5B:66:E5:F1	00:0D:ED:A5:51:70	00:0D:65:C9:32:76	6	11.0	96	PING Reply
27	00:09:5B:66:E5:F1			6	11.0	14	802.11 Ack
28	00:09:5B:66:E5:F1			6	11.0	14	802.11 Ack
29	00:09:5B:66:E5:F1	00:0D:ED:A5:51:70	00:0D:65:C9:32:76	6	11.0	96	PING Reply
30	00:09:5B:66:E5:F1			6	11.0	14	802.11 Ack
31	00:09:5B:66:E5:F1			6	11.0	14	802.11 Ack
32	00:09:5B:66:E5:F1	00:0D:ED:A5:51:70	00:0D:65:C9:32:76	6	11.0	96	PING Reply
33	00:09:5B:66:E5:F1			6	11.0	14	802.11 Ack
34	00:09:5B:66:E5:F1			6	11.0	14	802.11 Ack
35	00:09:5B:66:E5:F1	00:0D:ED:A5:51:70	00:0D:65:C9:32:76	6	11.0	96	PING Reply
36	00:09:5B:66:E5:F1			6	11.0	14	802.11 Ack
37	00:09:5B:66:E5:F1			6	11.0	14	802.11 Ack
38	00:09:5B:66:E5:F1	00:0D:ED:A5:51:70	00:0D:65:C9:32:76	6	11.0	96	PING Reply
39	00:09:5B:66:E5:F1			6	11.0	14	802.11 Ack
40	00:09:5B:66:E5:F1			6	11.0	14	802.11 Ack

RF Interference

Wideband and narrowband RF interference sources can wreak havoc on a wireless LAN and on the analyst's interpretation of captures. Some analyzers do not list corrupted frames in the display at all. Other analyzers list the frames, but have a column dedicated to illustrating which frames were corrupted when they were received. Figure 11.11 illustrates frames corrupted due to RF interference. Notice the "Flag" column in the analyzer. "C" means CRC Error (corrupted data).

FIGURE 11.11 RF Interference Causing Corrupted Frames

Packet	Address 1	Address 2	Address 3	Flags	Channel	Data Rate	Size	Protocol
47	00:0D:ED:A5:51:70	00:09:5B:66:E5:F1	00:0D:65:C9:32:76	C	6	36.0	96	IP Fragment
48	00:0D:ED:A5:51:70	00:09:5B:66:E5:F1	00:0D:65:C9:32:76	C	6	36.0	96	IP Fragment
49	00:09:5B:66:E5:F1			#	6	24.0	14	802.11 Ack
50	00:09:5B:66:E5:F1	00:0D:ED:A5:51:70	00:0D:65:C9:32:76		6	36.0	96	PING Reply
51	80:99:E2:A5:51:70	20:1F:CF:A4:51:70	FF:FF:FF:FF:FF:FF	*CW	6	1.0	129	802.11 Assoc Req
52	00:0D:ED:A5:51:70	00:09:5B:66:E5:F1	00:0D:ED:A5:51:70		6	36.0	28	802.11 Data
53	00:09:5B:66:E5:F1			#	6	24.0	14	802.11 Ack
54	00:09:5B:66:E5:F1	00:0D:ED:A5:51:70	00:0D:65:C9:32:76	C	6	36.0	96	PING Reply
55	00:0D:ED:A5:51:70	00:09:5B:66:E5:F1	00:0D:ED:A5:51:70		6	36.0	28	XNS
56	00:09:5B:66:E5:F1			#	6	24.0	14	802.11 Ack
57	00:0D:ED:A5:51:70	00:0D:ED:A5:51:70	FF:FF:FF:FF:FF:FF	*C	6	1.0	129	802.11 Reassoc Rsp
58	09:27:B4:67:AD:5E	E6:39:E4:04:AB:99	8D:61:C1:AB:7C:1E	C	6	36.0	96	802.11 Frag
59	00:09:5B:66:E5:F1	00:0D:ED:A5:51:70	00:0D:65:C9:32:76		6	36.0	96	PING Reply
60	00:09:5B:66:E5:F1			#	6	24.0	14	802.11 Ack
61	00:09:5B:66:E5:F1			#	6	24.0	14	802.11 Ack
62	00:09:5B:66:E5:F1	00:0D:ED:A5:51:70	00:0D:65:C9:32:76	C	6	36.0	96	LSAP-41
63	00:09:5B:66:E5:F1			#	6	24.0	14	802.11 Ack
64	00:09:5B:66:E5:F1			#	6	24.0	14	802.11 Ack
65	00:09:5B:66:E5:F1	00:0D:ED:A5:51:70	00:0D:65:C9:32:76		6	11.0	96	PING Reply
66	00:09:5B:66:E5:F1	00:0D:ED:A5:51:70	00:09:5B:66:E5:F1	*C	6	11.0	98	802.11 Disassoc
67	00:09:5B:66:E5:F1	00:0D:ED:A5:51:70	00:0D:65:C9:32:76		6	5.5	96	PING Reply
68	00:29:5A:66:E5:F1			#C	6	24.0	14	802.11 Ack
69	00:09:5B:66:E5:F1	00:0D:ED:A5:51:70	00:AD:7A:10:3A:7B	C	6	36.0	96	LSAP-32

Co-channel and Adjacent Channel Interference

When access points in close proximity to each other are transmitting at high power with high-gain antennas, there is likely to be co-channel and/or adjacent channel interference. For this reason, a high-quality site survey is essential in avoiding this type of interference. Co-channel and adjacent channel interference (both of which are the same type of interference) may cause a large degradation of throughput due to transmission retries. Stations in a BSS will honor duration values of received transmissions from adjacent cells that are on the same channel. This situation equates to two or more adjacent basic service sets sharing bandwidth, and severely degrades overall network throughput.

With a proper site survey and implementation, transmissions on one channel are barely heard, if at all, on other channels. When an access point is transmitting on channel 1, a wireless LAN protocol analyzer might be able to successfully (no CRC errors) receive some transmissions on channels 2 and 3 when at close range to the access point. When the analyzer is positioned at a reasonable range, most of these frames will be shown as corrupted and having low received signal strength. Figure 11.12 illustrates this scenario.

FIGURE 11.12 Capturing on an Adjacent Channel

Packet	Address 1	Address 2	Address 3	Flags	Channel	Data Rate	Size	Protocol
26	FF:FF:FF:FF:FF:FF	00:0D:ED:A5:51:70	00:0D:ED:A5:51:70	*	3	1.0	128	802.11 Beacon
27	FF:FF:FF:FF:FF:FF	00:0D:ED:A5:51:70	00:0D:ED:A5:51:70	*	3	1.0	128	802.11 Beacon
28	00:0D:ED:A5:51:70	00:09:5B:66:E5:F1	00:0D:ED:A5:51:70	*	3	1.0	74	802.11 Assoc Req
29				#C	3	11.0	30	802.11 Control
30	FF:FF:FF:FF:FF:FF	00:0D:ED:A5:51:70	00:0D:ED:A5:51:70	*	3	1.0	128	802.11 Beacon
31				CW	3	11.0	30	802.11
32	FF:FF:FF:FF:FF:FF	00:0D:ED:A5:51:70	00:0D:ED:A5:51:70	*	3	1.0	128	802.11 Beacon
33	FF:FF:FF:FF:FF:FF	00:0D:ED:A5:51:70	00:0D:ED:A5:51:70	*	3	1.0	128	802.11 Beacon
34	FF:FF:FF:FF:FF:FF	00:0D:ED:A5:51:70	00:0D:ED:A5:51:70	*	3	1.0	128	802.11 Beacon
35				CW	3	11.0	30	802.11
36	05:B3:20:D3:4F:33	BC:DE:24:B8:B8:B8	05:B3:20:D3:4F:33	*CW	3	11.0	24	802.11 Assoc Req
37				C	3	11.0	30	802.11
38	FF:FF:FF:FF:FF:FF	00:0D:ED:A5:51:70	00:0D:ED:A5:51:70	*	3	1.0	128	802.11 Beacon
39				CW	3	11.0	30	802.11
40	23:18:EB:2C:D6:CC	AE:97:BA:D9:D9:D9	00:0D:ED:A5:51:70	C	3	11.0	24	802.11 Data
41	FF:FF:FF:FF:FF:FF	00:0D:ED:A5:51:70	00:0D:ED:A5:51:70	*	3	1.0	128	802.11 Beacon
42	FF:FF:FF:FF:FF:FF	00:0D:ED:A5:51:70	00:0D:ED:A5:51:70	*	3	1.0	128	802.11 Beacon
43	FF:FF:FF:FF:FF:FF	00:0D:ED:A5:51:70	00:0D:ED:A5:51:70	*	3	1.0	128	802.11 Beacon
44	4F:3B:84:F1:45:90	CF:65:87:FA:FA:FA	00:0D:ED:A5:51:70	C	3	11.0	24	802.11 Data
45				#CW	3	11.0	30	802.11 Control
46	FF:FF:FF:FF:FF:FF	00:0D:ED:A5:51:70	00:0D:ED:A5:51:70	*	3	1.0	128	802.11 Beacon
47	FF:FF:FF:FF:FF:FF	00:0D:ED:A5:51:70	00:0D:ED:A5:51:70	*	3	1.0	128	802.11 Beacon

Figure 11.13 illustrates frames successfully captured on a channel adjacent to the channel on which they were sent. The access point was transmitting beacons on channel 1 at high power while the wireless LAN protocol analyzer was listening for frames on channel 2.

FIGURE 11.13 Frames Captured on an Adjacent Channel – AiroPeek NX

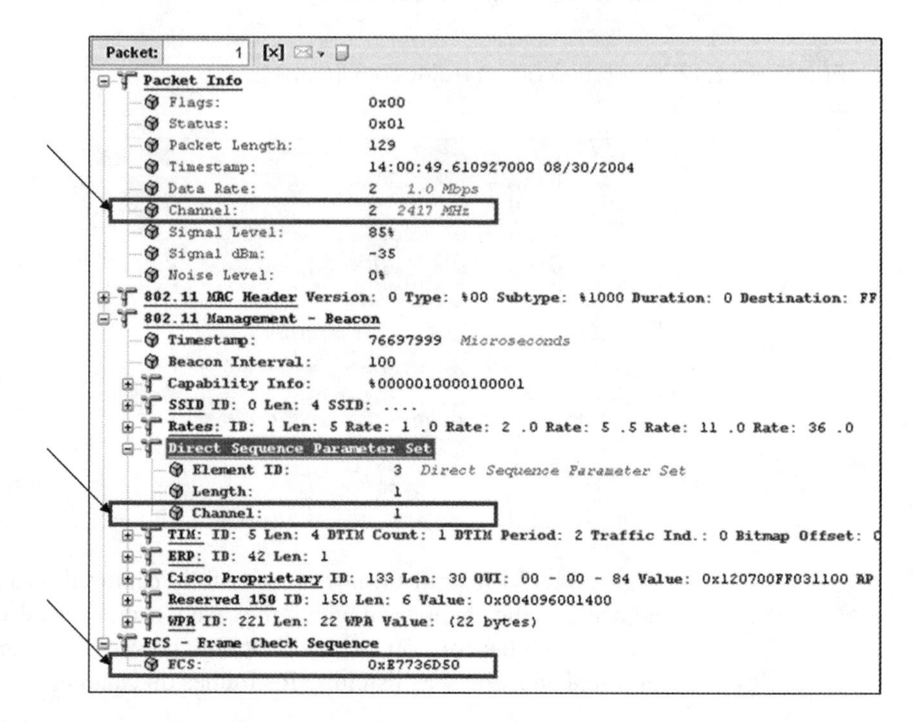

When frames can be successfully captured across a large channel gap, such as when they are transmitted on channel 1 and can be successfully captured at a reasonable distance on channel 6, it is an indication of too much output power on the access point.

FIGURE 11.14 Frames Captured on an Adjacent Channel – AiroPeek NX

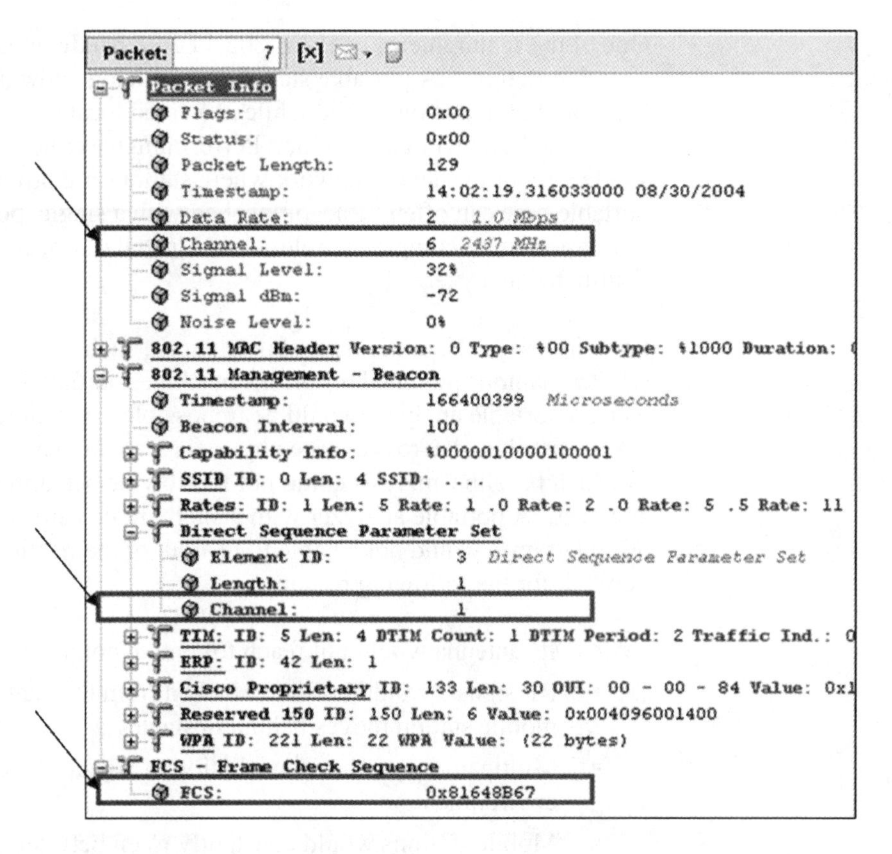

This "cross-channel interference" can lead not only to degraded throughput because of the honoring of duration values by stations, but also because of automatic enabling of protection mechanisms by access points. Some vendors implement protection mechanisms on access points in such a way that when one access point enables *Use_Protection* in its beacons, nearby access points that can hear these beacons also enable *Use_Protection* in their beacons. This implementation becomes a chain reaction throughout the ESS. This "ripple effect" can cross channels when an access point on channel 1 can hear beacons from an access point on channel 6 due to excessive output power. The site surveyor should take care to space the cells appropriately, use only the necessary output power and antenna gain, and properly reuse non-overlapping channels.

Mobile vs. Portable Impacts

One of the requirements of IEEE 802.11 is to handle *mobile* as well as *portable* stations. A portable station is one that is moved from location to location, but that is only used while at a fixed location. Mobile stations access the LAN while in motion. Performing baseline analysis within a single BSS or in a small network where stations are primarily of the portable type can often be accomplished with a single, portable analyzer. For larger, more complex wireless LAN installations, consider a distributed analyzer.

For wireless LANs utilizing mobile stations, such as a warehouse with wireless laptops mounted to forklift machines, defining baseline usage using a portable analyzer would be impossible. Additionally, wireless LANs distributed across a large physical area would also pose a situation in which baseline analysis could not feasibly be performed by a portable analyzer. A portable analyzer with a single radio card for capturing 802.11 frames would not be able to hear all of the traffic across the entire network for the following reasons:

- Its antenna would not reach to remote points of the network
- RF signals would become blocked from the analyzer when a mobile station moves behind obstacles
- Multipath may cause cancellation of signals depending on the RF environment
- Mobile stations would constantly roam between access points on different channels, and the analyzer can only capture all of the frames in a single area on a single channel at a time

In cases like this, using distributed analyzers (or the distributed analysis function integrated into the wireless infrastructure devices) that can track statistics across the entire network simultaneously is necessary. Distributed analyzers utilize multiple sensors that report to a single console. The console can then extrapolate and interpret data collected from multiple sources in order to give the administrator useful information.

Mobile stations are becoming more prevalent in the wireless LAN market with PDAs, VoWiFi IP Phones, etc. In addition to performing baseline analysis, it is often necessary to track these devices in real-time, keep records of roaming and authentication, and even be able to decode traffic such as VoIP for troubleshooting. Often protocol analysis and positioning are mutually exclusive, but that has recently been a movement toward merging these technologies.

Wireless LAN positioning systems (also called wireless LAN tracking systems) have traditionally been standalone applications, but manufacturers have recently begun integrating this functionality into wireless LAN switches, EWGs, WIDS, and other infrastructure devices. Figure 11.15 illustrates a standalone wireless LAN positioning system application. Figures 11.16 – 11.18 illustrate positioning systems integrated into the wireless LAN infrastructure equipment.

FIGURE 11.15 Standalone Wireless LAN Positioning System Application - Ekahau

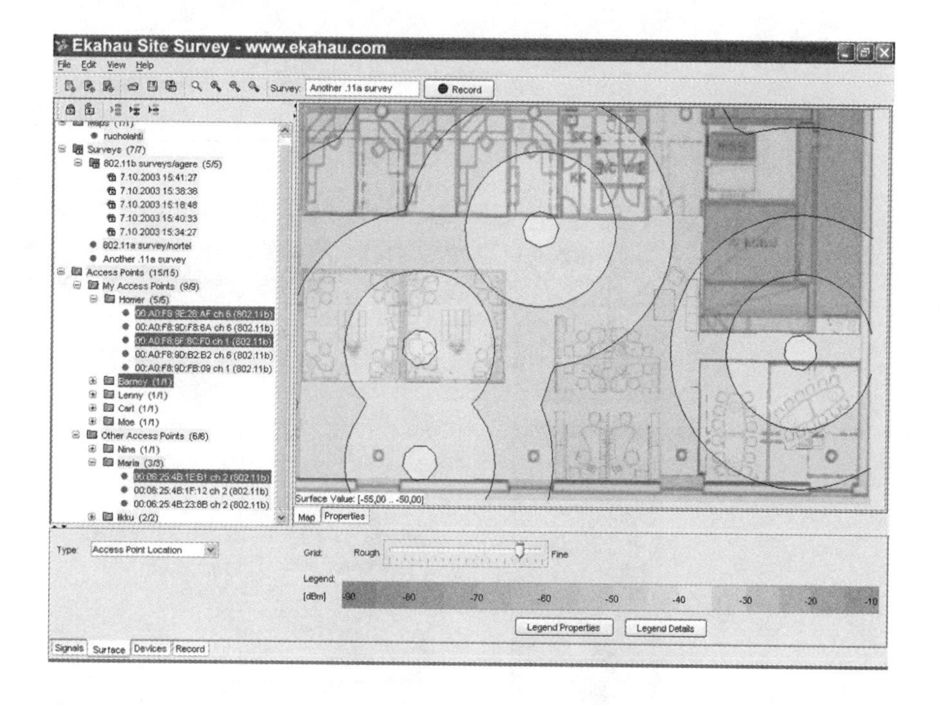

FIGURE 11.16 Integrated Wireless LAN Positioning System – Aruba Networks

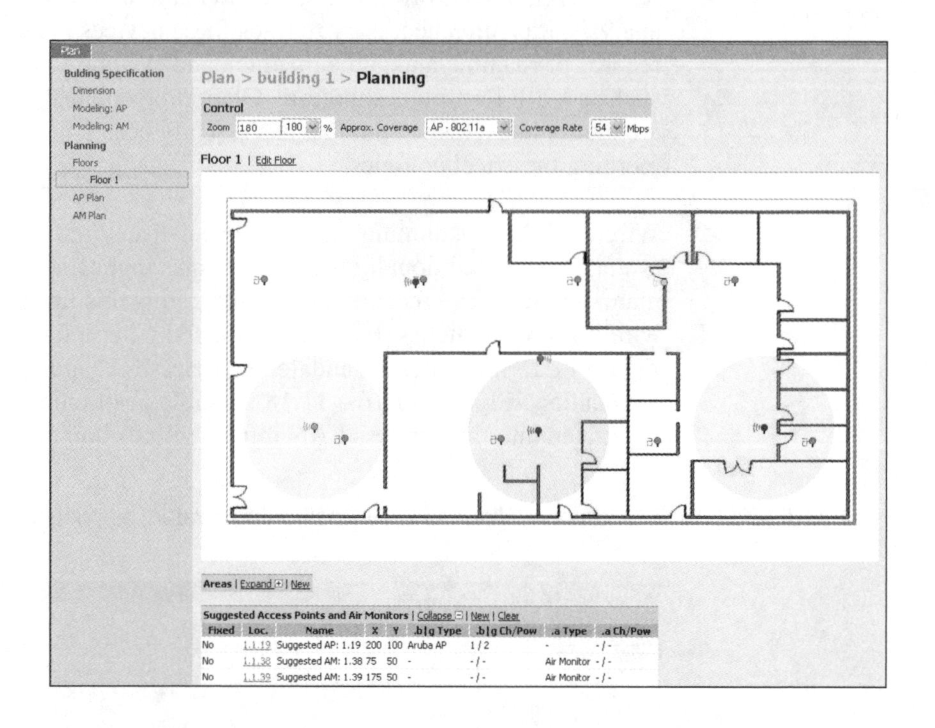

FIGURE 11.17 Integrated Wireless LAN Positioning System – Airespace Networks

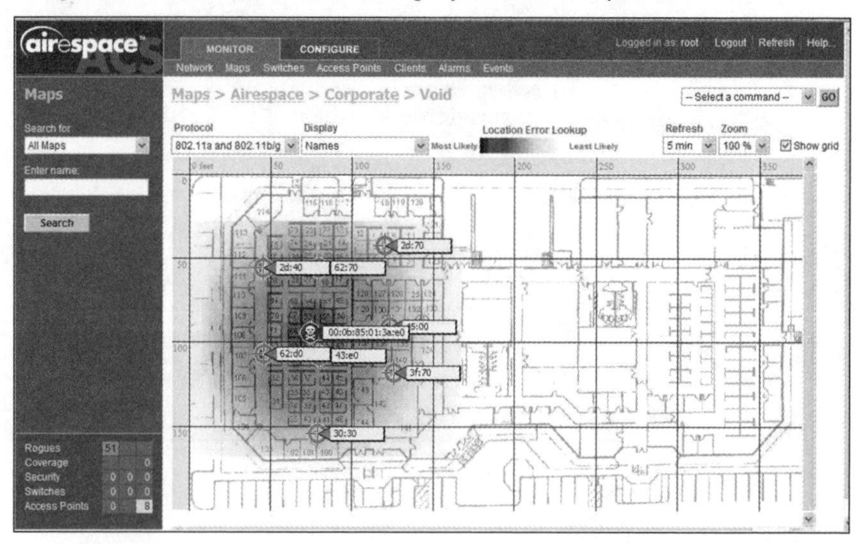

FIGURE 11.18 Integrated Wireless LAN Positioning System – Trapeze Networks

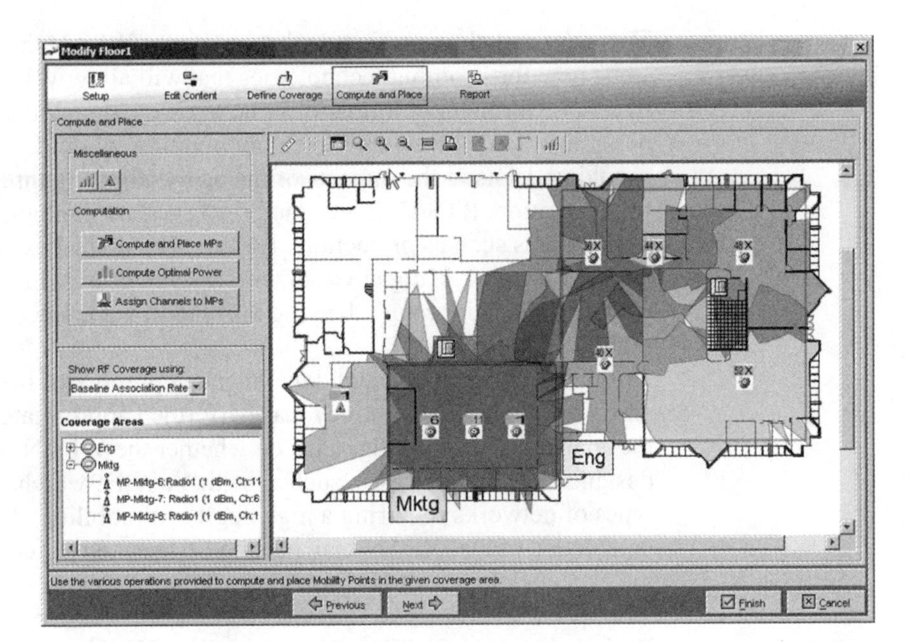

Installing a greater number of access points at lower power forms "micro cells" and offers the following advantages in an enterprise wireless LAN:

- Higher throughput per access point because fewer users would be attached to an access point, and client devices would only connect at higher data rates since they would be closer to the access point
- Since access points also often act as WIDS sensors, tracking client devices in real time would be much more precise with more access points in place.

Summary

There are a plethora of factors that may limit WLAN throughput and some new forthcoming technologies that will allow WLANs to both overcome throughput limitations due to external factors and boost the possible WLAN throughput by a minimum of twofold. The analyst should understand the impacts of manually adjusting thresholds such as fragmentation, RTS/CTS, etc., and the impacts of automatically invoked mechanisms such as protection. It's not always the best idea to have a mixed mode 802.11b/g environment. In some cases, higher throughput may be obtained through denying 802.11b client connectivity in an ESS.

An analyst should understand the intended purpose of the wireless LAN and assure that the site survey was performed appropriately. From there, the analyst can make judgments on whether the WLAN is performing as designed. There are cases, such as with VoWiFi, warehousing, and other types of networks requiring a high degree of mobility where certain parameters of the WLAN may need to be tweaked or redesigned to give the expected degree of performance.

With the advent of RFID tags designed for 802.11 networks and WLAN positioning systems, it is now possible to track mobile devices and people wearing 802.11 RFID tags in real time to a specific location within the premises. This adds to the security and productivity of a wireless environment.

Key Terms

Before taking the exam, you should be familiar with the following terms:

adjacent channel interference

beacon rate

co-channel interference

CRC Error

cross-channel interference

DTIM rate

fragmentation threshold

hidden nodes

MIMO

multipath

retransmissions

VoWiFi

WLAN positioning systems

Review Questions

1. When 802.11 frames are captured on a channel adjacent to the channel on which they are transmitted, they generally exhibit what two characteristics?

2. When an analyzer displays an abnormally large number of frame retransmissions relative to the baseline performance test, what are two things to check?

3. Stations in adjacent basic service sets must honor what parameters of each other's transmissions if they can hear each other?

4. What is the minimum fragmentation threshold value according to the 802.11 standard?

5. Name two distinct effects that decreasing the beacon rate will have on the BSS.

6. Name one distinct effect of decreasing the DTIM rate.

CHAPTER

12

Additional Information

In This Chapter

What's Missing?

Some topics invariably get left out of every book, every whitepaper, and every conversation. A good example is layers 5 and 6 of the OSI model. Every good network engineer understands most of layers 1-4 and 7. What happened to layers 5 & 6? Most engineers will tell you to leave those layers to the programmers. That same thing happens to other topics that are relevant, but often avoided, due to a lack of thorough knowledge of the topic. The LLC sublayer, the SNMP protocol, and wireless middleware all fall into this category of "Left Out for Unspecified Reasons." The LLC sublayer is directly related to wireless because the protocol analyzers decode the LLC layer for the analyst to observe and understand. SNMP is being used to manage most every kind of wireless device today. Wireless middleware is becoming more and more common and affects how networks perform – especially in the face of problems such as RF interference.

802.2 Logical Link Control (LLC)

The 802.2 LLC is the highest layer of the IEEE 802 Reference Model and provides functions similar to the traditional data link control protocol, which operates above the 802.11 and 802.3 protocols. ISO/IEC 8802-2 (ANSI/IEEE Standard 802.2), dated May 7, 1998, specifies the LLC sublayer. The purpose of the LLC is to exchange data between end users across a LAN using an 802-based MAC controlled link. The LLC provides addressing and data link control, and it is independent of the topology, transmission medium, and medium access control technique. As a result, 802.2 provides interoperability between users operating over different 802 protocols (e.g., 802.3 and 802.11) and corresponding 802.11 data frames containing 802.2 packets commonly flow over the wireless LAN, which makes it imperative that the wireless LAN analyst understand their behavior.

What is the LLC Sublayer?

Logical Link Control, or LLC, is IEEE's 802.2 standard way for multiple higher layer protocols to share the data link. For lack of agreement, the IEEE defined multiple implementations of LLC. The group of people favoring a best-effort connectionless datagram data link model got LLC-1. The group of people favoring a reliable connection-oriented protocol on top of the basic datagram got LLC-2. Initially, both models took 3 (or 4) octets from the frame body of LAN frames to carry the identity of higher layer protocols at the data link source and destination, and, in the case of LLC-2, to carry sequence, windowing, and acknowledgment information. Today LLC-1 is the more popular implementation, and a protocol analyst may never even see LLC-2.

The 802.3 committee, responsible for standardizing Ethernet, imagined that the LLC destination and source service access point fields (DSAP and SSAP), one octet each, would take the place of the two-octet Ethernet type field for identifying upper layer protocols. Three bits of each SAP field were devoted to other purposes, leaving 5 bits to uniquely identify up to 32 higher layer protocol clients (such as IP) running on each machine on a data link. While 32 higher layer protocols is enough for most any practical machine, it is not enough for every possible protocol client to be assigned a globally approved SAP value in the way of the registry for Ethernet types.

Advocates of keeping the type field approach lobbied for and won acceptance of yet another LLC standard: SNAP (sub network access protocol). One of the 32 global SAP values, “AA”, was assigned for this use. In a given frame body with an LLC header, if both SAPs are “AA,” then following the normal three-octet LLC header is a five octet SNAP header, the last two octets of which are the original Ethernet type field.

To add insult to injury, all those responsible for how IP uses Ethernet simply declined to use LLC, with or without SNAP. Still later, the IEEE legitimized the use of the Ethernet type field by adding its description to the IEEE 802.3 standard, ending forever the need to distinguish between “Ethernet” and “IEEE CSMA/CD”. Thus, today on Ethernet, the IP protocol does not use LLC at all. Other protocols on Ethernet use the type field, LLC-1 without SNAP, or LLC-1 with SNAP.

How Does the LLC Sublayer Relate to 802.11?

While LLC is optional on 802.3 Ethernet, it is required by the 802.11 standard. LLC is used primarily to carry the higher layer protocol identity, either in the SAP fields (without SNAP) or more likely in the SNAP type field.

When multiple similar data link segments are bridged together to deliver data units between two end stations, the Ethernet type field and LLC headers are operational only at the two end stations and not on the bridges. In other words, while the protocol information is carried in the frame header (Ethernet) or frame body (other IEEE 802 LANs) it is not acted upon by the bridges. Instead, the bridges operate at the MAC layer, making forwarding and filtering decisions based on MAC addresses found in the frame header.

This same model holds for 802.11 access points as they distribute frames within or between basic service sets using a distribution system. The LLC and SNAP headers are neither acted upon nor changed. When multiple dissimilar data link segments are bridged together, such as an 802.11 access point with integrated Ethernet, the situation becomes much more complicated. LLC may need to be removed from or added to frame bodies as part of the creation of a new frame before it is transmitted on the next data link. The IEEE has a recommended practice, IEEE 802.1H, for bridges that addresses the LLC translation problem. Figure 12.1 shows how frames are transmitted from station to station across access points, both across the wireless medium and from wireless to Ethernet. You can see in this figure how the LLC sublayer comes into play on wireless stations, but not on wired stations.

FIGURE 12.1 LLC Layer Illustration

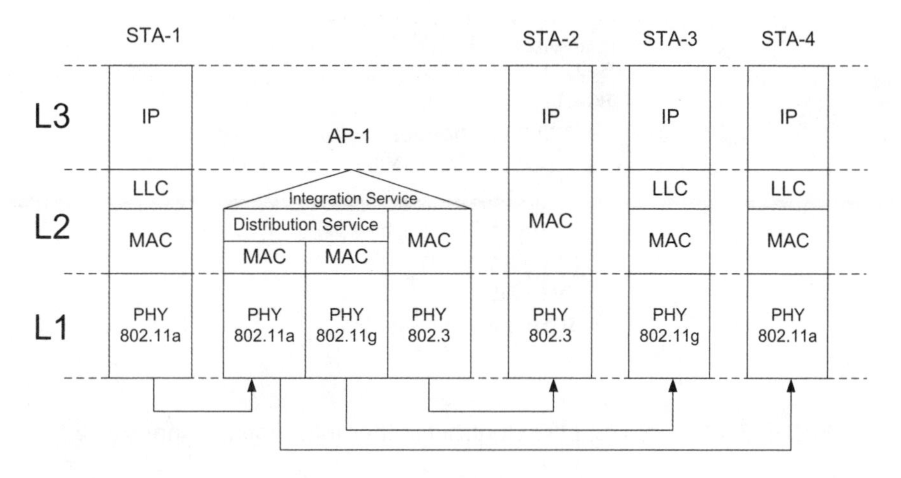

Higher layers, such as TCP/IP, pass user data down to the LLC. The LLC in turn appends a control header, which includes addressing information, creating an LLC protocol data unit (PDU).

FIGURE 12.2 802.2 LLC PDU

DSAP 8 bits	SSAP 8 bits	Control 8 or 16 bits	Data

Before transmission, the LLC PDU is handed down through the MAC service access point (SAP) to the MAC layer, which appends control and FCS information at the beginning and end of the packet, forming a MAC frame. Keep in mind that the LLC control header is part of the MSDU, and is therefore encrypted, along with other upper layer protocol information. The LLC control header is illustrated in the protocol analyzer decodes in Figures 12.3 and 12.4.

FIGURE 12.3 802.2 LLC Control Header Information (Analyzer #1)

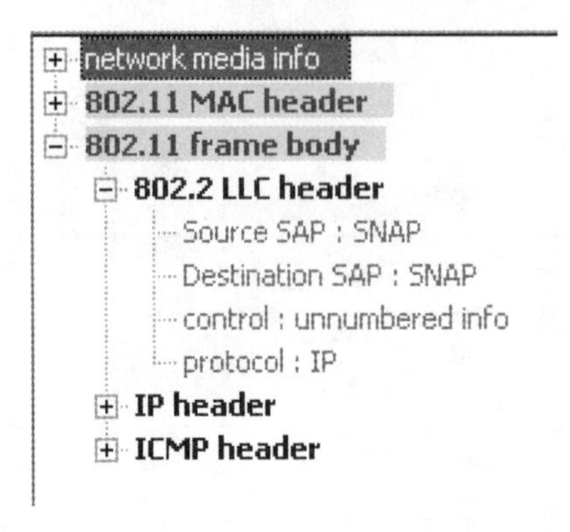

FIGURE 12.4 802.2 LLC Control Header Information (Analyzer #2)

```
802.2 Logical Link Control (LLC) Header
    Dest. SAP:              0xAA    SNAP
    Source SAP:             0xAA    SNAP
    Command:                0x03    Unnumbered Information
    Vendor ID:              0x000000
    Protocol Type:          0x0800   IP
IP: Ver: 4 HLen: 5 TLen: 1028 ID: 1352 Offset: 0 TTL: 1:
ICMP - Internet Control Messages Protocol ICMP Type: 8
```

802.2 LLC Services

The LLC provides the following three services for a Network Layer protocol:

- Unacknowledged connectionless service
- Connection-oriented service
- Acknowledged connectionless service

These services apply to the communication between peer LLC layers; meaning one layer located on the source station and one layer located on the destination station. All three LLC protocols employ the same PDU format, which consists of four fields. The Destination Service Access

Point (DSAP) and Source Service Access Point (SSAP) fields each contain 7-bit addresses that specify the destination and source stations of the peer LLCs. One bit of the DSAP indicates whether the PDU is intended for an individual or group station(s). One bit of the SSAP indicates whether it is a command or response PDU. The Data field contains the information from higher-layer protocols that the LLC is transporting to the destination. The Control field has bits that indicate whether the frame is one of the following types:

- Information - Used to carry user data.
- Supervisory - Used for flow control and error control.
- Unnumbered - Various protocol control PDUs.

Unacknowledged Connectionless Service

The unacknowledged connectionless service is a datagram-style service that does not involve any error-control or flow-control mechanisms. This service does not involve the establishment of a data link layer connection (such as between peer LLCs). This service supports individual, multicast, and broadcast addressing. The unacknowledged connectionless service simply sends and receives LLC PDUs with no acknowledgement of delivery. Because the delivery of data is not guaranteed, a higher layer must deal with reliability issues.

The unacknowledged connectionless service offers advantages in the following situations:

- If higher layers of the protocol stack provide the necessary reliability and flow-control mechanisms, then it would be inefficient to duplicate them in the LLC. In this case, the unacknowledged connectionless service would be appropriate. TCP, for example, already provides the mechanisms necessary for reliable delivery.
- It is not always necessary to provide feedback pertaining to successful delivery of information. The overhead of connection establishment and maintenance can be inefficient for applications involving the periodic sampling of data sources, such as monitoring sensors. The unacknowledged connectionless service would best satisfy these requirements.

Connection-Oriented Service

The connection-oriented service establishes a logical connection that provides flow control and error control between two stations needing to exchange data. This service involves the establishment of a connection between peer LLCs by performing connection establishment, data transfer, and connection termination functions. The service can connect only two stations; therefore, it does not support multicast or broadcast modes. The connection-oriented service offers advantages mainly if higher layers of the protocol stack do not provide the necessary reliability and flow-control mechanisms, which is generally the case with terminal controllers.

Flow control is a protocol feature that ensures that a transmitting station does not overwhelm a receiving station with data. With flow control, each station allocates a finite amount of memory and buffer resources to store sent and received PDUs. Networks, especially wireless networks, suffer from induced noise in the links between network stations that can cause transmission errors. If the noise is high enough in amplitude, it causes errors in digital transmission in the form of altered bits. Such errors will lead to inaccuracy of the transmitted data, and the receiving network device may misinterpret the meaning of the information.

The noise that causes most problems with networks is usually Gaussian and impulse noise. Gaussian noise, also known as "white noise," is the natural noise which occurs when electricity is passed through a conductor and is due to the random vibration of electrons in the conductor. In the case of RF, it is the “background” radiation level caused by various sources in the environment. Theoretically, the amplitude of Gaussian noise is uniform across the frequency spectrum, and it normally triggers random single-bit independent errors. Impulse noise, the most disastrous, is characterized by long quiet intervals of time followed by high amplitude bursts. This noise results from lightning and switching transients. Impulse noise is responsible for most errors in digital communication systems and generally provokes errors to occur in bursts.

To guard against transmission errors, the connection-oriented and acknowledged-connectionless LLC use error control mechanisms that detect and correct errors that occur in the transmission of PDUs. The

LLC automatic repeat request mechanism recognizes the possibility of the following two types of errors:

- Lost PDU - A PDU fails to arrive at the other end or is damaged beyond recognition.
- Damaged PDU - A PDU has arrived, but some bits are altered.

When a frame arrives at a receiving station, the station checks to determine whether there are any errors present by using a Cyclic Redundancy Check (CRC) error detection algorithm. In general, the receiving station will send back a positive or negative acknowledgement, depending on the outcome of the error detection process. In case the acknowledgement is lost in route to the sending station, the sending station will retransmit the frame after a certain period of time. This process is often referred to as Automatic Repeat Request (ARQ).

Overall, ARQ is best for the correction of burst errors because this type of impairment occurs in a small percentage of frames, thus not invoking many retransmissions. Because of the feedback inherent in ARQ protocols, the transmission links must accommodate half-duplex or full-duplex transmissions. If only simplex links are feasible, then it is impossible to use the ARQ technique because the receiver would not be able to notify the transmitter of bad data frames.

The following are two approaches for retransmitting unsatisfactory blocks of data using ARQ.

- Continuous ARQ: (often called a sliding window protocol) the sending station transmits frames continuously until the receiving station detects an error and transmits a negative ACK.
- Stop-and-wait ARQ: the sending station transmits a frame, then stops and waits for an acknowledgment from the receiver on whether a particular frame was acceptable or not. If the receiving station sends a negative acknowledgment, the frame will be sent again. The transmitter will send the next frame only after it receives a positive acknowledgment from the receiver.

Acknowledged Connectionless Service

As with the unacknowledged connectionless service, the acknowledged connectionless service does not involve the establishment of a logical connection with the distant station. The receiving stations with the acknowledged version do confirm successful delivery of datagrams. Flow and error control is handled through use of the stop-and-wait ARQ method.

The acknowledged connectionless service is useful in several applications. The connection-oriented service must maintain a table for each active connection for tracking the status of the connection. If the application calls for guaranteed delivery, but there are a large number of destinations needing to receive the data, then the connection-oriented service may be impractical because of the large number of tables required.

Examples that fit this scenario include process control and automated factory environments that require a central site to communicate with a large number of processors and programmable controllers. In addition, the handling of important and time-critical alarm or emergency control signals in a factory would also fit this case. In all these examples, the sending station needs an acknowledgment to ensure successful delivery of the data; however, the urgency of transmission cannot wait for a connection to be established.

LLC/MAC Layer Service Primitives

The 802.2 LLC layer communicates with its associated MAC layer (such as 802.11 or 802.3) through the following specific set of service primitives:

- MA-UNITDATA.request: The LLC layer sends this primitive to the MAC layer to request the transfer of a data frame from a local LLC entity to a specific peer LLC entity or group of peer entities on different stations. The data frame could be an information frame containing data from a higher layer or a control frame (such as a supervisory or unnumbered frame) that the LLC generates internally to communicate with its peer LLC.

- MA-UNITDATA.indication: The MAC layer sends this primitive to the LLC layer to transfer a data frame from the MAC layer to the LLC. This primitive occurs only if the MAC has found that a frame it has received from the Physical layer is valid and has no errors and the destination address indicates the correct MAC address of the station.
- MA-UNITDATA-STATUS.indication: The MAC layer sends this primitive to the LLC layer to provide status information about the service provided for a previous MA-UNITDATA.request primitive.

SNMP

Simple Network Management Protocol (SNMP) traffic is very common over wireless LANs, especially in large enterprises. In some cases, the presence of SNMP can cause performance and security problems, so it is important that the wireless LAN analyst have at least a basic understanding of SNMP.

SNMP is an application layer protocol that facilitates the exchange of management information among network devices. SNMP is part of the TCP/IP family of protocols and generally runs over User Datagram Protocol (UDP), a connectionless transport layer protocol for sending data between applications. Thus, SNMP can be utilized for just about any operating system and hardware platform. SNMP may provide a means for managing access points, client devices, or infrastructure devices such as enterprise wireless gateways.

A typical SNMP scenario includes a network management station that centrally monitors and controls network devices. In order to provide a standardized environment, an SNMP-based network management system includes the following elements:

- **Network management station**. A network management station runs software that monitors and controls managed devices via the SNMP protocol.
- **Managed device**. A managed device is an IP-addressable node on the network capable of using the SNMP protocol.

- **Management information base (MIB)**. The MIB is a database that stores management information on the managed device. The network management station monitors and updates values within the MIB, which contains managed objects having standardized object identifiers. Each object provides a specific characteristic about the managed device.
- **Agent**. The agent is software (or integrated firmware) residing on board the managed device. Through the use of the SNMP protocol, agents update local management information in the MIB with the network management station.

The first version of SNMP (SNMPv1) includes a very basic set of messages that the agent and network management station use to communicate. The following list provides a brief overview of the SNMPv1 message types.

- **GET**. The network management station uses the GET message to request object information from a particular MIB on a managed device. This information could be anything that the network administrator feels is important to support effective operational support, such as polling for the presence of all infrastructure devices.
- **GET-NEXT**. If the object has multiple instances contained within a table, the network management station can issue a GET-NEXT message in order to retrieve additional information related to the table.
- **GET-RESPONSE**. After receiving a GET or GET-NEXT message, the agent issues a GET-RESPONSE message containing the requested information. If the network management station requests invalid information, the agent will respond with an error indication and an applicable explanation.
- **SET**. The network management station can send a SET message to the agent when needing to change the value of an object within the managed device. For example, a SET message could change an object that changes the transmit power level of a particular access point's 802.11g radio. The agent responds to the SET message with a SET-RESPONSE, indicating that the change has been made.

- **TRAP**. The agent is the only element that can generate a TRAP message, which is spontaneously sent to the network management station and signifies a particular event. The agent, for example, could send a TRAP when a GPS-enabled client device roams outside a particular area. TRAPs conserve the limited bandwidth of wireless networks by pushing information to the network management station only when necessary.

This set of messages is simple, and it is not enough for some applications. Additional versions of SNMP build upon SNMPv1 by including efficiency and security features. An issue with SNMPv1 is that moving multiple instances of data related to one particular object is inefficient. For example, some applications may poll a hundred or more wireless client devices every ten minutes. To accomplish this task, the management station would repeatedly issue GET-NEXT messages in order to obtain the corresponding data from the device MIB tables. The overhead of GET-NEXT messages, however, can put a staggering forty to fifty percent reduction in throughput capacity.

SNMPv2 is now available, which includes the existing set of SNMPv1 message types and solves the overhead issue by adding the GET-BULK message. A network management station can send a GET-BULK message to the managed device in order to retrieve large blocks of data from a particular object's table. The agent sends a single corresponding response message that contains the requested information. The use of the GET-BULK message saves on overhead, which improves the performance of the network. As a result, strongly consider the use of GET-BULK when often extracting tables from large numbers managed devices.

SNMPv2 also includes another message, INFORM, that enables a network management station to send TRAP information to another management station. The INFORM message is useful for a number of applications. A user roaming to a remote area, for example, can send a TRAP to the local management station which in turn sends the TRAP to the home management station.

A problem with SNMPv2 is that it is incompatible with SNMPv1. SNMPv2 messages have a different header and protocol data unit format. Of course SNMPv2 also includes two additional message types. In order to solve this problem, consider using an SNMP proxy agent. When an SNMP proxy agent is employed, a SNMPv2 agent acts as a proxy agent for SNMPv1.

A SNMPv2 network management station can then issue messages targeted for a SNMPv1 agent. The SNMPv2 proxy agent forwards GET, GET-NEXT, and SET messages to the SNMPv1 agent. The proxy converts GET-BULK messages into a series of GET-NEXT messages to achieve interoperability. The proxy agent also maps SNMPv1 TRAP messages to ones compliant with SNMPv2 TRAP messages before sending them to the network management station.

SNMPv1 and SNMPv2 lack authentication and encryption, which makes the use of these protocols over wireless networks extremely vulnerable to hackers. It is possible to modify the MIB and make client software react erratically and even deny service. In addition, you can alter messages in a way that causes the network management station to log inaccurate accounting records. These vulnerabilities are an issue for public wireless network operators because a hacker can exploit them and take advantage of free access to the network by fooling the accounting system with bogus management data.

With no encryption on wireless networks, monitoring of the transmission of SNMP packets is a way to effortlessly discover management information. You can watch for TRAPs using a wireless analyzer, for example, which provides critical information regarding events happening within an organization. In addition, a hacker can purposely send unauthorized GET messages in order to extract valuable management information from managed devices. As a result, strongly consider disabling GET message responses when integrating SNMPv1 and SNMPv2 into mobile wireless applications.

To achieve a secure network management system for mobile wireless applications, utilize SNMPv3, which includes authentication and encryption mechanisms. Using SNMPv3 will protect your system from unscrupulous activity. SNMPv3 comprises certain security features in addition to all of the SNMPv2 message types.

SNMP adds overhead to the network, which requires you to carefully weigh the benefits of making use of various management objects residing on mobile devices. If a network management station continuously polls devices for magnitudes of data, then there may not be much throughput left for application data. As a result, include the expected number and frequency of SNMP frames when planning and allocating bandwidth.

GET-BULK messages are relatively large and experience approximately twenty percent higher retransmission rates due to the elevated probability of being hit by an error as compared to GET-NEXT messages. However, the savings in overhead with GET-BULK messages increases efficiency to the point that significantly counters the higher retry rate. So, you should still consider using the GET-BULK process when retrieving large amounts of data from managed devices in the presence of marginal radio conditions, such as when signals levels are low or interference exists.

Something else to consider is using UDP versus TCP as the transport protocol with SNMP. TCP requires considerable time and overhead to establish a connection before management data can be sent. TCP will typically decrease throughput capacity by as much as twenty percent. TCP may be feasible for extremely large GET-BULK transfers, but the larger number of TCP header overhead and packet exchanges also increases the probability of bit errors and retransmissions.

Wireless Middleware

As with wired network solutions, a mobile wireless client interfaces directly with a database or application on a central system. The problem is that wireless networks offer a less than perfect environment for corresponding communications. Traditional connectivity mechanisms such as terminal/host and client/server may satisfy wireless application requirements, but they can fall short in providing a satisfactory user experience.

As a result, some companies are deploying wireless middleware for curing the trials and tribulations of wireless connectivity. An effective middleware solution enables application developers to focus more on the application itself without worrying about the underlying network. As a result, mobile applications operate more smoothly, which significantly reduces support costs.

Wireless Issues Overview

Wireless systems are difficult to install in a way that provides complete coverage where users operate their mobile client devices. Obstacles such as buildings, walls, and even people offer varying amounts of attenuation to the propagation of radio waves. The result is somewhat spotty coverage that periodically interrupts connectivity over the wireless network as users roam.

Interference from microwave ovens and cordless phones causes decreased performance and sometimes connectivity loss for an indefinite period of time. This somewhat erratic disturbance makes communications over the network slower and even unavailable if the interference is severe enough. This problem becomes more significant as the number of active users on the network increases.

Because wireless systems allow users to roam, end users will run into problems when crossing subnets and different wireless networks. Wireless network technologies provide roaming among radio cells, but roaming across different networks requires software developers to integrate special roaming utilities. For example, a client device using a wireless LAN card for network connectivity will not be able to roam to another subnet within the same facility without special provisions. Nor will the client be able to natively roam from the wireless LAN to a 3G cellular system.

Terminal/Host Connectivity

Some mobile wireless applications need to interface directly with application screens running on a host computer, such as an IBM AS/400, Unix box, or 3270 mainframe. In this case, terminal emulation software runs on the client device, which communicates with application software running on the host. For example, VT220 terminal emulation corresponds with applications running on Unix and 5250 terminal emulation works with AS/400-based systems.

An advantage of terminal emulation is that it has low initial cost and is somewhat of a plug-and-play solution. Keep in mind that wireless systems using terminal emulation may not be able to maintain adequate connections with legacy applications because legacy applications may have timeouts entrenched throughout the application software that functions better with the more reliable wired networks. Users may run into a coverage hole that causes a loss of connectivity for an unacceptable amount of time. The timeouts automatically disconnect a communications session if they do not sense activity within a given time period, causing an unpleasant experience for the users of mobile devices.

Consequently, the corporate MIS staff spends much of their time responding to end-user complaints of dropped connections and the associated issues of incomplete data transactions. The implementation of terminal emulation can have a significant detrimental effect on long-term support costs, diminishing the benefits that the mobile applications offer the enterprise.

In addition, terminal emulation applications tend to push significant unnecessary data over the wireless network, which consumes valuable throughput. Because the application is actually running on the host, all input screens, print streams, and other correspondence with the application must traverse the wireless network. As a result, uninformed companies that begin their attempt into wireless LAN technology using wireless terminal emulation eventually search for better solutions in order to reduce support issues.

Client/Server Connectivity

Scores of mobile applications today follow the client/sever model, in which thin or thick client application software interfaces with a database residing on a server. In such cases, the application communicates with the database using proprietary database commands or Open Database Connectivity (ODBC) protocols. The software on the client device with this configuration provides nearly all application functionality.

This approach provides flexibility when developing mobile applications, mainly because the programmer has complete control over the functions that are implemented and is not constrained by applications running on a host system. The problem with client/server connectivity is that it operates over TCP/IP, which is cumbersome over wireless networks. TCP/IP uses relatively large headers and adds significant delays when re-establishing connections while the user traverses coverage holes.

The use of web browsers over a wireless network loosely fits the client/server model, with the browser acting as a thin client with the data and much of the application software residing on a server. Many web server applications can tolerate wireless impairments, but be careful when implementing applications that require immediate responses between the client and the server. In some cases, your application may not be able to recover from loss of connectivity in the transactions that have dependencies.

Advantages of Wireless Middleware

Wireless middleware is software offering intermediate communications between user client devices and the application software or databases located on a server. The middleware, which generally runs on a dedicated gateway platform attached to the wired network, processes the packets that pass between the wireless client and the wired network. In general, middleware offers efficient and reliable communications over the wireless network, while maintaining appropriate connections to application software and databases on the server via the more reliable wired network. Figure 12.5 below illustrates the general wireless middleware network model.

FIGURE 12.5 Wireless Middleware Model

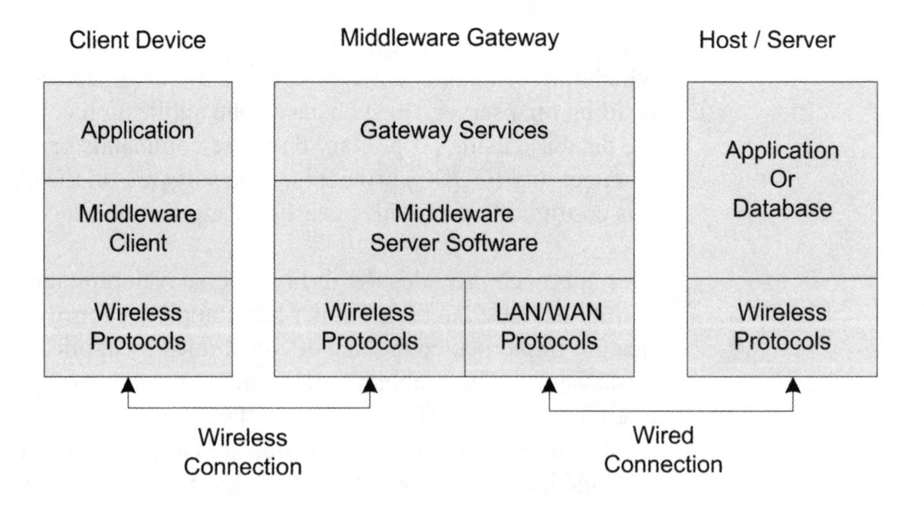

Connection Management

Wireless middleware has store-and-forward messaging that queues traffic to ensure delivery to client devices that lose connections with the network. The middleware is able to act as a proxy of the client. In fact, most applications have no idea that the middleware server is maintaining the connection with the host. Once the client gains wireless connectivity again, the middleware lets the client resume communications. This is especially beneficial when servers and applications can't tolerate the loss of connectivity.

If transmissions are unexpectedly cut at midstream, a middleware recovery mechanism known as "intelligent restart" detects the premature end of a transmission. When the connection is reestablished, the middleware resumes transmission from the break point instead of at the beginning. This functionality not only enables wireless clients to easily reestablish connectivity, but it also helps with improving performance and reducing battery drain because there will be fewer packet retransmissions.

Performance Optimization

Some middleware products include data compression at the transport layer to help minimize the number of bits sent over the wireless link. Some implementations of middleware use header compression, where mechanisms replace traditional packet headers with a much shorter bit sequence before transmission. Using a shorter bit sequence helps free up needed throughput for supporting higher performance applications, such as video streaming and large numbers of active users.

Middleware packages are often capable of combining smaller data packets into a single large packet for transmission over the wireless network, which can improve performance and help lower transmission service costs of wide area networks. Since most wireless data services charge users by the packet, data bundling results in a lower aggregate cost.

Middleware products sometimes offer a development environment that allows developers to use visual tools to "scrape" and "reshape" portions of existing application screens to more effectively fit on the smaller displays of mobile devices, such as PDAs and cell phones. This is a good alternative as compared to using existing larger screens offered by legacy terminal/host connections, which would generally require users having small screens to spend most of their time scrolling. Developers can pick and choose parts of several legacy screens in order to make the user interface more efficient and streamlined for traversing the wireless network.

Roaming Support

To facilitate roaming, a few middleware products offer home and foreign agent functions to support the use of Mobile IP protocols, which seamlessly handles changes in a client's point of attachment to the Internet. Mobile IP technology enables wireless users to roam across different subnets, which is crucial when deploying mobile applications that span a large company.

In addition, some middleware offers proprietary solutions for roaming across different wireless network types, which generally requires special client software installed on the client device. For example, middleware makes it possible for users to roam from a wireless LAN to a cellular system, assuming the client device has both network adapters. Mobile phone manufacturers are making this type of cross-protocol roaming a reality, however, with the phones having built in 802.11 interfaces. Integrated chipsets are now available to facilitate 802.11/cellular roaming.

Multi-Host Support

Middleware can also provide multi-host support, which is invaluable when integrating mobile applications into complex end system environments. An example might be a mobile application that requires wireless client devices to retrieve pricing data from an IBM mainframe (using 3270 terminal emulation) and record data entered by the user into a database located on a Windows 2003 server (through ODBC). Such integration would be difficult to implement with terminal emulation and direct database connectivity.

The use of middleware provides a centralized component that molds both the terminal/host and client/server worlds together. This might be especially useful if the pricing data (from the example above) is going to be moved over to the Windows 2003 sever. The middleware makes it possible to develop a common interface for the users that doesn't need to change after a significant change such as moving the pricing data over to the Windows 2003 server.

Centralized Development Environment

A strong advantage from a development perspective is the consistent API that many wireless middleware tools offer. Developers can use the wireless middleware as a development environment for integrating their applications to various types of end systems, regardless of the underlying network.

Developers can also centrally implement and manage the wireless connections through the middleware platform. For example, changes made to the solution can easily be made available to all clients because of the centralized nature of the middleware server. The server also sees all client traffic, making it possible to identify the presence of users, track user activity, and detect rogue users and potential denial of service attacks.

Summary

Wireless stations have an LLC layer to identify upper layer protocols in use over the wireless network. This same LLC layer is not found in Ethernet clients using IP. This creates a situation where access points, functioning as a translator between 802.3 Ethernet and 802.11 must either insert LLC layers, translate them, or remove them, depending on where the traffic is originating and is destined. Analysts will see the LLC sublayer in protocol decodes, and it is important to understand why it is there and what the values mean.

The SNMP protocol is used to manage most wireless infrastructure devices on the market today. Understanding the underlying functionality of the SNMP protocol, the differences between versions, and how it is implemented in the wireless infrastructure to ease the burden of managing numerous devices is crucial from design, troubleshooting, and security perspectives. Analysts may be able to improve network performance without every enabling a protocol analyzer by verifying proper configuration of SNMP features.

Wireless middleware is a unique type of product that functionally sits between the wireless client application and the backend server application. Middleware allows proper operation of a wireless application under poor connectivity conditions. Additionally, middleware can ease the burden of migrating between backend systems such that clients are unaware of the migration.

Key Terms

Before taking the exam, you should be familiar with the following terms:

Automatic Repeat Request (ARQ)

client/server

flow control

Logical Link Control (LLC)

Management Information Base (MIB)

primitives

Service Access Point (SAP)

Simple Network Management Protocol (SNMP)

SNMP agent

SNMP community string

terminal/host

wireless middleware

Review Questions

1. What is the purpose of the LLC sublayer?

2. What are two networking services provided by the LLC sublayer?

3. Which wireless network devices have LLC sublayers?

4. Does the LLC sublayer header get encrypted when TKIP is used on the wireless LAN?

5. Which version of SNMP provides for encrypted community strings?

CHAPTER

13

Case Studies

No text on wireless LAN analysis would be complete without some example troubleshooting scenarios. The abbreviated case studies below highlight simple but common troubleshooting scenarios in three topical areas: security, performance, and general fault finding. While some real-world scenarios can be complex and take hours or even days to troubleshoot down to the root cause, these scenarios will give the analyst an idea of what troubleshooting wireless LANs using protocol analysis is like.

Case Study 1 - Security

Westcott Telecom's management team has decided that it is time to implement wireless LAN technology in order to improve productivity in their Field Services division. Before haphazardly deploying wireless LAN technology, they request that their best engineer, Mr. Jones, oversee security solution testing in their research and development lab. Mr. Jones makes a list of items that he wants to have in his security solution, and another list of items that he wants to avoid having in his security solution. As Mr. Jones tests each security solution, he compares them to the lists he has prepared. One of the items on his AVOID list is transmission of the wireless client's user name in clear text across the RF medium since this may lead to dictionary attacks.

After configuring his protocol analyzer to look for a string called *user* in all captured frames, Mr. Jones begins building and testing security solutions. Mr. Jones is using a test user called user1 in his sample configurations. He analyzes each authentication process and reviews the analyzer's findings. Figures 13.1 and 13.2 illustrate configuration of a filter and a notification when the filter is matched.

FIGURE 13.1 Filter Configuration – Pattern Matching – Commview for WiFi

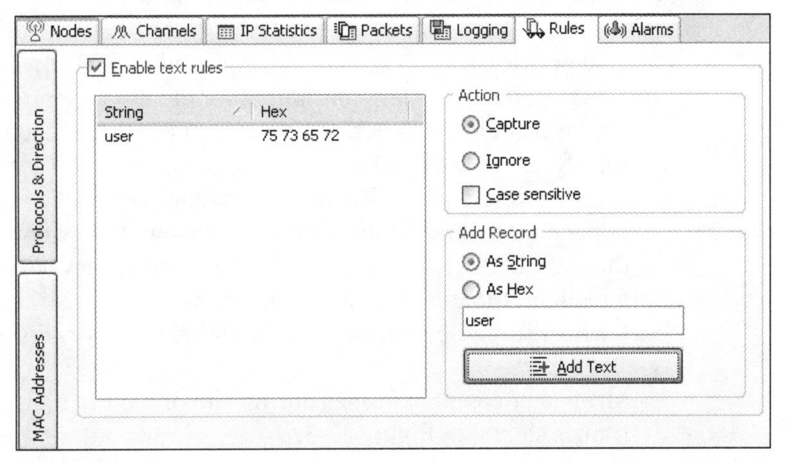

FIGURE 13.2 Filtered Capture – Commview for WiFi

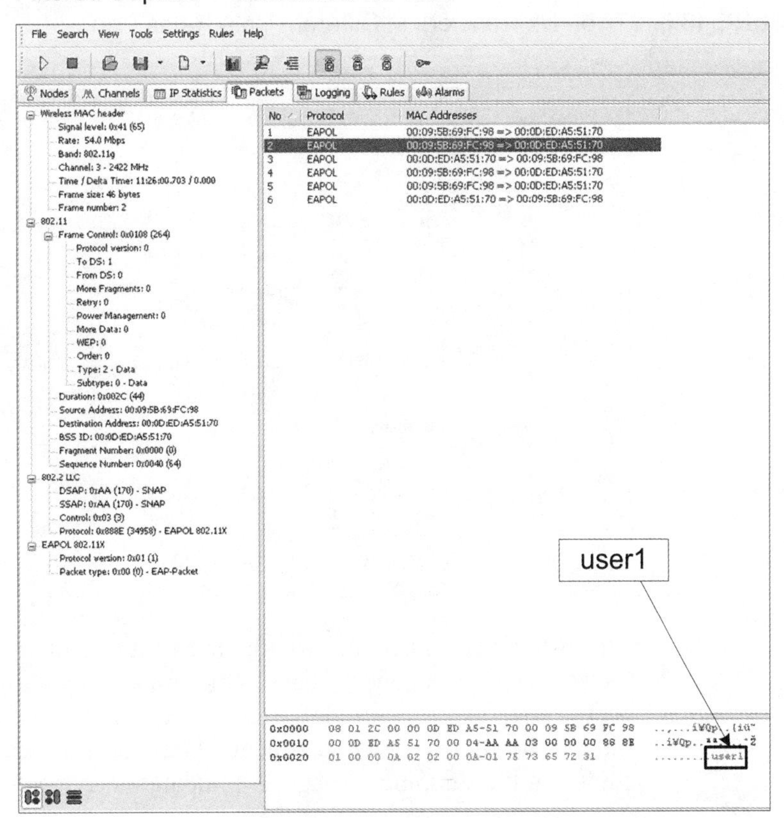

Case Study 2 - Performance

Coleman Research Ltd. has implemented an 802.11a solution, and all access points were left configured for default supported rates when deployed. The engineering department often uses the wireless network to transfer large CAD drawings to the company's file server. These file transfers often take a few minutes to complete. The engineering staff has noticed that while the files always transfer at 54 Mbps data rates due to the short distance between stations and the access point, the speed at which the file transfers complete varies greatly. Mr. Coleman has asked for help from the company's wireless analyst, Mr. Fisher.

Mr. Fisher captures frames during one of the file transfers, and notices the frames shown in Figure 13.3.

FIGURE 13.3 Two File Transfers at Different Data Rates

Source Physical	Dest. Physical	BSSID	Channel	Signal	Data Rate	Size	Protocol
00:E0:B8:5C:5B:CF	00:09:5B:66:E6:80	00:0D:65:B8:61:63	36	70%	9.0	1336	FTP Data
00:09:5B:66:E6:80	00:0D:65:B8:61:63		36	27%	6.0	14	802.11 Ack
00:E0:B8:5C:5B:CF	00:09:5B:66:E6:80	00:0D:65:B8:61:63	36	70%	9.0	864	FTP Data
00:09:5B:66:E6:80	00:0D:65:B8:61:63		36	30%	6.0	14	802.11 Ack
00:09:5B:66:E6:80	00:E0:B8:5C:5B:CF	00:0D:65:B8:61:63	36	20%	6.0	76	FTP Data
00:0D:65:B8:61:63	00:09:5B:66:E6:80		36	68%	6.0	14	802.11 Ack
00:E0:B8:5C:5B:CF	00:09:5B:69:FC:98	00:0D:65:B8:61:63	36	52%	54.0	1100	FTP Data
00:09:5B:69:FC:98	00:0D:65:B8:61:63		36	75%	24.0	14	802.11 Ack
00:E0:B8:5C:5B:CF	00:09:5B:69:FC:98	00:0D:65:B8:61:63	36	54%	54.0	864	FTP Data
00:09:5B:69:FC:98	00:0D:65:B8:61:63		36	74%	24.0	14	802.11 Ack
00:09:5B:69:FC:98	00:E0:B8:5C:5B:CF	00:0D:65:B8:61:63	36	80%	48.0	76	FTP Data
00:0D:65:B8:61:63	00:09:5B:69:FC:98		36	54%	24.0	14	802.11 Ack
00:E0:B8:5C:5B:CF	00:09:5B:69:FC:98	00:0D:65:B8:61:63	36	55%	54.0	1100	FTP Data
00:09:5B:69:FC:98	00:0D:65:B8:61:63		36	78%	24.0	14	802.11 Ack
00:E0:B8:5C:5B:CF	00:09:5B:69:FC:98	00:0D:65:B8:61:63	36	55%	54.0	1336	FTP Data
00:09:5B:69:FC:98	00:0D:65:B8:61:63		36	78%	24.0	14	802.11 Ack
00:09:5B:69:FC:98	00:E0:B8:5C:5B:CF	00:0D:65:B8:61:63	36	78%	48.0	76	FTP Data
00:0D:65:B8:61:63	00:09:5B:69:FC:98		36	70%	24.0	14	802.11 Ack
00:09:5B:66:E6:80	00:E0:B8:5C:5B:CF	00:0D:65:B8:61:63	36	27%	6.0	76	FTP Data
00:0D:65:B8:61:63	00:09:5B:66:E6:80		36	70%	6.0	14	802.11 Ack
00:09:5B:66:E6:80	00:E0:B8:5C:5B:CF	00:0D:65:B8:61:63	36	25%	6.0	76	FTP Data
00:0D:65:B8:61:63	00:09:5B:66:E6:80		36	70%	6.0	14	802.11 Ack
00:E0:B8:5C:5B:CF	00:09:5B:69:FC:98	00:0D:65:B8:61:63	36	70%	54.0	864	FTP Data
00:09:5B:69:FC:98	00:0D:65:B8:61:63		36	70%	24.0	14	802.11 Ack
00:09:5B:69:FC:98	00:E0:B8:5C:5B:CF	00:0D:65:B8:61:63	36	78%	48.0	76	FTP Data
00:0D:65:B8:61:63	00:09:5B:69:FC:98		36	70%	24.0	14	802.11 Ack
00:E0:B8:5C:5B:CF	00:09:5B:69:FC:98	00:0D:65:B8:61:63	36	70%	54.0	864	FTP Data
00:09:5B:69:FC:98	00:0D:65:B8:61:63		36	75%	24.0	14	802.11 Ack

Mr. Fisher notices that along with the 54 Mbps data frames captured from the engineer's file transfer, data frames transmitted through the same access point at 6 Mbps and 9 Mbps are coming from another computer outside the engineering department. After some investigation, Mr. Fisher finds that the shipping manager's computer is connecting at data rates as low as 6 Mbps due to the distance to the engineering department's access point. After speaking with Mr. Anderson, the shipping manager, Mr. Fisher discovers that Mr. Anderson occasionally uploads documentation to the company's file server over the wireless LAN. The site survey shows that no clients should connect to access points at data rates slower than 36 Mbps.

Mr. Anderson's computer should not be connecting to the engineering department's access point at all, so another investigation ensues. Each access point at Coleman Research is configured for an individual VLAN ID that corresponds to the department that uses it. The shipping department uses SSID 103, which is mapped to VLAN 103. The engineering department uses SSID 101, which is mapped to VLAN 101. Mr. Anderson's computer is configured with SSID 103, but after analyzing the wireless authentication process between Mr. Anderson's computer and the shipping access point, Mr. Fisher notices that Mr. Anderson's computer is being disassociated after a successful authentication. Figure 13.4 shows the captured frames. Notice the disassociation and deauthentication frames at the end of the authentication/association frame exchanges.

FIGURE 13.4 Captured Frames, Shipping Access Point

Packet	Address 1	Address 2	Address 3	Flags	Channel	Data Rate	Size	Protocol
30	00:0D:ED:A5:51:70	00:09:5B:66:E6:80	00:0D:ED:A5:51:70	*	1	1.0	34	802.11 Auth
31	00:09:5B:66:E6:80			#	1	1.0	14	802.11 Ack
32	00:09:5B:66:E6:80	00:0D:ED:A5:51:70	00:0D:ED:A5:51:70	*	1	11.0	34	802.11 Auth
33	00:0D:ED:A5:51:70			#	1	11.0	14	802.11 Ack
34	00:0D:ED:A5:51:70	00:09:5B:66:E6:80	00:0D:ED:A5:51:70	*	1	1.0	83	802.11 Assoc Req
35	00:09:5B:66:E6:80			#	1	1.0	14	802.11 Ack
36	00:09:5B:66:E6:80	00:0D:ED:A5:51:70	00:0D:ED:A5:51:70	*	1	54.0	94	802.11 Assoc Rsp
37	00:0D:ED:A5:51:70			#	1	24.0	14	802.11 Ack
38	00:0D:ED:A5:51:70	00:09:5B:66:E6:80	00:0D:ED:A5:51:70		1	54.0	40	EAPOL-Start
39	00:09:5B:66:E6:80	00:0D:ED:A5:51:70	00:0D:ED:A5:51:70		1	54.0	82	EAP Request
40	00:0D:ED:A5:51:70			#	1	24.0	14	802.11 Ack
41	00:0D:ED:A5:51:70	00:09:5B:66:E6:80	00:0D:ED:A5:51:70		1	54.0	40	EAPOL-Start
42	00:09:5B:66:E6:80			#	1	24.0	14	802.11 Ack
43	00:09:5B:66:E6:80	00:0D:ED:A5:51:70	00:0D:ED:A5:51:70		1	54.0	82	EAP Request
44	00:0D:ED:A5:51:70			#	1	24.0	14	802.11 Ack
45	00:0D:ED:A5:51:70	00:09:5B:66:E6:80	00:0D:ED:A5:51:70		1	54.0	53	EAP Response
46	00:09:5B:66:E6:80			#	1	24.0	14	802.11 Ack
47	00:0D:ED:A5:51:70	00:09:5B:66:E6:80	00:0D:ED:A5:51:70		1	54.0	53	EAP Response
48	00:09:5B:66:E6:80			#	1	24.0	14	802.11 Ack
49	00:09:5B:66:E6:80	00:0D:ED:A5:51:70	00:0D:ED:A5:51:70		1	54.0	82	EAP Request
50	00:0D:ED:A5:51:70			#	1	24.0	14	802.11 Ack
51	00:0D:ED:A5:51:70	00:09:5B:66:E6:80	00:0D:ED:A5:51:70		1	54.0	80	EAP Response
52	00:09:5B:66:E6:80			#	1	24.0	14	802.11 Ack
53	00:09:5B:66:E6:80	00:0D:ED:A5:51:70	00:0D:ED:A5:51:70		1	54.0	82	EAP Success
54	00:0D:ED:A5:51:70			#	1	24.0	14	802.11 Ack
55	00:0D:ED:A5:51:70	00:09:5B:66:E6:80	00:0D:ED:A5:51:70		1	54.0	64	EAP Request
56	00:09:5B:66:E6:80			#	1	24.0	14	802.11 Ack
57	00:09:5B:66:E6:80	00:0D:ED:A5:51:70	00:0D:ED:A5:51:70		1	54.0	82	EAP Response
58	00:0D:ED:A5:51:70			#	1	24.0	14	802.11 Ack
59	00:09:5B:66:E6:80	00:0D:ED:A5:51:70	00:0D:ED:A5:51:70		1	54.0	84	EAPOL-Key
60	00:0D:ED:A5:51:70			#	1	24.0	14	802.11 Ack
61	00:09:5B:66:E6:80	00:0D:ED:A5:51:70	00:0D:ED:A5:51:70	*	1	54.0	30	802.11 Disassoc
62	00:0D:ED:A5:51:70			#	1	24.0	14	802.11 Ack
63	00:0D:ED:A5:51:70	00:09:5B:66:E6:80	00:0D:ED:A5:51:70	*	1	1.0	30	802.11 Deauth
64	00:09:5B:66:E6:80			#	1	1.0	14	802.11 Ack

Instead, Mr. Anderson's computer is being successfully authenticated and associated through the engineering access point since the engineering and shipping departments are located adjacent to each other. Mr. Fisher checks the group assignment for Mr. Anderson's user in the RADIUS server and finds that his user is misconfigured to be part of the engineering group, which is assigned to VLAN 101. This mistake is verified by a wired capture between the engineering access point and the RADIUS server as shown below.

FIGURE 13.5 Captured Frames, Engineering Access Point

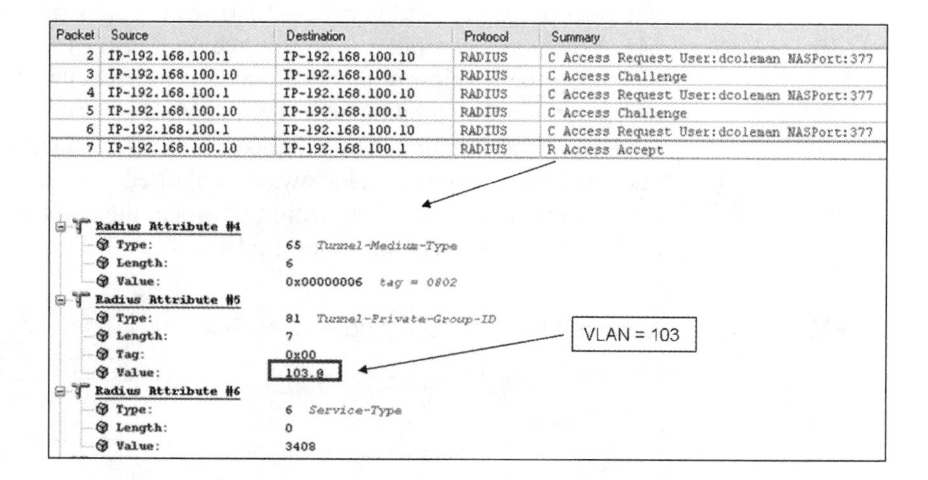

Packet	Source	Destination	Protocol	Summary
2	IP-192.168.100.1	IP-192.168.100.10	RADIUS	C Access Request User:dcoleman NASPort:377
3	IP-192.168.100.10	IP-192.168.100.1	RADIUS	C Access Challenge
4	IP-192.168.100.1	IP-192.168.100.10	RADIUS	C Access Request User:dcoleman NASPort:377
5	IP-192.168.100.10	IP-192.168.100.1	RADIUS	C Access Challenge
6	IP-192.168.100.1	IP-192.168.100.10	RADIUS	C Access Request User:dcoleman NASPort:377
7	IP-192.168.100.10	IP-192.168.100.1	RADIUS	R Access Accept

A simple change to the RADIUS server configuration alleviated a performance problem caused by a security misconfiguration.

Case Study 3 – General Fault Finding

Donohue's Aircraft Repair Depot recently upgraded their single-access point wireless LAN from 802.11b to 802.11g so that more technicians working in the hangar could simultaneously communicate with the file server located in the parts department. While using the 802.11b network that was previously installed, they were unhappy with the performance and thought that 802.11b technology was simply slow. After upgrading, performance has not improved as expected, so they have contacted an analyst, Mr. Parsons, to determine the root cause of the problem(s).

Mr. Parsons begins by placing an analyzer close to the access point. He notices that Donohue's access point is using channel 11, the RF medium is not very busy, and a relatively low amount of frames are being transmitted even during the busiest part of the work day. Figure 13.6 illustrates the low RF medium usage.

FIGURE 13.6 Low RF utilization

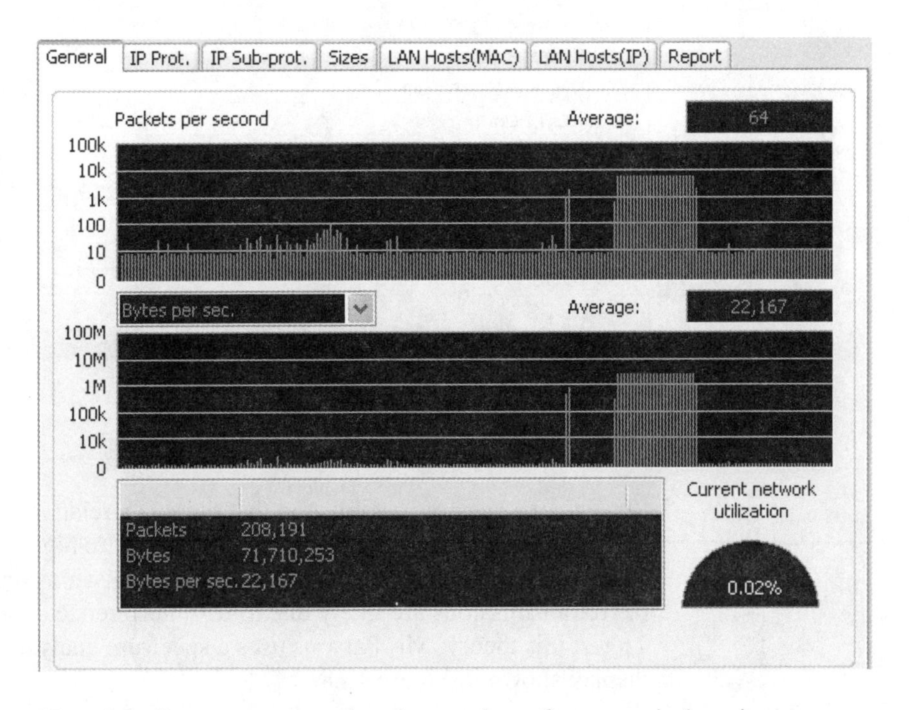

Next, Mr. Parsons notices that the number of retransmissions is abnormally high, considering that "normal" is less than 3% and "high" is around 10% on average. The reason for the retransmissions is that many frames are being corrupted. Figure 13.7 illustrates an analyzer displaying retransmission percentages.

FIGURE 13.7 Abnormally High Retransmission Count

Statistic	Current
⊞ General	
⊞ Errors	
⊞ Counts	
⊞ Size Distribution	
⊞ Wireless	
⊟ 802.11 Analysis	
Average Signal Strength	73.726
Average Signal dBm	-0.149
Average Noise	0.000
Average Noise dBm	0.000
802.11 Data	30.655%
802.11 Management	22.024%
802.11 Control	43.750%
Local	65.774%
From DS	17.560%
To DS	13.095%
DS-DS	0.000%
Retry	21.131%
WEP	30.655%
WEP ICV Errors	0.000%
Order	0.000%

Mr. Parsons has now established that there is a relatively low amount of traffic on the RF medium, but retransmissions are high. Since the CSMA/CA protocol typically leads to retransmissions under a heavy load, the retransmissions are likely due to RF interference or severe multipath. To test this theory, Mr. Parsons uses a spectrum analyzer and sees the display shown in Figure 13.8.

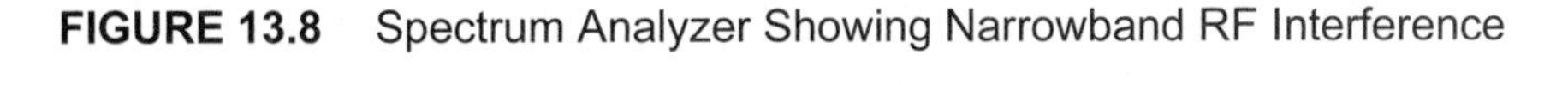

FIGURE 13.8 Spectrum Analyzer Showing Narrowband RF Interference

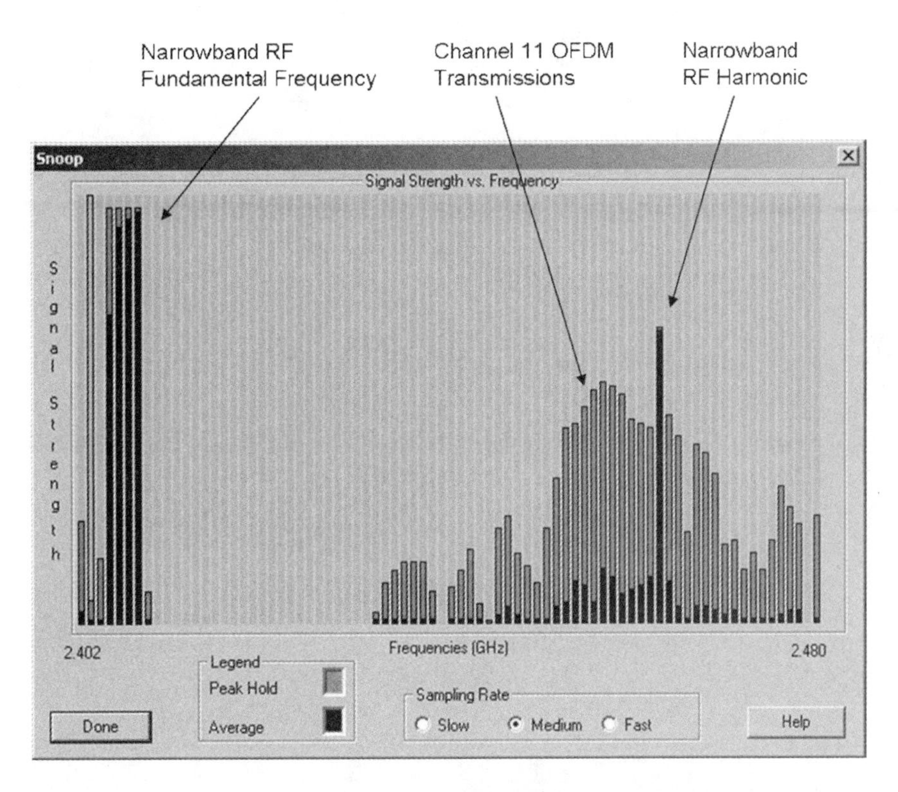

After finding that channel 1 has a significant source of narrowband RF interference with harmonics that fall into the channel 11 series of frequencies, Mr. Parsons tries to find the source. He finds that the source of this RF interference is a power tool that is used repeatedly throughout the day to repair aircraft engines. Since this tool must be used, the alternative to discontinued use is to move the access point to channel 6. This change gives the client devices and access point the best chance of receiving uncorrupted frames. After moving the access point to channel 6, performance improves dramatically, and the problem is solved.

B

C

D

E

K

L

M

N

O

P

Q

R

S

INTERNATIONAL CONTACT INFORMATION

AUSTRALIA
McGraw-Hill Book Company Australia Pty. Ltd.
TEL +61-2-9900-1800
FAX +61-2-9878-8881
http://www.mcgraw-hill.com.au
books-it_sydney@mcgraw-hill.com

CANADA
McGraw-Hill Ryerson Ltd.
TEL +905-430-5000
FAX +905-430-5020
http://www.mcgraw-hill.ca

GREECE, MIDDLE EAST, & AFRICA
(Excluding South Africa)
McGraw-Hill Hellas
TEL +30-210-6560-990
TEL +30-210-6560-993
TEL +30-210-6560-994
FAX +30-210-6545-525

MEXICO (Also serving Latin America)
McGraw-Hill Interamericana Editores S.A. de C.V.
TEL +525-117-1583
FAX +525-117-1589
http://www.mcgraw-hill.com.mx
fernando_castellanos@mcgraw-hill.com

SINGAPORE (Serving Asia)
McGraw-Hill Book Company
TEL +65-863-1580
FAX +65-862-3354
http://www.mcgraw-hill.com.sg
mghasia@mcgraw-hill.com

SOUTH AFRICA
McGraw-Hill South Africa
TEL +27-11-622-7512
FAX +27-11-622-9045
robyn_swanepoel@mcgraw-hill.com

SPAIN
McGraw-Hill/Interamericana de España, S.A.U.
TEL +34-91-180-3000
FAX +34-91-372-8513
http://www.mcgraw-hill.es
professional@mcgraw-hill.es

UNITED KINGDOM, NORTHERN, EASTERN, & CENTRAL EUROPE
McGraw-Hill Education Europe
TEL +44-1-628-502500
FAX +44-1-628-770224
http://www.mcgraw-hill.co.uk
computing_neurope@mcgraw-hill.com

ALL OTHER INQUIRIES Contact:
Osborne/McGraw-Hill
TEL +1-510-549-6600
FAX +1-510-883-7600
http://www.osborne.com
omg_international@mcgraw-hill.com